MW00843342

Emergence of Pico- and Nanosatellites for Atmospheric Research and Technology Testing

Emergence of Pico- and Nanosatellites for Atmospheric Research and Technology Testing

Edited by
Purvesh Thakker
Program Manager, ION Cubesat
University of Illinois at Urbana-Champaign, Urbana, Illinois

Wayne A. Shiroma
Professor, Department of Electrical Engineering, and
Co-Director, Hawaii Space Flight Laboratory
University of Hawaii at Manoa
Honolulu, Hawaii

Volume 234
PROGRESS IN
ASTRONAUTICS AND AERONAUTICS

Frank K. Lu, Editor-in-Chief
University of Texas at Arlington
Arlington, Texas

Published by
American Institute of Aeronautics and Astronautics, Inc.
1801 Alexander Bell Drive, Reston, Virginia 20191-4344

American Institute of Aeronautics and Astronautics, Inc., Reston, Virginia

1 2 3 4 5

ISBN 978-1-60086-768-2

Progress in Astronautics and Aeronautics

A Note from the Series Editor

I would like to echo the sentiments voiced in the Foreword by Bob Twiggs, Cubesat pioneer, on the exciting potential of pico- and nanosatellites for research and also for exploitation of space. Like many human endeavors, what starts out as imagination or vision is transformed into reality through a collective sharing of that imagination or vision. The chapters within this volume focus on all of these elements. They point to future potential—sometimes beyond our immediate imagination. One can imagine, for example, a versatile, autonomous pico- and nanosat swarm that, working together, exceeds the capabilities of existing satellites. One can imagine how such satellites can create new launch vehicles. One can think about opening up space to individuals because of the reduced cost. I applaud the editors for putting together material that challenges our imagination and opens up new ideas.

Foreword

The picosat is generally accepted as the smallest size satellite, and its inception is due to university programs. The nano- or cubesat standard was proposed to meet the picosat size of one kilogram. From the beginning of the cubesat in 1998, there seems to have been a transition in the small satellite world that continues to bring new surprises to space exploration. Initially, the cubesat was scoffed at by major government laboratories and aerospace businesses as a "toy"—too small to be of any practical use.

The cost of launching, coupled with reduced budgets for space and the current flood of commercial, unique, small low-power devices with exceeding capability, is making this toy a choice of means to do both science and military experiments previously not thought possible. We see the National Science Foundation actively funding cubesats for space weather science. The U.S. Army recently announced experiments with cubesats.

What will the future bring with pico- and nanosats? It is not the size, the low launch cost, or the low development costs that will bring the greatest return. These satellites are opening the door to students with a real opportunity to dream, to innovate, to take to space, to bring success to space experiment, and to make operational failure a learning success. Students can now experiment and be allowed to fail, but make giant leaps in technology.

Can we predict where this access to space will take us? That is the most exciting part; we don't know, but a young engineer named Steve Wozniak opened a new vista that seems to exceed new bounds at an alarmingly fast rate. Where will the pico/nanosat go?

Bob Twiggs
Professor of Astronautics
Space Science Center
Morehead State University
Morehead, Kentucky

Table of Contents

Preface

As Bob Twiggs expressed in the Foreword, pico- and nanosatellites have attracted a growing amount of attention over the past decade. This book arose from the need to have a single-volume reference that could not only describe to the research community what pico- and nanosatellites are capable of, but also convey to the academic community a valuable set of lessons learned for those wishing to start their own programs.

The book is organized into six parts, each consisting of one or more contributed chapters on related topics. Each chapter is a stand-alone unit that can be read independently of the others. In fact, some of the chapters are reprints of papers that were previously published in conference proceedings. This format allows the reader to investigate a topic of interest without having to refer to other portions of the book. The fact that each part contains chapters from different institutions also provides different perspectives on the same topic.

Part I: Overview of University Programs discusses the implementation of cubesat programs at two universities. Chapter 2 describes a formal design course at the University of Illinois. Chapter 3 overviews the University of Hawaii's Small-Satellite Program. Both chapters provide valuable lessons learned for those interested in starting a new university program.

Part II: Missions contains five chapters that describe satellite missions that either investigate some scientific phenomenon or validate some engineering mechanism. Chapter 4 provides an overview of atmospheric and other research using small university satellites. This is followed by a more detailed discussion of the mission of the Illinois Observing Nanosatellite (ION) in Chapter 5. Chapter 6 describes a cubesat mission to characterize ionospheric scintillations. Chapter 7 presents an interesting picosatellite architecture for collecting solar power. Chapter 8 describes an experiment that investigates degradation of tether structures in the low-Earth-orbit environment.

Part III: Attitude Determination and Control Systems covers the not-required but essential attitude determination and control systems (ADCS). Chapter 9 discusses models, algorithms, and hardware, in the context of ION ADCS system. The following two chapters present novel thruster technologies that can be used for both propulsion and attitude control. Chapter 10 overviews several microthruster propulsion techniques and presents a novel microthruster based on sublimating naphthalene. Chapter 11 describes a vacuum arc thruster in which high-velocity ions are ejected as a result of a vacuum arc.

Part IV: Electrical Power Systems details the electrical power system (EPS), which is responsible for the generation, storage, and distribution of power to the payload and other subsystems. Chapter 12 provides a collection of techniques and recommendations that address fundamental power system design considerations, based on experience with ION. Chapter 13 explains the concept of a power system architecture trade study that compares energy storage vs solar generation with energy storage, and describes unregulated vs regulated bus configurations.

Part V: Command and Data Handling and Telecommunication Systems includes one chapter on the command and data handling (C&DH) system and three chapters on the telecommunication system. Chapter 14 discusses lessons learned with ION software system, focusing on system scheduling, boot sequence, and communication protocol. Chapter 15 is a wide-ranging overview of cubesat radio communications, including aspects of propagation, modulation, link budgets, and antennas. Chapter 16 is about the construction of a mobile ground station. Chapter 17 describes a novel concept for self-steering satellite cross-links.

The appendix contains an end-to-end, comprehensive case study that integrates all of the elements of the preceding sections—mission, requirements, system and subsystem design, testing, and project management.

The idea for this book was originated by Gail Klein of the Jet Propulsion Laboratory and Jordi Puig-Suari of CalPoly San Luis Obispo, but editorial responsibilities were later turned over to the undersigned. Over the three years spent assembling this book, the institutional affiliations of many of the authors have changed—for example, many of the student authors are now employed in industry—but we decided to keep the affiliations that were in place at the time of original manuscript submission.

Developing this book was a long, yet rewarding process that involved numerous interactions with the contributing authors and editorial staff. We wish to acknowledge all of their assistance and contributions, without which this book would never have materialized. We also acknowledge Justin Akagi of the University of Hawaii, who reorganized the content of the book in its current form.

Purvesh Thakker
Wayne A. Shiroma
September 2010

I. Overview of University Programs

Chapter 1

Introduction

Purvesh Thakker* and Gary Swenson†
University of Illinois at Urbana-Champaign, Urbana, Illinois

PICO- and nanosatellite activity (<30 kg) has expanded greatly in the last 10 years thanks in large part to activity within the university satellite community. This book describes the current state of this technology and includes a number of detailed examples. Although much of the activity has been based around educational objectives, a number of programs have taken the extra step to apply this technology to real application problems. These application problems include obtaining atmospheric measurements and performing technology experiments for later use on larger satellites. In fact, the National Science Foundation has already recognized the potential of these satellites for atmospheric science and has initiated a program to fund their development [1]. Recently, industry has been getting involved as well. This book contains two industry publications from Aerospace Corporation.

Before delving into the details of individual satellite programs, this chapter first provides some background information about pico- and nanosatellites including recent trends, advantages, and launch information. In addition, this chapter provides some systems background information comparing these satellites to their larger cousins.

I. Pico- and Nanosatellite Background

A. Trends

A number of recent developments have enabled the rapid expansion of pico- and nanosatellites for the past nine years. These developments include coordination among pico- and nanosatellite programs [2], a significant increase in the number of pico- and nanosatellite programs, demonstrations showing that

*Program Manager, ION Cubesat, Department of Electrical and Computer Engineering.

†Professor, Department of Electrical and Computer Engineering.

pico- and nanosatellites can obtain valuable measurements [3–10], improvements in small-satellite technology, the advent of shared cluster launches [11], the introduction of low-cost "responsive" launch vehicles, changes in the way launch contracts are written, and a willingness to fly standardized cubesats (see Sec. I.D) as secondary payloads. As a result of these developments, 32 cubesats have already launched, and 18 have made it to space. In addition, a number of nanosats have launched.

B. Advantages

Pico- and nanosatellites offer a number of advantages over the traditional approach utilizing large government satellites. The most obvious benefit is lower development costs (see Sec. I.A) and launch costs (see Sec. I.D). In addition, many different satellites with different instruments can fulfill the need for more atmospheric measurements [1]. Constellations of many small satellites can also simultaneously obtain measurements around the globe. Unlike ground instruments, a single satellite can offer global coverage. Satellites provide another major benefit over ground instruments in that they provide a space perspective for remote sensing and can provide in situ space measurements that often cannot be made any other way. Disposable satellites would further allow more perilous orbits for shorter durations of time with more experimental missions. Finally, the atmospheric research community would benefit greatly if it had direct control over instrument selection. Although atmospheric researchers have been able to propose satellites for development through NASA, they have not had the resources to directly fund their own satellite program because of the size of these programs. However, the atmospheric sciences community does fund and develop many ground-based instruments including large phased-array antennas.

In addition to atmospheric missions, these satellites offer the opportunity to fly experimental technology and small test missions prior to flying larger, more expensive missions. Without such flight heritage, this technology may never be considered for larger systems.

C. Pico- and Nanosatellite Programs

Pico- and nanosatellites have been developed within a number of organized programs such as the Student Explorer Demonstration Initiative (STEDI) program, the University-Class Explorer (UNEX) program [12], the nanosat program [13], and the cubesat program [2]. In addition to these programs, a few individual universities have been developing custom satellites and arranging custom launches on their own for some time. Although the STEDI and UNEX [14] programs (~$10 million) have shown significant results, their measurements could likely be accomplished with today's nanosats (~$100,000 for parts). The nanosat and cubesat programs have engaged large numbers of universities, which each develop their own independent satellites. These universities have developed nanosats and cubesats for largely educational reasons with very limited support. In doing so, some universities have demonstrated the potential of these pico- and nanosatellites for addressing real application problems. Although this work is relevant to all pico- and nanosatellites, it refers primarily to the cubesat and nanosat programs. Most current pico- and nanosatellite activity resides in these programs, and their engineering and management guidelines are unique because funding levels are much lower. The STEDI/UNEX programs operate more according to government/industry funding levels and norms, and so they are placed in a category between university satellites and government/industry satellites.

The Air Force Research Laboratory (AFRL) began the nanosat (30 kg) program, which consists of a two-year competition between selected universities. Universities submit proposals to AFRL, which selects participants based on the strength of the proposals and their relevance to AFRL objectives. Universities receive approximately $100,000 in parts support and attend some organized training events. AFRL also hosts reviews at major milestones, such as at the critical design review, in order to assist the students and award points towards selection of a winner. The winner is awarded a launch as a secondary piggyback payload using a deployer called the Lightband.

Although many universities can participate in the nanosat program, only a small number of participants receive a launch (approximately 8%). This is in contrast to the cubesat (1 to 3 kg) program, where any organization can launch as part of a cubesat cluster as long as it can pay the $40,000 piggyback launch cost for a 1-kg, 10-cm cube. So far, the cubesat program has facilitated the launch of 32 satellites with 18 currently in space. Moreover, developers have engineered and launched these satellites without any formal funding program. The workforce has been primarily the university communities. All organizations have been raising their own development and launch funds through internal programs, grants, or donations. The key document in the cubesat program is the CubeSat Spec, where the interface between cubesats and their deployers is defined (data available online at http://cubesat.atl.calpoly.edu/pages/documents/developers.php). It was introduced in 1999 by Stanford and Cal-Poly Universities and allows smooth, efficient integration of many independently developed pico- and nanosatellites. The Spec places an emphasis on protecting the launch vehicle, and 80 organizations have registered intentions to develop satellites conforming to it. The most popular deployer is the PPOD developed by Cal-Poly Pico/Nano [11], which can hold up to three cubes. A new six-pack version of the PPOD will be able to support up to six cubes, optionally as part of a single satellite.

Although cubesats are significantly smaller than nanosats, a few cubesat developers have demonstrated that these satellites can offer real utility in addition to their effective cluster deployment techniques. As a result, the cubesat program offers the potential to affordably launch large numbers of pico- and nanosatellites for making atmospheric measurements. As atmospheric scientists require larger spacecraft, developers can apply the same cluster launch and other concepts from the cubesat community to nanosats and new intermediate class satellites between cubesats and nanosats. In recognition of this potential, the National Science Foundation's Atmospheric Sciences group has recently begun a program for funding cubesats and nanosats [1]. Even when fully funded, these satellites may cost under $1 million.

D. Launch Background

A number of barriers have traditionally made it difficult for pico- and nanosatellites to access space, as listed here:

- High launch-vehicle costs
- Difficulty adding secondary payloads
- Integration costs
- Integration hassles
- Disparate schedules
- Complex customer interface
- Launch-vehicle safety

Universities have addressed immediate launch needs by flying as secondary payloads and efficiently sharing launches, such as with cluster launches. For example, one launch included fourteen independent cubesats [2] and three additional pico- and nanosatellites as secondary payloads on a single Russian DNEPR launch vehicle.

The barriers include a high minimum cost to launch satellites that would dominate smaller program budgets if they flew as primary payloads. The recently introduced SpaceX Falcon-1 launch vehicle brings the minimum launch cost in the United States down to $7.9 million [15]. Entrepreneurial activity and DARPA's "Responsive Space" program should continue to provide innovations. Note that satellites developed using U.S. taxpayer dollars must use a U.S. launch provider because the U.S. government views access to space as a strategic resource. Second, many barriers have made it difficult for satellites to launch as secondary payloads. Launch contracts have traditionally given a single customer rights to an entire launch vehicle. Neither the customer nor the launch provider can modify the contract unilaterally to add secondary payloads later. However, in some recent contracts, such as those for the Falcon-1 and the DNEPR, the launch provider supplies a ride to space rather than a launch vehicle. The customer providing the primary payload still controls the launch parameters, but the launch provider can add secondary payloads in unutilized space.

Even with these new contracts, small satellite developers must overcome significant barriers to launch as secondary payloads including integration costs, integration hassles, and disparate schedules. In addition, launch providers want to interface with few customers, and so very small satellite developers must cooperate through an intermediary. The cubesat community overcomes these barriers with a standardized deployer interface defined in the cubesat specification. Since its introduction in 1999, over 80 organizations have registered intentions to develop cubesats according to this specification. With this simple specification, each organization can independently develop and launch satellites without having to develop a deployer or arrange for a launch. An intermediary launch integrator, such as Cal-Poly, collects the satellites from the independent universities and installs them into deployers, such as the PPOD. Cal-Poly then integrates a number of PPODs with three cubesats each into unutilized space on the launch vehicle, perhaps in the same space with each subsequent launch. Because the launch provider must only interface with Cal-Poly and its PPODs, it experiences few integration hassles for all secondary payloads. With all satellites conforming to the same spec, it also becomes possible to swap satellites if a particular satellite cannot make the flight.

In addition to providing a simple interface between cubesats and their deployers, the requirements in the CubeSat Spec protect the launch vehicle and primary payloads. For example, the satellite must wait in the PPOD for deployment without power. Also, the deployer completely encloses the satellite, and the satellite must go through vibration and bake-out procedures prior to launch. With these requirements, cubesats present few flight risk concerns to the launch provider and primary payload even with little information provided about the design of the satellites themselves. As a result, the launch provider can deploy large numbers of cubesats in cluster launches such as with the DNEPR launch containing fourteen cubesats. Thanks to this growing secondary payload launch infrastructure, universities and other organizations have been able to fly their satellites. As launch costs continue to fall and satellite numbers continue to increase, it might not be long before universities can launch clusters as primary payloads.

II. Systems Background

A. Components

Because of their limited space, mass, power, and budgets, pico- and nanosatellites differ from larger satellites in many ways. The primary difference is that pico- and nanosatellites utilize commercial components instead of space grade components. These commercial components are more readily available, cheaper, newer, smaller, better performing, better documented, and easier to use. Because they are not designed to operate in a radiation environment, pico- and nano satellites are relegated to short-term (<1 yr) low-Earth-orbit missions where radiation levels remain low. Next, because larger satellites cost more and are expected to last much longer, their components must be engineered with end-of-life specifications and with knowledge of permanent failure probabilities. Pico- and nanosatellites have a short lifetime requirement, so that they can be engineered more with beginning-of-life specifications and the permanent failure probabilities of most components can be ignored. For example, solar cells on large satellites degrade over years, and so the power generation of larger satellites decreases at the end of their life. Components also fail permanently as a result of accumulated radiation, and so failure probabilities need to estimated, and backup systems need to be implemented to achieve specified reliability goals.

B. Power

Satellite power systems utilize two basic architectures, direct energy transfer (DET) and power point tracking (PPT) [16]. DET designs do not place any series component in between the solar cells and the system, whereas PPT designs always operate the solar panels at the point where they transfer maximum power to the load. DET designs include the fully regulated bus, where the bus voltage variation is 2 to 5% of the nominal voltage, and the sun-regulated bus, where the bus is only regulated when the satellite is in sunlight. The design for the sun-regulated bus is less complex than the fully regulated bus. PPT designs typically place a switching regulator in series with the solar panels and the load. PPT designs are advantageous when the panels do not actively track the sun, as with all pico- and nanosatellites to date. They are also advantageous when solar radiation and array temperature vary over a wide range, thereby varying the array voltage. Pico- and nanosatellites sometimes utilize a very lean DET architecture in which the solar array, battery, and loads are permanently connected in parallel. However, it can be challenging to match solar array and battery voltages for such small satellites.

For power sources, spacecraft have four options: static, dynamic, fuel cell, and photovoltaic [17]. Static power sources generate power from temperature gradients caused by a radioactive source. Dynamic power sources use a heat source and a heat exchanger to drive an engine. The heat source can be concentrated solar energy, radioisotopes, or a controlled nuclear reaction. Fuel cells convert the chemical reaction of an oxidation reaction to electricity. These are self-contained generators and are used on the space shuttle. The longer the mission is, the larger the reactant tanks required. Photovoltaic sources are the best choice for pico- and nanosatellites because of the small mass required and simplicity of implementation. However, the amount of power available is very limited because the surface area on which cells can be placed is very limited. Panels can be deployed to a certain extent, but at some point the panel mass and volume become too great. As a result, pico- and nanosatellites face strict power limitations as compared to larger satellites. These power restrictions limit instrument operation times and communication capabilities. The panels can be made of lower-cost silicon solar cells, or higher-efficiency gallium-arsenide solar cells. Another major difference is that larger spacecraft often use tracking and pointing mechanisms to ensure that solar cells are properly pointed towards the sun, whereas pico- and nanosatellite programs have cells that are fixed on the sides of the satellite or to fixed deployed panels.

For batteries, larger spacecraft generally use NiCd or NiH_2 batteries, which have more flight and testing heritage. Pico- and nanosatellites, however, often utilize Li-Ion batteries [6,7,18,19], which are found in portable devices such as cell phones. Li-ion batteries offer a 65% volume and 50% mass advantage. They require safety circuits and a special charging profile. The batteries go through some qualification tests, such as thermal vacuum tests, to give them the best chance of success.

Another difference is in the spacecraft bus. Larger spacecraft generally distribute power throughout the satellite using a 28 Vdc bus. The required voltages are converted off of this bus including both ac and dc regulated voltages. Pico- and nanosatellites, however, use much newer and lower power components including 3V components. As a result, the bus typically runs at lower voltages and uses only dc regulators. Larger satellites also often use a more distributed distribution system, where the required voltages are converted at each load separately. Because there are fewer components, pico- and nanosatellites typically use a more centralized distribution system where the bus directly supports all loads with regulated voltages.

C. Attitude Determination and Control

Pico- and nanosatellites sometimes do not have any attitude control capability. Cubesats, in particular, have just begun to offer attitude control capability [20, 21]. However, for many remote-sensing missions, the instrument must be pointed in the direction of its target. Once pointed, the satellite is subject to disturbance torques in orbit and sometimes needs to perform slew maneuvers [17]. Larger satellites might require the attitude control system to maintain attitude during orbit insertion, where lateral movement is required. Because the larger motors required for lateral movement can create larger disturbances, this can increase the requirements for the attitude control system actuators. Pico- and nanosatellites rely on the launch vehicle to perform orbit insertion.

The satellites can use passive or active techniques to control attitude [17]. Passive techniques include magnets, gravity gradient, and spin stabilization. Magnets can be used for near equatorial orbits where the Earth's magnetic field direction does not change. Gravity gradient is a simple technique that uses an extended boom. Elongated objects tend to align their long axis through the Earth's center. Spinning spacecraft have gyroscopic stiffness that passively resist torque on two axes. Permanent magnets and gravity gradient provide limited five-degree accuracy, while spin stabilization can provide one-degree accuracy or higher [17].

Active techniques include thrusters, magnetic torquers, and reaction wheels. Thrusters are not currently used on pico- and nanosatellites because of the extra complexity required and the need for fuel. Fuel storage adds to satellite complexity and adds flight risk to the launch vehicle and primary payload. Pico- and nanosatellites have implemented attitude control systems with magnetic torquers and have begun to do so with cubesats. Reaction wheels are too large for cubesats and require too much power, but they can be used on nanosats. Reaction wheels store momentum internally and require an external control torque from thrusters or magnetic torquers to dump the momentum.

The satellite must be oriented to external references such as the sun, the Earth's IR horizon, the local magnetic field direction, and the stars [17]. Gyroscopes can also provide a short-term reference between external updates. Pico- and nanosatellites can easily implement a three-axis magnetic sensor, so that this is the sensor of choice. If magnetic sensors are used alone, then a Kalman filter is required. Sun sensors make a good complement to magnetic sensors. A camera is required to use stars as an external reference.

D. Thermal

The space environment can cause equipment to get as hot as 100°C and as cold as −130°C with the changes occurring in 10's of seconds or minutes [17]. Convection with ambient air does not exist in space as it does on the ground. Sunlight is the major source of environmental heating. Albedo is sunlight reflected off of a planet or moon and provides another heat source. All incident sunlight not reflected as albedo is absorbed by the Earth and eventually reemitted as Earth IR energy, or blackbody radiation. Larger satellites will often employ passive and active thermal systems to achieve desired temperatures. Passive systems include the use of materials, coatings, blankets, or second surface mirrors. Active systems make use of heaters or thermoelectric coolers. Pico- and nanosatellites generally utilize passive techniques [5,21,22].

E. Communications

The communications architecture for larger satellites can vary depending on the orbit. In geosynchronous orbit, the satellite requires directional antennas, and the ground station antenna does not need to actively track the location of the satellite because it stays stationary. Satellites in low Earth orbit pass over the ground station for a total of about thirty minutes each day. As a result, the ground antenna must actively track the satellite. The communication distance is also much shorter for low-Earth-orbit satellites (~650 km vs 36,000 km). Pico- and nanosatellites are relegated to low Earth orbit because they cannot tolerate the radiation levels of geosynchronous orbit. Pico- and nanosatellites also typically use broad antenna patterns, so that the satellite antenna does not need to be pointed towards the moving ground station. Finally, pico- and nanosatellites generally make use of commercially available ham radio equipment and the amateur bands [6,7,19, 22–25], whereas larger satellites generally use custom space equipment. Ham radio equipment is available at 1200 baud and 9600 baud data rates.

III. Management and Design Constraints

To the seasoned spacecraft engineer in government or industry, the strategies employed by university satellite developers might seem to overstep the bounds of conventional wisdom. However, student-based organizations must make choices given the environment at hand (Sec. III.A). Certain guidelines can ease the task for the inexperienced student designer by allowing universities to develop satellites with limited expertise and resources. These guidelines substantially reduce the complexity of the development task by taking reasonable design risks that the extremely risk-averse space industry normally would not take. In addition to reducing complexity, this approach also allows the satellite to be developed more quickly and more cheaply with widely available commercial technology. For example, small teams of less than ten students can develop satellites with this approach in less than two years.

A. University Environment Constraints

Universities provide a unique environment for developing satellites, which requires a unique approach. Students, particularly at the undergraduate level, have limited capabilities. These include technical skills as well as basic communication, teamwork, and leadership skills, which are developed as students participate on the project. In addition, the students working on a particular element of the project often change and must always juggle responsibilities with coursework. University programs also face financial, equipment, and advising limitations. Furthermore, universities often push the envelope of technology and do not generally test systems as thoroughly as industry. However, university satellite developers must thoroughly test their systems because the developers cannot recover the systems after deployment. In addition, developers must often choose the simplest, most effective solution for engineering decisions outside of their large research goals. The system as a whole is very complex, so that each component cannot individually be a complex research project. In addition to placing a great deal of emphasis on functional and operational testing, developers must account for the special environment that satellites operate in. This environment includes radiation, thermal, and magnetic effects. To ensure that the satellites can operate reliably in this environment, developers put the satellites through thermal-vacuum, vibration, and other environmental testing procedures.

Universities also possess advantages. Unlike the extremely risk-averse aerospace industry, students are very willing to try new things. University students are not guided as much by conventional wisdom. Students enjoy learning about new technologies and have less of a fear of failure because they learn a great deal either way. Faculty advisors also know that the projects always succeed in their educational mission [14] and are willing to take some reasonable risks with more long-term research missions than industry can tolerate (data also available online at http://cubesat.atl.calpoly.edu/pages/documents/developers.php). Universities are well known for their ability to perform cutting-edge research for these and other reasons. In addition, the low cost of these university programs makes them much more risk tolerant than their larger counterparts allowing for more experimental missions. This risk tolerance allows universities to take on more innovative challenges. With additional innovations, such as with thrusters, atmospheric researchers could make multipoint or multi-angle measurements. With improved small satellite thruster technology, disposable satellites could also dive into lower Earth orbits to take measurements. University satellites can also serve as an important preliminary step in the realization of larger spacecraft missions by testing new technologies, such as thrusters [15], or by performing preliminary measurements to verify anticipated signal and noise levels [9].

B. Goal and Mission Constraints

The first step to keeping the project manageable is to limit the number of missions. Cubesats should normally focus on one primary mission, whereas larger satellites can accommodate more. Although it sounds easy to limit the number of missions, it is a lesson that is easily ignored in the excitement of proposing a new satellite. The project is always much more complex than anyone realizes at the beginning, and later project participants bear the burden of implementation. For example, the ION project with its five missions was delivered two years late after a tremendous effort. The project barely fit within space, mass, and power limitations after much finagling. Fortunately, the launch date was also postponed by two and a half years.

The next important step to keeping the project manageable is to select missions in low Earth orbit (LEO). LEO satellites still receive a great deal of radiation protection from the Earth's atmosphere (Fig. 1.1). By sticking to LEO, the satellite can utilize commercial components that are not designed to operate in a radiation

Fig. 1.1 TID vs shielding for ION.

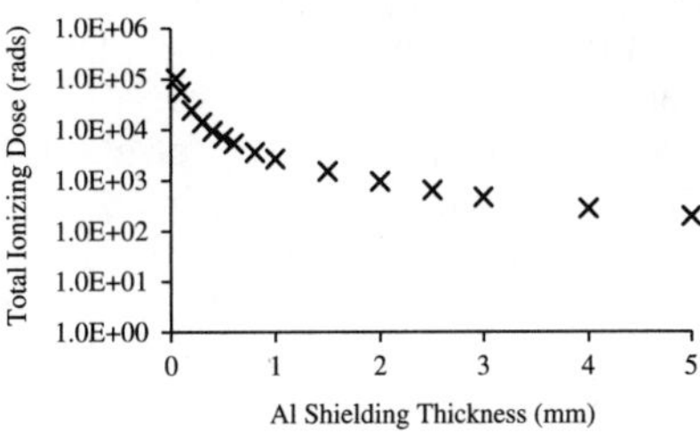

environment. The satellite also can avoid the use of complex fault-tolerance techniques opting instead for only the simplest, most effective techniques.

The final important step to keeping the project manageable is to define a short lifetime requirement, such as six months, to keep the total ionizing dose (TID), or total radiation dosage, low. This is important because the components degrade with radiation exposure. With ION's shielding level and orbit, it receives 2.7 krads of radiation dosage after six months. Because commercial complementary metal-oxide-semiconductor (CMOS) components begin to fail at around 10 krads, the project can largely ignore permanent failure of components caused by radiation. Note that commercial CMOS components can sometimes tolerate megarads of radiation before failing, and so it is possible that the satellite will last much longer than the required lifetime. Also note that the system design must still account for single event upsets, which are temporary bit flip errors caused by radiation.

C. Risk Constraints

NASA defines four risk classification levels for its payloads [16]. Class A payloads generally have a lifetime of greater than five years. They are high-priority payloads with no reflight opportunities. All practical methods are taken to minimize risk to mission success, and the highest assurance standards are used. Critical single-point failures are not permitted without a formal waiver. The heritage of previously used materials is verified, and all new or changed materials are qualified. Testing includes full formal qualification and acceptance testing and integrated end-to-end testing at all hardware and software levels. Class B payloads have a medium lifetime of two to five years. They are high-priority payloads with few or no reflight opportunities. Stringent assurance standards are used with only minor compromises to maintain a low risk to mission success. Critical single-point failures might be permitted, but are minimized and mitigated by use of high-reliability parts and additional testing. Previously tested/flown materials are used, and new materials are qualified. Testing is the same as Class A for hardware and might use a combination of qualification and protoflight hardware. Qualified software simulators can be used to verify software and system. Class C payloads have a short lifetime of less than two years. They are medium-priority payloads with some or few reflight opportunities. Critical single-path failures may be permitted, but are mitigated by use of high-reliability parts and additional testing. Previously tested/flown materials are used, and new materials are qualified. Testing includes limited qualification testing for new aspects of the design plus full acceptance testing. Testing is also required for verification of safety compliance and interface compatibility. Class D payloads also have a short lifetime of less than two years. They are low-priority payloads with significant reflight opportunities. Critical single-path failures may be permitted, but are mitigated by use of high-reliability parts and additional testing. Materials requirements are based on applicable safety standards and are assessed for life limits. Testing is required only for verification of safety compliance and interface compatibility. The program is acceptance tested for critical performance parameters.

Note that even NASA's Class D programs can cost two to three orders of magnitude more than pico/nano satellites. As a result, pico- and nanosatellites can tolerate much higher risk levels. Pico- and nanosatellites have short lifetime requirements of less than one year. They are low-priority piggyback payloads that can easily be rebuilt and flown piggyback on a subsequent flight. It is not unusual for there to be many critical single-path failures, and mass market commercially available parts are utilized. Universities generally avoid projects with complex safety standards that might place the primary payload at risk. Testing is required for interface compatibility, and the program is acceptance tested for critical performance parameters.

The satellite must be designed with a certain level of fault tolerance to minimize risk because the satellite cannot be recovered for repairs after deployment and it is subject to single event upsets. However, it is very easy to get carried away with fault tolerance. ION did not employ much in the way of redundancy because space could just accommodate a single unit of each component. The ION project did include many of the following possible techniques. Power techniques such as watchdog timers, ground-controlled reset capability, resettable fuses, large power margin, and a sleep mode can be effectively implemented. Radiation techniques such as aluminum shielding and tantalum tape on components can be effectively implemented. A variety of software resilience techniques including a NOP sled, termination of misbehaving applications, a special reset mode, periodic satellite resets, and activity logging can be implemented.

D. Configuration Constraints

The configurations available for pico- and nanosatellites are largely dictated by the launch opportunities available. Pico- and nanosatellites launch as secondary payloads because the cost of flying as a primary is too high. As a result, the orbits available to pico- and nanosatellites are limited to the flight opportunities available near the completion date of the satellite. CubeSats are the most readily available configuration because anyone can purchase launch space. However, space and mass are fixed to specific dimensions and are very limited. Other configurations are possible, many of which utilize the lightband deployer used with nanosats. However, these often require custom launch arrangements as a piggyback payload. In addition, the launch often must be donated because the cost of these large configurations often exceeds pico- and nanosatellite budgets.

The small size also limits the surface area of any lenses used in remote sensing, which in turn limits the light gathering power and therefore the sensitivity of the instrument. Similarly, cubesats face strict power limitations. Even the largest cubesats using state-of-the-art solar cells and extended solar panels cannot support more than a few watts of power. It is challenging to develop an efficient power system that operates a satellite at just 1W. For communications, it is challenging to communicate over distances of 3000 km with the limited power available. It is also challenging to achieve high data rates that would allow cubesats to send large amounts of data back down. Cubesats typically use commercially available hardware from the ham radio community. Faster rates might be possible, but currently data rates generally do not exceed 9600 baud resulting in a maximum download per day of about 1MB, for a single ground station. It is also challenging to achieve attitude stabilization onboard cubesats because of their limited space for attitude hardware and limited processing capability for attitude software. ION has a pointing accuracy goal of five degrees.

E. Team Constraints

The major challenge with teams on university satellites is dealing with the inexperience of the students and the constant turnover. Documentation and training materials make it easier for new students to get up to speed quickly. Turnover is a reality that university programs must face. Turnover can be reduced by offering incentives such as credit, leadership titles, assistantships, thesis work, hourly positions, and opportunities to learn and explore. Most importantly, turnover can be reduced with quality leadership. Create a fun work environment with solid team unity behind exciting, but realistic goals. Encourage student ownership of the work while facilitating and assisting. In addition, the students can be given overlapping work assignments so that new students learn from veteran ones who are likely leaving soon. A shared network drive on all computers allows students to archive all of their work in one place no matter which computer they are working from. This makes transitions much easier because all material is in one place.

F. External Communication Constraints

University satellite programs communicate with external programs much more openly than in industry, where processes and methods are often kept proprietary. Cubesat workshops are held twice a year, once as an independent event and once in conjunction with the Small Satellite Conference. These workshops present an excellent opportunity for new programs or new students in existing programs to interact with more experienced members of the community. The students are eager to pass on and share whatever they learn. This happens through formal publications, online documentation, and a willingness to talk about designs at community meetings. This open communication accelerates the learning for all participants allowing all to develop an understanding of what works and what doesn't based on the entire community's experience.

G. Cost Constraints

Because the budgets for university satellites are much smaller, they must operate in a manner different. First, they cannot afford to fly as primary payloads, and so they must either fly as cubesats or make custom arrangements. Next, the programs must purchase low-cost commercial components that are available in mass quantities. Fortunately, these components are also newer and better documented. In the case of communications systems, university satellites often purchase the entire radio as a unit because suitable equipment is available from the ham radio community. Next, universities often do not have clean rooms for storing the satellite parts. Fortunately, clean room requirements are payload dependent, and most university satellite payloads do not require clean rooms. The satellite optics on ION was cleaned immediately before integration and delivery, and the satellite was stored in a clean room at Cal-Poly while awaiting launch. Another consequence of small budgets is that they make it more difficult to make backups. These backups allow for mistakes and reduce stress particularly during integration when a long series of steps must occur in a specific order. The ION project did not have backups of many components. This saved some time, but the result was that a second copy of the satellite could not be built after the first satellite was lost.

References

[1] "CubeSat-Based Science Missions for Space Weather and Atmospheric Research," *Program Solicitation*, D. f. G. D. o. A. S. National Science Foundation, Arlington, VA, 2007.

[2] Lan, W., Brown, J., Toorian, A., Coelho, R., Brooks, L., Suari, J. P., and Twiggs, R., "CubeSat Development in Education and into Industry," *Proceedings of AIAA Space*, AIAA, Reston, VA, 2006, also AIAA Paper 2006-7296.

[3] Flagg, S., Bleier, T., Dunson, C., Doering, J., DeMartini, L., Clarke, P., Franklin, L., Seelback, J., Flagg, J., Klenk, M., Safradin, V., Cutler, J., Lorenz, A., and Tapio, E., "Using Nanosats as a Proof of Concept for Space Science Missions: QuakeSat as an Operational Example," *18th Annual AIAA/USU Conference on Small Satellites*, Paper SSC04-IX-4, AIAA, Reston, VA, 2004.

[4] Rysanek, F., and Hartmann, J. W., "MicroVacuum Arc Thruster Design for a CubeSat Class Satellite," *16th Annual AIAA/USU Conference on Small Satellites*, Paper SSC02-I-2, Logan, Utah, Aug. 2002.

[5] Voss, D. L., Kirchoff, A., Hagerman, D. P., Zapf, J. J., Hibbs, J., Dailey, J., White, A., Voss, H. D., Maple, M., and Kamalabadi, F., "TEST—A Modular Scientific Nanosatellite," *Space 2004 Conference and Exhibit*, AIAA Paper 2004-6121, Sept. 2004.

[6] Holmes, W. C., Bryson, J., Gerig, B., Oehrig, J., Rodriguez, J., Schea, J., Schutt, N., Voss, D., Voss, J., Whittington, D., Bennett, A., Fennig, C., Brandle, S., Dailey, J., and Voss, H. D., "TU Sat 1—A Novel Communications and Scientific Satellite," *16th Annual AIAA/USU Conference on Small Satellites*, Paper SSC02-I-1, Aug. 2002.

[7] Obland, M., Hunyadi, G., Jepsen, S., Larsen, B., Klumpar, D. M., Kankelborg, C., and Hiscock, W. A., "The Montana State University NASA Space Grant Explorer-1 Science Reflight Commemorative Mission," *15th Annual AIAA/USU Conference on Small Satellites*, Paper SSC01-III-2, 2001.

[8] "The Montana Nanosatellite for Science, Engineering, and Technology for the AFRL/NASA University Nanosat Program," *17th Annual AIAA/USU Conference on Small Satellites*, Logan, UT, Paper SSC03-IX-4, Aug. 2003.

[9] Waydo, S., Henry, D., and Campbell, M., "CubeSat Design for LEO-Based Earth Science Missions," *IEEE Aerospace Conference Proceedings*, Vol. 1, IEEE, Piscataway, NJ, 2002, pp. 1-435–1-445.

[10] Gregorio, A., Bernardi, T., Carrato, S., Kostadinov, I., Messerotti, M., and Stalio, R., "AtmoCube: Observation of the Earth Atmosphere from the Space to Study 'Space Weather' Effects," *Proceedings of Recent Advances in Space Technologies*, Aeronautics and Space Technologies Inst., Istanbul, 2003, pp. 188–193.

[11] Nason, I., Puig-Suari, J., and Twiggs, R., "Development of a Family of Picosatellite Deployers Based on the CubeSat Standard," *IEEE Aerospace Conference Proceedings*, Vol. 1, 2002, IEEE, Piscataway, NJ, pp. 1-457–1-464.

[12] "Announcement of Opportunity Explorer Program: Small Explorers and Missions of Opportunity," N. A. a. S. Administration, Ed., 2007.

[13] Hunyadi, G., Ganley, J., Peffer, A., and Kumashiro, M., "The University Nanosat Program: An Adaptable, Responsive and Realistic Capability Demonstration Vehicle,"

IEEE Aerospace Conference Proceedings, Vol. 5, IEEE, Piscataway, NJ, 2004, p. 2858.

[14] Marchant, W., Hurwitz, M., Sholl, M., and Taylor, E. R., "Status of CHIPS: A NASA University Explorer Astronomy Mission," *IEEE Aerospace Conference Proceedings, 2002*, Vol. 7, IEEE, Piscataway, NJ, 2002, pp. 7-3527–7-3533.

[15] "Falcon 1 Overview," SpaceX, 2008, http://www.Spacex.com/falcon1.php.

[16] Patel, M. R., *Spacecraft Power Systems*, CRC Press, Boca Raton, 2005.

[17] Larson, W. J., and Wertz, J. R., *Space Mission Analysis and Design*, 3rd ed., Microcosm Press, El Segundo, CA, 1999.

[18] Obland, M., Klumpar, D. M., Kirn, S., Hunyadi, G., Jepsen, S., and Larsen, B., "Power Subsystem Design for the Montana EaRth Orbiting Pico-Explorer (MEROPE) CubeSat-Class Satellite," *IEEE Aerospace Conference Proceedings*, Vol. 1, IEEE, Piscataway, NJ, 2002, pp. 1-465–1-472.

[19] Young-Keun, C., Byoung-Young, M., Ki-Lyong, H., Soo-Jung, K., and Suk-Jin, K., "Development of the HAUSAT-2 Nanosatellite for Low-Cost Technology Demonstration," *Recent Advances in Space Technologies, 2005. RAST 2005*, Aeronautics and Space Technologies Inst., Istanbul, 2005, pp. 173–179.

[20] Scholz, A., "COMPASS-1 Hands-on Experience for the Space Engineers of Tomorrow," *Proceedings of Recent Advances in Space Technologies*, Aeronautics and Space Technologies Inst., Istanbul, 2003, pp. 55, 56.

[21] Young-Keun, C., Je-Hong, P., Young-Hyun, K., Byoung-Young, M., and Myung-Il, M., "Design and Development of HAUSAT-1 Picosatellite System (CubeSat)," *Proceedings of Recent Advances in Space Technologies*, Aeronautics and Space Technologies Inst., Istanbul, 2003, pp. 47–54.

[22] Ubbels, W. J., Bonnema, A. R., van Breukelen, E. D., Doorn, J. H., van den Eikhoff, R., Van der Linden, E., Aalbers, G. T., Rotteveel, J., Hamann, R. J., and Verhoeven, C. J. M., "Delfi-C3: A Student Nanosatellite as a Test-Bed for Thin Film Solar Cells and Wireless Onboard Communication," *Proceedings of 2nd International Conference on Recent Advances in Space Technologies*, Aeronautics and Space Technologies Inst., Istanbul, 2005, pp. 167–172.

[23] Hunyadi, G., Klumpar, D. M., Jepsen, S., Larsen, B., and Obland, M., "A Commercial off the Shelf (COTS) Packet Communications Subsystem for the Montana EaRth-Orbiting Pico-Explorer (MEROPE) CubeSat," *IEEE Aerospace Conference Proceedings*, Vol. 1, IEEE, Piscataway, NJ, 2002, pp. 1-473–1-478.

[24] Twiggs, B., and Kuroki, S., "BioExplorer Bus—Low Cost Approach [Satellite Design]," *IEEE Aerospace Conference Proceedings*, Vol. 1, IEEE, Piscataway, NJ, 2002, pp. 1-427–1-434.

[25] "Tokyo Tech CubeSat—CUTE-I—Design & Development of Flight Model and Future Plan," AIAA Paper 2003-2388, 2003.

Chapter 2

Management and Implementation of a Cubesat Interdisciplinary Senior Design Course

Purvesh Thakker* and Gary Swenson†

University of Illinois at Urbana-Champaign, Urbana, Illinois

I. Introduction

DESIGNING and operating space systems requires close interaction between various fields of engineering and science. The interdisciplinary design aspects contained within the Illinois Observing Nanosatellite (ION) project [1,2] serve as an incubator for developing system engineers with hands-on expertise. Students develop real satellites that launch into space as part of the CubeSat Program [3]. Figure 2.1 shows ION, a double cubesat, with its Russian launch vehicle.

A cubesat is a cube-shaped satellite measuring 10 cm on a side. The CubeSat Spec (data available online at http://cubesat.atl.calpoly.edu/pages/documents/developers.php), developed by Stanford and Cal-Poly Universities, defines a simple interface between cubesats and their deployers. These deployers are integrated into unutilized space on existing launches as secondary payloads. Each deployer can hold three cubes including three individual satellites or combined double or triple satellites. The satellites can be independently developed at different universities. As a result, universities are able to launch real student spacecraft for a reasonable cost. The spacecraft are often developed as educational exercises, but they can also perform real missions as demonstrated by the ION program and others. These missions include performing atmospheric measurements [4–7] and testing new space technology [1,8]. Satellites that perform real missions often provide a better educational experience because they are more motivating for all to participate in.

As Illinois' first cubesat, ION launched in July of 2006. Unfortunately, the launch vehicle crashed back to Earth with its payload of 18 satellites including ION. Nevertheless, the project trained over 100 engineers from seven disciplines.

---

*Program Manager, ION Cubesat, Department of Electrical and Computer Engineering.

†Professor, Department of Electrical and Computer Engineering.

Fig. 2.1 Completed ION (double cubesat) with launch vehicle.

More than a dozen unique papers on different elements of the system are being published. The project also laid the foundations for future spacecraft and high-altitude balloon developments at Illinois. ION's missions include measuring oxygen airglow emissions from the Earth's mesosphere using a photometer, testing new MicroVacuum Arc Thrusters (μVAT), testing a new processor board for small satellites, performing Earth imaging with a CMOS camera, and demonstrating ground-based attitude stabilization.

A. Course Overview

The course objectives include training students to identify, formulate, and solve engineering problems as part of a large multiteam project. The interdisciplinary design class that houses ION (ENG 491) is open to senior-level students from all departments within the College of Engineering, which has an enrollment of approximately 6000 undergraduates across 13 engineering disciplines. The cubesat section primarily attracts students from electrical and computer engineering and aerospace engineering (80% combined), with other students coming from computer science, mechanical engineering, general engineering, and theoretical and applied mechanics. In addition to undergraduates, some graduate students also participate through teaching assistantships, independent studies, and thesis work. The class size is typically 15–20 students.

Before students join cubesat, most of their coursework involves individual work in a classroom setting. The large team activities in cubesat are much closer to the professional work environment that students will experience in industry after graduation. The educational experience includes project specification, design, simulation, fabrication, assembly, functional testing, and environmental testing. In addition, students develop communication skills through informal and formal, written, and oral communication activities. Students also develop leadership skills as they move into team lead positions and train incoming students. The practical system engineering, communication, and leadership training obtained is highly

Fig. 2.2 Programming the satellite.

valued by academia and industry alike, as is illustrated by the heavy recruitment of participants by graduate schools and industry. Figure 2.2 shows members of the software team programming ION, which can be seen standing on the table.

Student-developed satellites present many challenges beyond technical issues as a result of limited expertise, limited equipment, limited budgets, and the students' short-term involvement in the project. As a two-semester course, half of the students change every semester, and all must balance participation with other classes. In addition, Illinois faced many startup challenges with ION. Many cubesat programs have mentors with space systems' experience that help programs get started. For example, G. Swenson's experiences at NASA and at Lockheed Palo Alto provided some invaluable space systems' experience. The university environment provides an advantage in some respects as well. As a student project with limited financial investment, the organization can tolerate a high degree of risk allowing it to do things that a more expensive project cannot. Even if the satellite fails, students earn invaluable real hands-on engineering experience. Figure 2.3 gives some feel for the complexity of the project. It provides an overview of many of the software components on ION. A single processor includes system software for managing nine applications employing many device drivers.

B. Cubesat Community

Over 80 organizations have registered intentions to develop cubesats since the program began in 1999, and 32 have launched with 18 currently in space. Most of these organizations are universities, which are spread out all over the globe. Cubesat workshops are held twice a year, once as an independent event, and once in conjunction with the Small Satellite Conference. These workshops present an excellent opportunity for new programs or new students in existing programs to interact with more experienced members of the community.

In the cubesat community, programs communicate very openly, unlike in industry where processes and methods are often kept proprietary for good reasons. The students are eager to pass on and share whatever they learn. This happens through formal publications, online documentation, and a willingness to talk about designs at community meetings. This open communication accelerates the learning for all participants allowing all to develop an understanding of what works and what does not based on the entire community's experience.

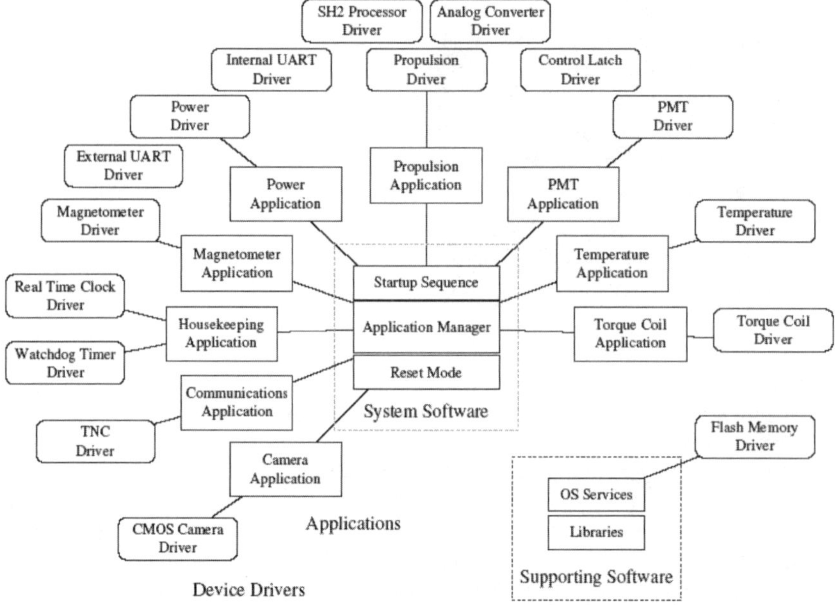

Fig. 2.3 Software overview with most satellite components.

II. Organization

A. Organizational Structure

The organization typically consists of about six teams with three to five students each (Fig. 2.4). One or two graduate students serve as program managers (PM) in conjunction with three faculty advisors. These people together are the course instructors who are responsible for administering the class and evaluating student work.

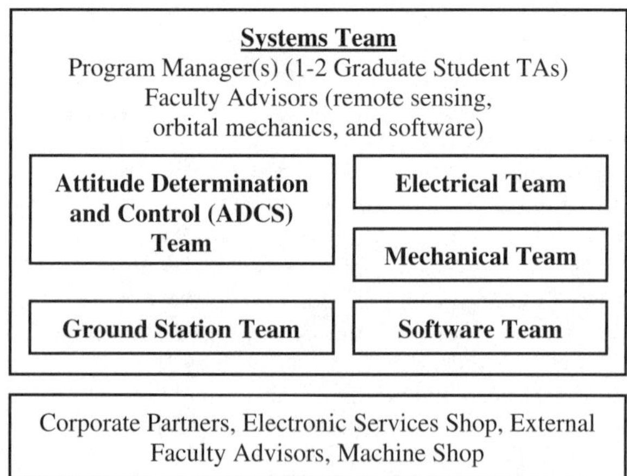

Fig. 2.4 Typical team structure.

Numerous other faculty provide technical advice in their specialty areas as required. The PM carries most of the responsibility for administering the course and managing the project. The PM sets up the teams at the beginning of the semester, runs systems meetings and reviews, and coordinates logistical details such as ordering parts. The PM is integrally involved in all major programmatic and technical decisions, and the PM sometimes develops part of the spacecraft directly. The PM provides a communication interface between all project participants including students, team leads, faculty advisors, external faculty advisors, and parties external to the university. The PM also keeps the team focused on its highest priorities as the semester progresses. These priorities are established at the beginning of the semester through discussions with veteran students and faculty advisors and can change as the semester progresses. The best PMs are often former team leads who are already familiar with the details of the project. They can appropriately prioritize work at a much more detailed level and can drill down deeper with questions at group meetings.

Work is broken down along team lines, and each team has its own team lead. The team leads typically consist of veteran students who are instrumental in coordinating all activities within their team. They explain existing designs and previously made decisions to new team members, which is critical to avoid starting over from scratch each semester as students change. The team leads are involved in most internal team decisions, and they provide an interface with other teams and the course instructors. They assign individuals to specific work areas within the team's realm of responsibility and carefully manage the completion of tasks at group work sessions.

It is a challenge to operate a long-term project within a semester-based course structure. The project must be organized around specific time schedules with individual and group tasks defined around specific end-of-semester objectives. Because half the organization changes each semester, a special effort must be made to maintain continuity. This is primarily achieved through veteran students, TAs, and faculty advisors that carry over to the next semester. In addition, students who are leaving the team are asked to formally document their work and train students that are staying with the team. Oftentimes, students are also asked to write up a single large document that captures the current state of the entire project. New students can use this document as a starting point to get familiar with the work.

B. Facilities

The facilities consist of the cubesat lab and the ground station. The cubesat lab is located in the aerospace building. It contains four lab benches with test equipment and six computers. The computers can largely be shared because files are stored on a shared network drive called the class drive. The class drive serves as a working directory for each of the teams. Each team has a folder on the class drive where it can save its work or download files. This way, the team is not tied to any individual computer in the lab. Also, when the student leaves, all of his work is readily accessible to new students. Before the class drive was implemented, files were scattered over many computers, both at school and at homes. A special effort had to be made to ensure that students leaving the course made their files available, but files were frequently lost in spite of these efforts. The shared drive made transitions between semesters much smoother. The shared drive and the computers can also be accessed remotely, so that students can work from other

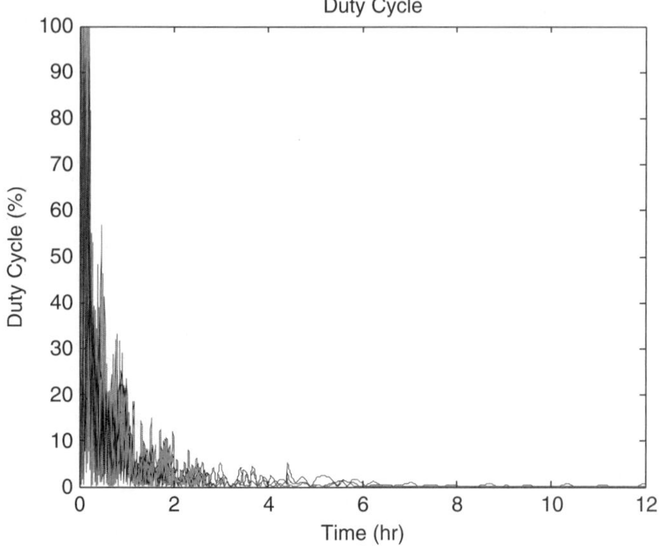

Fig. 2.5 Attitude stabilization simulation.

locations. Figure 2.5 shows simulation results for attitude stabilization of the satellites. Some of this attitude stabilization work was completed remotely while a student was off campus. The lab also has a thermal vacuum chamber for environmental testing of components (Fig. 2.6).

The ground station is located in the electrical and computer engineering building. It has a tracking antenna on its roof with computers and communication

Fig. 2.6 Thermal vacuum testing.

equipment inside. All students are given a key to the cubesat lab, so that they can work at their own convenience. Students that require access to the ground station are given a key to that room as well. The cubesat team relies heavily on the machine shop to help fabricate parts. The experienced personnel in the machine shop can fabricate components with minimal direction. The parts shop also provides a resource for electrical equipment and parts that the team commonly needs. Finally, vibration testing is done off campus with the assistance of a nearby company that specializes in the procedure.

III. Guidelines

At the onset, a set of guidelines is required to set the overall program philosophy. These guidelines ease the task for the inexperienced student designer. To the seasoned spacecraft engineer in government or industry, these might seem to be overstepping the bounds of conventional wisdom, but student-based organizations must make choices given the environment at hand. The following list summarizes some of the major design guidelines for the University of Illinois Cubesat Program.

A. Management Guidelines

1. Project Definition

The first question that everyone asks is "What does the satellite do?" Satellites that perform real missions are better respected and are more motivating to work on. Also, the people defining the project will have many competing ideas and often do not grasp the full complexity of their proposals. The program's leadership must take responsibility for keeping the project manageable for those who will carry the burden of implementing the ideas. The leadership must also prioritize ideas based on scientific merit and simplicity of implementation. Overcomplicating the project at this stage can lead to a great deal of frustration later and will likely require the team to descope a portion of the project that some students have worked very hard on. Adding missions is easier than removing them. It is best to begin with a plan for completing one good mission on time and within budget.

2. Risk Management

Use commercial technology. Commercial technology is much more affordable than space technology. It is also much more widely available, better documented, and newer. This is particularly true for electronic components and sensors, which are continually getting smaller and more efficient. The young engineer, for the time given, already faces an enormous challenge in developing a fault-tolerant, robust, engineered product, tested for thermal vacuum and launch vibration loads. The tradeoff against using commercial components is that cubesat programs must accept some extra parts lifetime risk.

3. Mission Life

We have taken the tact to define a short six-month lifetime and plan for system reliability accordingly. This lower lifetime goal mitigates the lifetime risk of using commercial parts and simplifies the satellite's development. Most components are likely to last for six months in LEO orbits below 600 km with little extra effort.

4. *Student Ownership*

The satellite is a student-lead and student-built project. Faculty advisors give advice, but students make the decisions. The students have the best technical information for making the decisions, and leadership development is an important educational objective.

5. *Testing*

Test, test, test. Perform thorough functional and environmental tests. At the beginning, there was no infrastructure or unique test equipment at Illinois for environmental testing. This infrastructure evolved over the first few years with the emphasis on testing. There are always bugs remaining, but extensive testing and simulation minimize the risk. Because the purpose of testing is to find bugs, new tests must continually be developed. Also, establish an easily repeatable test suite to check hardware on a regular basis, such as after environmental tests that can damage hardware.

6. *Schedule*

The schedule plan must be flexible. The design team is primarily senior undergraduate level, and the skills and capabilities vary from year to year. Goals are set on a semester-by-semester basis, and some semester groups accomplish more than others. Also, avoid signing up for a launch date until the end is in sight. Students have many outside commitments to juggle and do not have the experience to predict how long things will take. Hopefully, launch opportunities for cubesats will increase in the coming years, so that launch opportunities are available on shorter notice.

B. Engineering Guidelines

1. *Clean Rooms*

Many cubesat teams do not have or do not use clean rooms. The clean rooms can protect the flight unit and keep optics clean, but most cubesats do not face design requirements that require high cleanliness. Cleans rooms are used on larger satellite programs to reduce failure risks as a result of the extremely high cost of the programs. For small low-cost programs, like cubesats, the need for clean rooms is largely payload dependent.

2. *Radiation*

A few things can be done to improve radiation immunity with minimal effort. These include adding shielding, using tantalum tape on commercial components, and implementing watchdog timers. Latch-up circuitry can also help protect boards with commercial components. However, many radiation immunity improvements require a great deal of effort. Because cubesats have less than perfect first contact success, it is best to stay focused on making the satellite operational on the first day.

3. *Redundancy*

Use very little. Space, mass, and expertise are at a premium. Be smart with resilient designs, but keep systems simple.

4. *Modularity*

Although it is a small package, design the mechanical, software, and electrical systems to be as modular as possible. This way the teams can work independently, and no one person or group ends up having to know everything. Also, having more people does not help if the system is not designed to be modular, as there are too many interdependencies. Use simple microcontrollers to handle unique hardware requirements of power, ACS, health and monitoring, and payload subsystems. Allow the software team to focus on data handling and operations.

5. *Reuse Designs*

There will not be enough capable and committed people to design and develop an entire satellite from scratch in the time available. During development, the team will likely have only five to eight people who can handle major tasks, and the rest will be helping with little tasks in preparation for future leadership. Even if more of these people are available, they may not be able to help because the system is very tightly bundled and interdependent. Focus key people on the most important pieces, and live with old designs for the rest. These older systems can sometimes be upgraded once higher-priority items have been completed.

6. *Legacy Designs*

Resist throwing out good work from previous students. It is difficult to pick up someone else's work, but starting over unnecessarily will prevent the team from moving forward. However, be prepared to begin anew if previous designs are overly complex, and new students have particular skills and motivation that will allow them to catch up quickly. Also, select designs that can easily be transitioned to new students. For example, select processors that are taught in previous classes or are easy to learn.

7. *Backups*

Plan to purchase extra components, and make extra boards. Hardware breaks during testing or because of mistakes. Progress on the project, especially during integration, requires that a series of many small events occur in a certain order. Having backups can save frustration and a great deal of time off of the project schedule. Also, sometimes multiple people need to work with the same hardware, and giving them each their own unit eliminates one more interdependency. Performance can be compared against the backup if there is a suspicion that a part is not operating properly. Finally, risks can be taken with the primary unit if a backup is available allowing more testing and less cautious development.

IV. Course Implementation

A. Systems Meetings

The PM runs the weekly systems meetings, which all participants are invited to attend. These systems meetings begin with any major announcements from the faculty advisors or PM. Then, the PM polls each team in turn to discuss the previous week's activity and plans for the upcoming week. The team leads usually provide a report, but all team members are welcome to speak about their activities. The PM, faculty advisors, and other students often ask clarifying questions as they seek to better understand the work and how it affects others on the project. The teams also get the opportunity to ask each other for any information or assistance that they might require. The weekly systems meetings provide an important forum for keeping the organization moving on a week-to-week basis along its prioritized paths. Figure 2.7 shows solar-panel testing in progress during the semester. Reports for small activities are like these are given at the weekly systems meetings.

B. Setting up the Teams

Each semester, there are a number of veteran students returning from earlier semesters and a number of new students. The semester begins with the PM touching base with the veteran students and course instructors to identify reasonable goals for the upcoming semester. If necessary, the team structure is updated to accommodate these new goals. For example, a new initiative might best be addressed by a new team that is independent of existing teams continuing their work from the previous semester. Next, the responsibilities and semester objectives are broken down clearly by team. The semester begins with an informational meeting that is open to all students, veteran and rookie. At this meeting, the PM

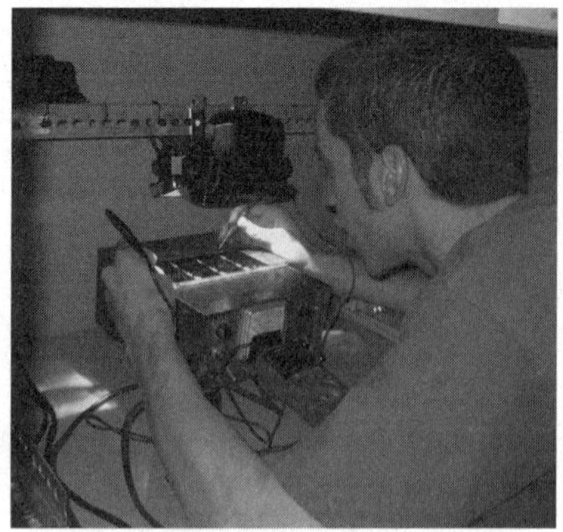

Fig. 2.7 Solar-panel testing.

provides a brief introduction to the project and the course. In addition, the PM outlines the team structure for the semester including each team's responsibilities, semester goals, desired size, and desired makeup. Next, a signup sheet is passed around allowing each student to sign up for a time to speak one-on-one with the PM.

For veteran students, the one-on-one meeting primarily serves as an opportunity for the student to update the PM with information about his or her team, especially including desired new students. In addition, this meeting serves as an opportunity to find out how many credits the veteran student is taking the class for and how many more semesters he or she plans to remain with the team. Identifying transitional needs at this stage is very helpful because extra students can be assigned to a team that is going to require new leadership soon. For new students, the primary purpose of the one-on-one meeting with the PM is to identify a team for the student to join. The students often need additional information to decide which team they want to join. At this meeting, they can ask the questions that they need to ask, and the PM can find out more about the student's long-term plans and credit objectives for the semester. Much of the meeting is spent with the PM providing additional details about specific work that needs to be done so that the student can choose a specific assignment from the work remaining. In most cases, it is possible to accommodate the student's interests while still making progress on the organization's major goals. However, sometimes organizational goals must be deferred by a semester, or new work must be started a semester early to accommodate the students' interests and skill sets. It is very important that effort be made to match students with suitable work because the cubesat course is more akin to a volunteer effort than a work environment. Because there is no compensation, the primary motivation for doing good work is the experience gained. Team leads are also solicited at this time if a veteran student has not already been established to lead a particular team.

The teams are mostly staffed after the one-on-one meetings with the PM, although a few new students may join over the next few weeks. To the get the ball rolling, the PM next sends out an email announcement with the members of each team. This way the team leads can quickly set up a meeting with their team. At these first team meetings, the team lead reviews the team's plan for the semester and begins to identify individual work for each student on the team. The team lead also begins training the team members in their work areas and sets up some regular work hours for the team. Oftentimes, the most challenging part to completing work is simply getting everyone together, and so setting up regular work hours can be instrumental to a team's success.

C. Work Plan

To increase the commitment to the semester objectives and maintain a written record of the work plan, each student is asked to fill out a work plan. This form requests information such as related experience, the number of credits that the student is registering for, the team's objectives, and the individual's objectives. It also asks for a week-by-week schedule and explicitly asks the students when their group meetings are scheduled because this is vital to their success. The work plan is updated with results halfway through the semester and at the end of the semester. Time logs are also included as part of the work plan submissions.

At the third week of the semester, each of the teams presents its work plan at the systems meeting. The teams present their refined semester goals as a group, and each individual also discusses what he or she will be working on. It is important that each individual announces his or her own semester plans because these ideas might be different from what others are thinking. This individual presentation serves as an opportunity to clarify any lingering confusion and serves as the individual's formal acceptance of his or her semester goals. Oftentimes, individuals will hesitate to speak up because they are hesitant to sign up for work that they do not yet know how to perform. It is important that individuals who do not sign up for a work assignment at this meeting are continually questioned in subsequent meetings until their individual plans are clear. If the course does not get a student's attention early, then he or she is often lost for much of the semester. At this time, the course instructors also ensure that the team has established some regular work hours. The students are free to work at their own convenience, but teams that establish a regular time to meet each week find much more success.

Per university guidelines, each student is expected to put in 45 lab hours for each credit he or she is registered for. Because work hours are flexible, the students are required to log their hours on a time log. They submit this log at the midpoint of the semester and again at the end of the semester. Students are allowed to adjust the number of credits that they have registered for if they are doing more or less work than they anticipated. Sometimes, the students cannot put in as much time as they planned because of outside commitments, and sometimes they are unable to perform their work because they are waiting for others on the team. This is avoided to the extent possible, and supplemental assignments are often given out. However, the reality is that the project is highly integrated, and communication and coordination with other team members is a major constraint to completing individual work.

D. Training

General training of new students begins immediately at the beginning of the semester. Students are sent some reading materials about the project, such as published papers, that help them better understand the nature of the work. In addition, the students are given keys and computer access as soon as possible. The students can begin browsing archived files on the class drive. Team leads usually train students in their specific work areas once teams have been formed. In addition to general training and team specific training, new students must often be taught new skills and procedures. A special folder exists on the class drive with tutorials that have been developed over time. These include tutorials on making PC boards, programming PIC processors, and operating the thermal-vacuum equipment. Thanks to these written tutorials, the veteran team members can spend less time on training new team members and more time on their own individual work. Figure 2.8 shows vibration test results for ION. Vibration testing was outsourced after the team discovered the difficulties of developing internal capability.

E. Reviews

Midpoint reviews and end-of-semester reviews are scheduled halfway through the semester and at the end of the semester, respectively. These reviews give the

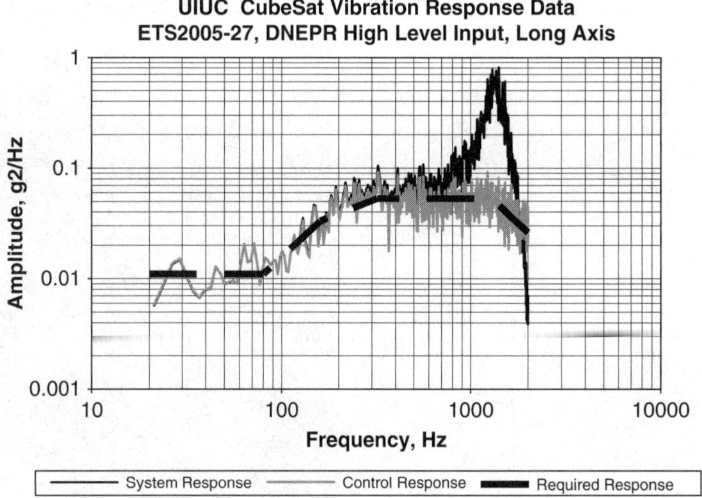

Fig. 2.8 Vibration response data.

students a milestone to work towards with a firm deadline. The reviews are held around a meeting table with the PM and faculty advisors present. Students from other teams are also encouraged to attend as their schedule permits. Each team is given 45 minutes to present its work in detail. They bring prepared documents with design information and demonstrate hardware or software, if possible. They also submit their time logs at these reviews. The reviews serve as important data points in establishing student grades because the students are able to present all of their work in a detailed organized fashion. Each student is expected to demonstrate a clear understanding of his own work and should demonstrate an ability to ask pertinent questions of other students. Pizza is provided after the midpoint and end-of-semester reviews as a small celebration of the work completed up to that point.

F. End of Semester

The semester comes to an end the week before final exams, so that students can finish strong without excessively impacting their other coursework. In addition to end-of-semester reviews, the students submit an end-of-semester form, which consists of their work plan with semester accomplishments, time logs, and check-out procedures. The document provides the students with an opportunity to remind the course instructors of the work that they have already reported at weekly meetings and reviews and during direct interaction with the PM and faculty advisors. The end-of-semester form is an important tool that course instructors use in their final grading discussions. The end-of-semester form also provides a checklist for students who are not returning for the next semester. It asks if the student has documented his work and trained members of the team who will be staying. It also asks if the student has returned all keys and borrowed equipment. Such details are easy to forget, but are very difficult to follow up on later. Figure 2.9 shows the

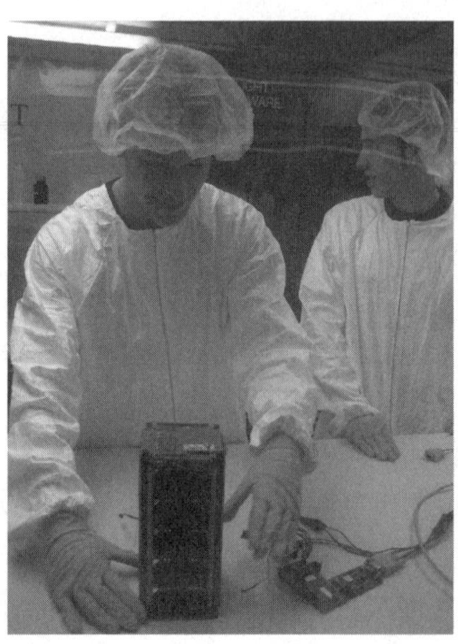

Fig. 2.9 Final satellite delivery.

completed satellite being delivered. It is being integrated with its deployer in a clean room in preparation for launch.

V. Dealing with Turnover

Student turnover is a major challenge to completing a long-term project, as students perform part-time work for a limited period of time. This issue can be addressed in a variety of ways.

A. Management

Management techniques for dealing with turnover include offering incentives to improve retention of key students. These incentives can include assistantships, thesis opportunities, hourly positions, leadership titles, or simply opportunities to explore and learn. In addition, students can be set up with overlapping work assignments, so that someone remaining on the team is familiar with the work as individual students leave. It is particularly important to identify the departure of key students at the beginning of their last semester, so that additional students can be assigned to their team. One or more successors can often be designated for the key student to train. Establishing a shared network drive for all documents downloaded or created by a team makes it easy for new students to quickly find and read through the relevant information.

Students can also be required to document their work and comment their code before leaving the team. It is important to note that documentation alone is not sufficient as it is not comprehensive. In one instance, a full element of work was fully documented in a Master's thesis and all working documents were burned to a DVD, but replacement students were still unable to continue the work. Documentation often does not include training in specific engineering skills, lessons in recognizing and dealing with specific engineering issues, and important details the previous engineer found obvious. Documentation also often focuses on the current status of the work without describing the full sequence of events and logic that lead to this status. As a result, the new engineer must guess at the reasoning behind decisions if he or she is not given adequate opportunity to question the previous engineer. A smooth transition often requires a full semester of overlapping work.

B. Leadership

In addition to implementing the management techniques just discussed, retention can be improved with quality leadership. The organization's leaders can create a fun work environment and can emphasize the significance of the work and the benefits to the students personally. The organization's leaders can also instill a sense of confidence that the work can be completed and can create a sense of being a team. Finally, the organization's leaders can approach their job as one of assisting and guiding students thereby creating a sense of student ownership.

In one example of the difference that leadership can make, nearly the entire ION team left after the second year. Previous students stated that, to a certain extent, they simply lost hope that they could complete the project. Recruiting efforts were weak, there were no software people for one year, there was only one good electrical person who was graduating, and the satellite was scheduled to be delivered in eight months. However, from years 3 to 5, retention in the organization was excellent. The primary difference was the presence of strong leadership. A few of the organization's new leaders had full confidence in their own abilities both to lead the team and to execute on the engineering work. As others saw that the team was making real progress, they began to believe in their own abilities, and new valuable people joined the team. Gradually, strong leaders were established across all elements of the organization creating the complete team necessary to follow through on all of the work. As these people graduated, they trained replacements of their own initiative. If they stayed in town after leaving the team, they would often continue attending meetings as advisors simply because they enjoyed being part of the group. Although turnover is a challenging problem to deal with for long-term projects in a student environment, the issue can be addressed in a variety of ways.

VI. Conclusion

The cubesat course provides students with the opportunity to work on an exciting large team project with students from other engineering disciplines. They develop valuable hands-on engineering skills, communication skills, and leadership skills in an environment that approaches what they will experience in industry. The management and implementation of the course has evolved over the course of its six years to the point where things now operate very smoothly. The insights from these successful practices can be utilized in other capstone courses that implement large interdisciplinary systems engineering projects. New projects, including high-altitude balloon flights and additional satellites, are now building on the success and infrastructure from the ION project. With a little luck, the next satellite will make it to orbit bringing back some value atmospheric data and flight-test data.

Acknowledgments

The authors would like to thank the University of Illinois, the College of Engineering, the Department of Electrical and Computer Engineering, the Department of Aerospace Engineering, and Alameda Applied Sciences Corporation for their support.

References

[1] Rysanek, F., and Hartmann, J. W., "MicroVacuum Arc Thruster Design for a CubeSat Class Satellite," *16th Annual AIAA/USU Conference on Small Satellites*, Paper SSC02-I-2, Aug. 2002.

[2] Swenson, G., Thakker, P., Kamalabadi, F., Frank, M., Coverstone, V., and Voss, H., "Optical Sensing Atmospheric Emissions from Cubesats and Nanosats," *Proceedings of the SPIE Sensors and Systems for Space Applications*, Vol. 6555, 2007, p. 655506.

[3] Lan, W., Brown, J., Toorian, A., Coelho, R., Brooks, L., Suari, J. P., and Twiggs, R., "CubeSat Development in Education and into Industry," *AIAA Space*, AIAA Paper 2006-7296, 2006.

[4] Holmes, W. C., Bryson, J., Gerig, B., Oehrig, J., Rodriguez, J., Schea, J., Schutt, N., Voss, D., Voss, J., Whittington, D., Bennett, A., Fennig, C., Brandle, S., Dailey, J., and Voss, H. D., "TU Sat 1—A Novel Communications and Scientific Satellite," *16th Annual AIAA/USU Conference on Small Satellites*, Paper SSC02-I-1, Aug. 2002.

[5] Gregorio, A., Bernardi, T., Carrato, S., Kostadinov, I., Messerotti, M., and Stalio, R., "AtmoCube: Observation of the Earth Atmosphere from the Space to Study 'Space Weather' Effects," *Proceedings of Recent Advances in Space Technologies*, Aeronautics and Space Technologies Inst., Istanbul, 2003, pp. 188–193.

[6] Obland, M., Hunyadi, G., Jepsen, S., Larsen, B., Klumpar, D. M., Kankelborg, C., and Hiscock, W. A., "The Montana State University NASA Space Grant Explorer-1 Science Reflight Commemorative Mission," *15th Annual AIAA/USU Conference on Small Satellites*, Paper SSC01-III-2, Aug. 2001.

[7] Flagg, S., Bleier, T., Dunson, C., Doering, J., DeMartini, L., Clarke, P., Franklin, L., Seelback, J., Flagg, J., Klenk, M., Safradin, V., Cutler, J., Lorenz, A., and Tapio, E., "Using Nanosats as a Proof of Concept for Space Science Missions: QuakeSat as an Operational Example," *18th Annual AIAA/USU Conference on Small Satellites*, Paper SSC04-IX-4, Aug. 2004.

[8] Simburger, E. J., Liu, S., Halpine, J., Hinkley, D., Srour, J. R., Rumsey, D., and Yoo, H., "Pico Satellite Solar Cell Testbed (PSSC Testbed)," *Conference Record of the 2006 IEEE 4th World Conference on Photovoltaic Energy Conversion*, Vol. 2, IEEE, Piscataway, NJ, 2006, pp. 1961–1963.

Chapter 3

Small Satellites 101: University of Hawaii Small-Satellite Program

Justin M. Akagi,* Tyler N. Tamashiro,† Wade G. Tonaki,‡ and Wayne A. Shiroma§

Hawaii Space Flight Laboratory, University of Hawaii at Manoa, Honolulu, Hawaii

THE United States launched Explorer 1 into space on 31 January 1958. The primary mission of this 91-cm-long × 15-cm-diameter cylindrical spacecraft was to test a cosmic ray detector, developed by Dr. James Van Allen and his University of Iowa graduate students, to measure the radiation environment of space. The results of this experiment led to the discovery of radiation belts trapped by Earth's magnetic field, later named the Van Allen Belts in honor of the researcher (data available online at http://www.jpl.nasa.gov/missions/past/explorer.html).

Since then, the average satellite has grown considerably in size (>100 kg) and cost (>$100 M per launch), thus limiting space opportunities for smaller, university-based research groups. However, recent technology advances have led scientists and space enthusiasts back into a field that only large government agencies could afford—satellite development. Nearly fifty years after the launch of the first U.S. satellite (a small satellite, of course), many universities have returned to the small-satellite field with renewed interest and vigor.

In this chapter, we provide an overview of the University of Hawaii's Small-Satellite Program focusing on small-satellite bus development for high-technology scientific missions. The chapter is organized as follows:

- University of Hawaii Small-Satellite Program Overview
- Phase I—*Mea Huaka`i* (2001–2004)

*Graduate Research Assistant; currently Industry/Government Outreach Coordinator, College of Engineering.

†Undergraduate Research Assistant; currently Launch Systems Integration Engineer, Launch and Operations Department, Northrop Grumman, Redondo Beach, CA.

‡Undergraduate Research Assistant; currently Graduate Research Assistant, Department of Electrical Engineering.

§Professor, Department of Electrical Engineering, and Co-Director, Hawaii Space Flight Laboratory.

- Phase II—*Hokulua* (2003–2005)
- Phase III—*Ho`okele* (2005–2006)
- Phase IV—*Ho`okia`i* (2005–2007)

Whether you are a part of a well-established small-satellite program or starting a new one, this chapter may provide some insight into both the technical and managerial aspects of small-satellite development.

I. University of Hawaii Small-Satellite Program Overview

The University of Hawaii's Small-Satellite Program was created with the intent of establishing a high-technology research and workforce-development infrastructure within the State of Hawaii. This multidisciplinary, student-driven effort provides an excellent opportunity for undergraduate students to work on (and lead) real-world engineering projects. Since 2001, over 150 undergraduate students in electrical, mechanical, and civil engineering have been involved in the development of four generations of nanosatellites with state-of-the-art scientific payloads. The experimental payloads developed by the undergraduate students in this program include thermal sensor units and thermal modeling software, geo-referenced imaging, solid-storage microthrusters, and retrodirective antenna array technologies. These projects have also provided the undergraduate students with great experience in collaborative work with the local scientific and high-tech industry.

The University of Hawaii (UH) began its Small-Satellite Program in 2001 with *Mea Huaka`i* ("*Voyager*" in Hawaiian), a standard 1-kg, 10 × 10 × 10 cm cubesat. Launched as part of the failed multi-university DNEPR mission in July 2006, the scientific payload for this first phase of our program was a set of temperature sensors to verify UH-designed thermal modeling software, but the actual mission was just learning to build a bus. In the next phase, *Hokulua* (*Twin Stars*, 2003–2005) was developed to test retrodirective antenna array technologies in space for possible use in distributed satellite networks. In the third phase (2005–2006), *Ho`okele* (*Way Finder*) was developed with geo-referenced imaging capabilities for a distributed satellite network to detect terrestrial-based natural disasters. The geo-referenced imaging payload consisted of a global positioning unit (GPS), nano-inertial measurement unit (nIMU), and a network camera.

The fourth phase of the program, *Ho`okia`i* (*Watchman*, 2005–2007), combined the knowledge and technologies developed in the previous three phases: satellite system bus design experience, retrodirective antennas, and distributed networking concepts. In addition, new microthruster technologies (Chapter 10) are undergoing development for the small-satellite platform for active stabilization. The design and realization of a mobile ground station for VHF/UHF communications has also been demonstrated (Chapter 16).

In the five years since the program's inception, over 150 engineering undergraduates have gone through this program. Seven students have gone on to pursue Ph.D. degrees, and many more have continued with M.S. degrees. A larger number have gone into the aerospace industry, including Boeing, Lockheed Martin, and Northrop Grumman. Two notable graduates of our program include national award-winning students Aaron Ohta and Blaine Murakami, who served as project directors for *Mea Huaka`i* and *Hokulua*, respectively. In 2003, Aaron Ohta received the Alton B. Zerby and Carl T. Koerner Award from the electrical engineering honor society Eta Kappa Nu, recognizing him as the most outstanding electrical/computer engineering student in the nation. Blaine Murakami received the same award in 2005. In the span of just three years, two of our Small-Satellite Program leaders were identified as the best in the nation.

II. *Mea Huaka`i (Voyager)*, 2001–2004

In 2001, UH began development of its first small satellite, *Mea Huaka`i* (*Voyager*). This was UH's first step in establishing an undergraduate-based small-satellite research program that eventually led to the foundation of the Hawaii Space Flight Laboratory. Over seventy undergraduates and nine faculty advisors participated in this first phase of the UH Small-Satellite Program, which involved the design, fabrication, and operation of a CubeSat: a cube-shaped satellite having a mass of no greater than 1 kg and a maximum volume of 1000 cm^3 (Fig. 3.1).

For over 30 years, the UH engineering curriculum followed the same pedagogical paradigm that is in place at most other engineering colleges. Students take calculus, physics, chemistry, and introductory engineering courses in the first two years, and discipline-specific engineering courses in the third year. Although this bottom-up teaching sequence is intended to provide a solid scientific and mathematical foundation for application-oriented coursework later in the curriculum (typically a capstone design project in the student's senior year), it is not necessarily the best one for providing motivation for students. This is particularly true for today's generation of students who want to see immediate applications of what they are learning. Faced with an alarming 35% retention rate at the UH College of Engineering, the faculty initiated a project-based education theme throughout the undergraduate curriculum [1].

This problem of retention is not specific to our university and is possibly the impetus for developing innovative freshman engineering courses (e.g., [2]). Providing early and sustaining motivation is just one aspect of a good engineering curriculum, however. The engineering education experience is further enhanced if it allows students to work on open-ended engineering projects that expose them to all of the *realities* of engineering: working to specifications, securing funding and keeping within budget, maintaining documentation, leading and working in multidisciplinary teams, conducting design reviews, and more. A real project, as opposed to just a canned laboratory exercise, also provides the benefit of allowing real-life issues—such as engineering ethics—to be experienced. At universities

Fig. 3.1 Part of the *Mea Huaka`i* team with the satellite housing.

such as ours, with faculty that are already overburdened and an engineering curriculum that is already bulging at its seams, real-life engineering issues such as ethics or proposal writing is best *experienced* within the confines of a real project rather than taught as a formal course.

Another motivating factor for curriculum change is providing students the opportunity to get involved with undergraduate research, especially at a Carnegie Doctoral/Research-Extensive University such as ours, where research often gets top billing. Integrating research into a project provides the sense of discovery-based education that steers students toward the pattern of lifelong learning.

All of this has led to a new paradigm—*project-based engineering education*, in which the CubeSat Program serves as an excellent example.

A. Project Organization

In keeping with the goal of project-based education, the two project leaders selected for *Mea Huaka`i* were undergraduate engineering students. Following a preliminary decision to maintain leadership continuity over the anticipated two-year span of the project and to limit the turnover associated with graduation, two juniors were selected by popular vote as project director (Aaron Ohta) and assistant project director (Michael Tamamoto). The responsibilities of the project directors were to coordinate the overall design effort, secure funds via proposal writing, organize design reviews, and deliver presentations to sponsors.

The $120,000 budget (Table 3.1) for this project was sponsored by grants from the Hawaii Space Grant Consortium, Northrop Grumman Space Technology, One Stop Satellite Solutions, Boeing, the UH College of Engineering, and many UH alumni and friends. A project management team was also assembled to manage the financial aspects of the project (fundraising, purchasing, reimbursements), coordinate public relations efforts with the media, conduct educational outreach to high school students, and maintain the project web page and file management system for archiving documentation.

B. Multidisciplinary Engineering

Satellite development is a multidisciplinary effort, involving aspects of all types of engineering, from software to telecommunications and structural analysis. Each satellite and payload subsystem team (shown in Fig. 3.2), consisting of 6–15 members, was led by an undergraduate student and advised by one or two faculty mentors. Although these teams were divided along the traditional civil, electrical,

Table 3.1 ***Mea Huaka`i*** **Budget**

Item	Cost
Launch, integration, and export licensing	$50,000
Parts	$15,000
Equipment	$10,000
Summer stipends for students	$45,000

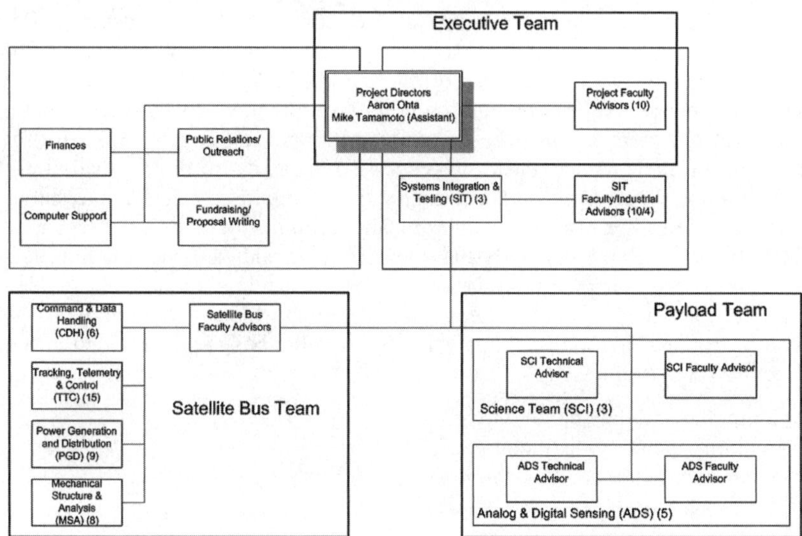

Fig. 3.2 ***Mea Huaka`i* organizational chart.**

and mechanical engineering disciplines, the nature of this project required a considerable amount of interaction and collaboration in weekly progress meetings and system integration sessions. This type of effort provided an invaluable learning experience for many of the students by building teamwork skills and teaching them the working dynamics of a multiteam engineering project.

1. Small-Satellite Development

Mission. When working on any engineering project, including small-satellite development, the first step is to define specific mission(s). The overall mission of the project drives all aspects of the project, from the organization of the team to the core bus design requirements. If the mission is to launch a simple cubesat with a beacon, the team may consist of five members. However, if the mission were as complicated as building a space station, five people would not be nearly enough. Similarly, although most of the bus may use a standardized architecture, the requirements of a particular mission may require slight to considerable reconfiguration.

When specifying a mission, it needs to be *purposeful, realistic, and realizable* [3]. The two primary goals of the *Mea Huaka`i* project were to 1) develop, test, and launch Hawaii's first small satellite as a collaborative, undergraduate-based effort, and 2) to perform thermal sensor measurements for verification of UH-designed thermal modeling software for space conditions. Because UH had never worked on a small-satellite project prior to *Mea Huaka`i*, the core bus development was a major endeavor in itself. The thermal modeling and sensor measurements addressed design concerns regarding effects of the space environment on small satellites in low-Earth orbit.

System bus. To achieve mission success, it is necessary to develop a bus capable of supporting any and all payload systems contained within the satellite. It must be capable of generating, storing, and producing sufficient power for its own processes, as well as the required power for any payload(s). It also requires an integrated system architecture and software algorithms capable of performing basic system maintenance, ensuring successful operation of the payload system, command and data handling, and establishing/maintaining reliable intersatellite communication. The wireless communication links should be able to support protocols for satellite commands and telemetry data and bandwidth requirements for scientific data. The structural housing for the satellite should be capable of ensuring a safe environment for storage and operation, which might require mission-specific structural and thermal analysis. The satellite bus can be divided into the following subsystems:

- Command and data handling (CDH)
- Tracking, telemetry, and control (TTC)
- Power generation and distribution (PGD)
- Mechanical structure and analysis (MSA)

Command and data handling. The *command and data handling* subsystem is responsible for scheduling the satellite's onboard processes and maintaining optimal operation throughout its mission lifetime. The satellite operates as an autonomous entity for a majority of the mission, with the exception of commands received from the ground station via a wireless communication link. As such, the primary microcontroller will make operational decisions based on satellite telemetry (as measured by various onboard electronics). These decisions, in addition to protecting the satellite from unsafe operation, should be designed to successfully achieve mission objectives. The satellite should perform system diagnostics regularly and store/process recent telemetry data. These telemetry data, provided by monitoring devices (such as battery monitors, thermal sensors, location and attitude measurement devices, etc.), are necessary to facilitate with program flow decisions. Based on the status of the satellite, the microcontroller should determine the system's current ability to perform mission procedures and operate accordingly. Some additional CDH subsystem responsibilities include establishing communication protocols, implementing distributed network architecture, and data processing for scientific payloads.

The diagram in Fig. 3.3 shows a breakdown of the basic building blocks for *Mea Huaka`i*'s CDH subsystem. These automated tasks are performed by a Rabbit Semiconductor RCM2000 microprocessor (data available online at http://www.rabbitsemiconductor.com). A programmable integrated circuit (PIC) chip is also included to aid the Rabbit processor in data routing and handling for the satellite. In addition, this unit can function as a backup processor in the event of main processor failure.

Tracking, telemetry, and control. The *tracking, telemetry, and control* subsystem is responsible for maintaining a reliable wireless link for all satellite communications. This includes satellite–ground station uplinks and downlinks and satellite–satellite links for distributed satellite networks. Upon initial deployment of the

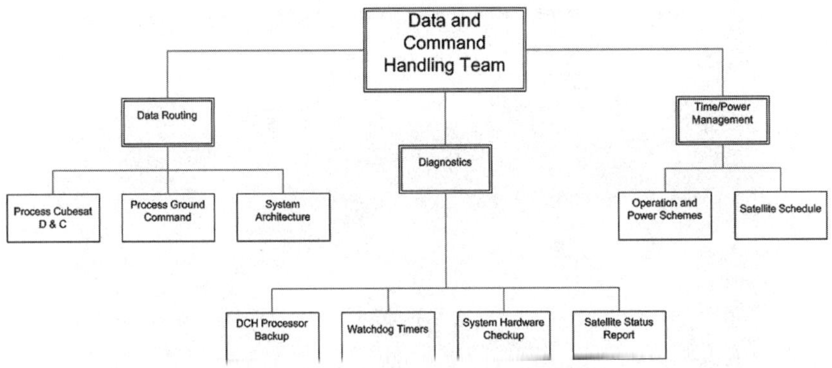

Fig. 3.3 CDH block diagram.

satellite, the electronics beacon the telemetry data to a receiver at the ground station. Upon successful contact, the transceiver/antenna configuration maintains the wireless communications link. To verify optimal operation, a link budget based on orbit and communications system parameters should be performed to ensure the reliability of the wireless link.

The communication protocol for *Mea Huaka`i* is initiated by data transmission from the onboard processor to the terminal node controller (TNC), which encodes the data using amplitude-frequency shift keying (AFSK) and a standard amateur radio protocol. The packets are then broadcast in the 70-cm UHF amateur radio band using a modified Yaesu VX-1R (data available online at http://www.yaesu.com), a commercial-off-the-shelf (COTS) handheld transceiver capable of half-duplex operation, which in turn is connected to a dual monopole antenna to provide an approximately isotropic radiation pattern. The deployed antenna configuration for *Mea Huaka`i* is shown in Fig. 3.4.

Power generation and distribution. The *power generation and distribution* subsystem is responsible for generating and supplying power for the satellite bus and payload. Energy received from the solar cells is used to power the system during sun-on periods and to recharge the battery pack for sun-off periods. During solar eclipse, the battery is used as the primary power source. The main power line (connected to the solar cells and battery) feeds into a number of dc-dc voltage converters, which provide the necessary supply voltages for the satellite's electronics. A battery monitor is also integrated into this subsystem to measure sustainable power consumption and facilitate in power management.

Solar power is collected and generated with Spectrolab 21%-efficiency gallium arsenide (GaAs) solar cells (data available online at http://www.spectrolab.com), providing about 1 W of continuous power. The solar cells charge lithium-ion batteries, which are used during the eclipse periods of the satellite orbit, or when the power demand exceeds the output of the solar cells. In addition, a Texas Instruments BQ2050 gas gauge was used to measure battery telemetry and detect any power systems problems.

The basic block diagram for *Mea Huaka`i*'s PGD subsystem is shown in Fig. 3.5.

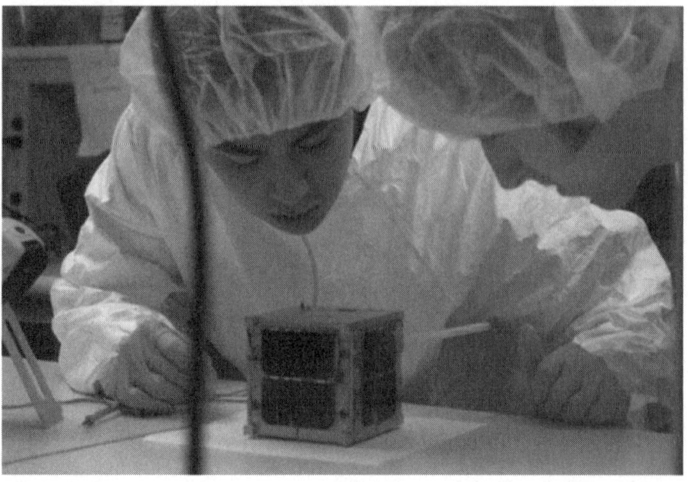

Fig. 3.4 UH student Monte Watanabe performing diagnostics on *Mea Huaka`i* in Cal Poly San Luis Obispo cleanroom.

Mechanical structure and analysis. The *mechanical structure and analysis* subsystem is responsible for the design and assembly of the satellite's mechanical structure. The structure must be designed to support both the satellite bus and accompanying payload(s). For flight models, structural analysis must be performed to ensure the satellite will survive launch and space conditions. In addition

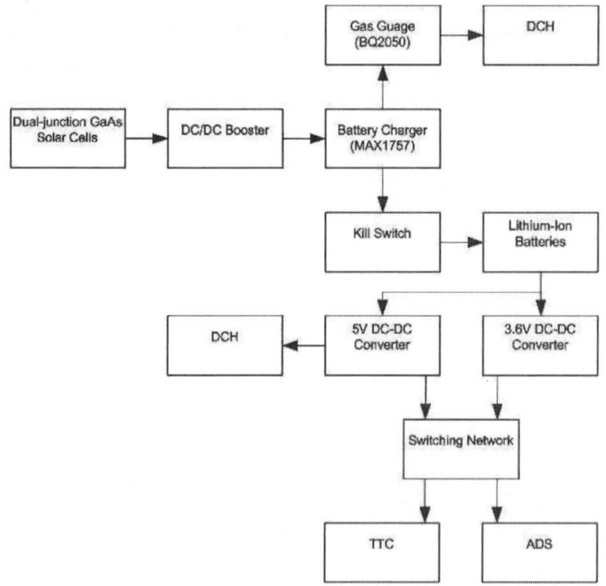

Fig. 3.5 PGD block diagram.

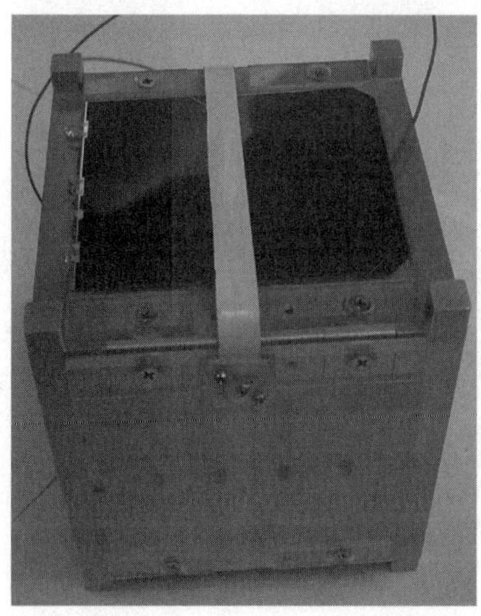

Fig. 3.6 ***Mea Huaka`i*, showing mountings for exterior components.**

to a robust mechanical structure capable of supporting the satellite's size/mass specifications, mounting brackets, connector slots, and other necessary mechanical considerations should be integrated into the design.

Structural considerations play an important role in the design of some of the subsystems. For example, the measuring-tape antenna (quarter-wavelength monopole) and corresponding mounting scheme, shown in Fig. 3.6, were affected by a number of practical concerns such as the small structure (finite ground plane), required elasticity of the antenna for deployment, and antenna feed specifications.

Some other relevant MSA concerns include the limited amount of space for internal electronics and thermal effects from the space environment. Thermal modeling and measurement comprise a portion of this subsystem; however, because it is a scientific mission for this project, it will be discussed in the following section.

System integration and testing. Although system integration and testing is technically not a subsystem, it is included in this section because of its importance in system bus development. It is responsible for the electrical and mechanical integration of all satellite bus subsystems and payload system(s). Although no actual hardware is designed or produced by this particular "subsystem," standardized connectors, PCB layout and testing procedures need to be established by the team. These processes and procedures should be fully documented to ensure repeatable subsystem verification results. In addition, SIT is responsible for ensuring that all electronic and mechanical components are well fitted (to minimize wasted space and loose wiring) to the satellite's mechanical structure.

Team members must be able to oversee the entire project development, from verifying hardware and software interface compatibility between subsystems to coordinating the mass, power, and size budgets. This team is also responsible for checking that the satellite meets all safety and environmental stress qualifications.

The SIT executive team consisted of a single student for *Mea Huaka`i*; however, in later phases of the UH Small Satellite Program, the SIT team generally consisted of the project leader(s) and a few members from the four core subsystem teams (CDH, TTC, PGD, MSA).

2. *Scientific Mission and Payload*

The primary scientific mission for this project was thermal measurement of the LEO space environment. *Mea Huaka`i* was equipped with six temperature sensors (one per cube face), with the purpose of determining the typical heat distribution of a cubesat in outer atmosphere thermal conditions. This knowledge is useful for designing future satellites, to ensure that the onboard electronics are kept within safe operating/storage temperature ranges.

These thermal data were collected to compare actual thermal data to the current thermal model (developed by UH undergraduate Cory Soon), based on assumed space conditions. The thermal model simulation, shown in Fig. 3.7, illustrates the three-dimensional heat distribution throughout the cube. As the shading implies, the sun illuminates the front left face, and the back right face is dark.

III. *Hokulua (Twin Stars)*, 2003–2005

In 1999, the Air Force Office of Scientific Research (AFOSR) and Air Force Research Laboratory Space Vehicle Directorate (ARFL/VS) established the University Nanosat Program (UNP), an initiative established to support universities within the United States develop, build, fabricate, and fly nanosatellite systems (data available online at http://www.vs.afrl.af.mil/UNP/index.html). UH was one of thirteen universities to win a $100,000 award for the third phase of the UNP.

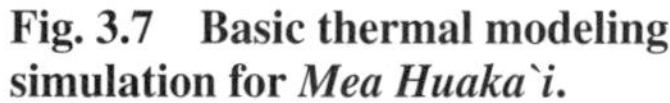

Fig. 3.7 Basic thermal modeling simulation for *Mea Huaka`i*.

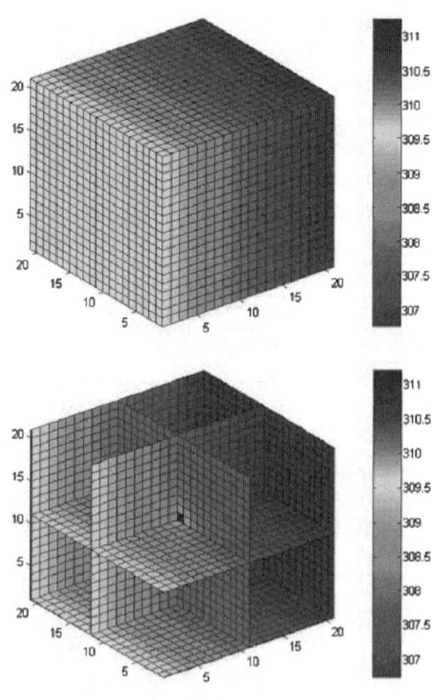

With strong experience in radio frequency (RF) and communications technology at UH, it was natural for our team to propose a satellite payload with a primary focus of communications technology research aimed at cooperative networks of small satellites.

UH is at the forefront of retrodirective antenna array research, a technology capable of establishing and maintaining high-directivity (narrow beamwidth), self-steering crosslinks for wireless communication. High-directivity antennas afford two major advantages over omnidirectional ones, which are typically used on small satellites. The first advantage is that a directive link is less susceptible to third-party eavesdropping because most of the radiated RF energy propagates in the direction of the target node (i.e., satellite or ground station). The second advantage of high directivity is the improvement of power transmission efficiency because less power is transmitted (and effectively wasted) in undesired directions. However, most small satellites have limited pointing/attitude control systems, meaning the highly directive beams are useless without a system capable of dynamically steering the beam in the direction of a desired satellite or ground station.

Retrodirective antenna array technology is particularly applicable to satellites without attitude control because it does not require prior knowledge of its target's position/orientation to establish a communication link, and unlike smart antenna technology, which can require complicated digital signal processing (DSP), dynamic beam steering is managed by analog circuitry. Figure 3.8 illustrates the advantages of retrodirective crosslinks in satellite-based wireless communication. Figure 3.8a shows omnidirectional crosslinks as the simplest means of communicating between satellites with limited attitude control. However, this method enables spy satellites or other third-party receivers to pick up the radiated signal. Moreover, the power radiated in directions other than the target satellite is effectively wasted. Figure 3.8b shows that retrodirective self-steering crosslinks are different from omnidirectional crosslinks as the radiation patterns of these communication links have higher gain (narrow beamwidth), concentrating radiating power in one main direction. This inhibits third-party eavesdropping, increasing network security while also increasing communication efficiency. Unlike conventional smart-antenna or phased-array architectures, these links are set up and maintained *autonomously*.

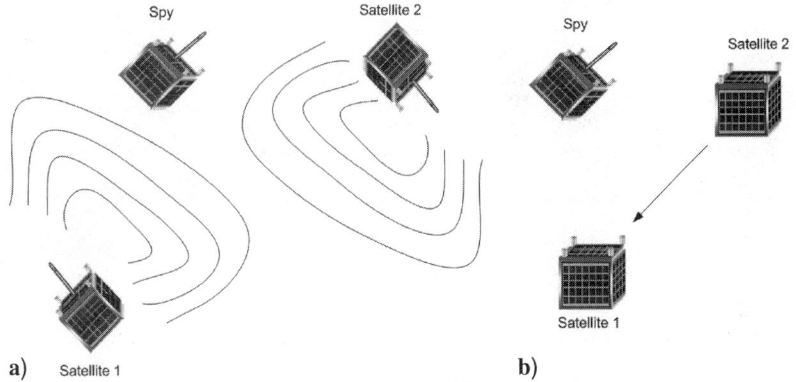

Fig. 3.8 Crosslinks, a) omnidirectional, and b) retrodirective "self-steering."

Although the idea of retrodirective antennas has been around since the 1960s, the application of retrodirective communication links between satellite nodes has never been implemented. With this in mind, UH's mission was as follows:

> To demonstrate and test the applicability of retrodirective antenna array technology for satellite-to-satellite communication within a distributed small-satellite network by testing and evaluating a retrodirective communication link on a pair of satellites on orbit.

A. Project Organization

The project organization for *Hokulua* was divided more strategically than the previous *Mea Huaka`i* project. The role of *project director* was replaced with a *project manager* and *project engineer*. The responsibilities of *project manager* (Scott Kawakami) included overseeing all non-technical issues (budget management, travel, meeting and design review scheduling, public relations and outreach, equipment/parts procurement, etc.). The responsibilities of *project engineer* (Blaine Murakami) included all high-level technical issues (advising, approving all designs, leading system integration work, acting as liaison between students and industrial advisory board, etc.).

The organizational chart for this phase is shown in Fig. 3.9. Most of the system bus teams remain the same, with the addition of a *thermal design and analysis* (TDA) subsystem.

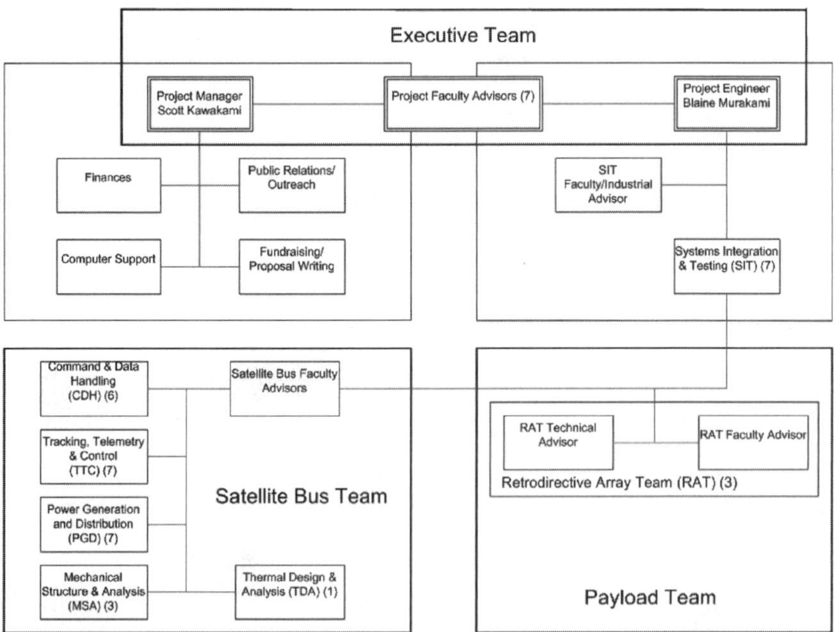

Fig. 3.9 ***Hokulua*** **organizational chart.**

Fig. 3.10 RCM3400 microcontroller.

1. Command and Data Handling

In this second phase of the program, the *CDH* subsystem upgraded its onboard microcontroller from the Rabbit Semiconductor RCM2000 to the newer RCM3400 (see Fig. 3.10; data available online at http://www.rabbitsemiconductor.com). This microcontroller was selected based on *Hokulua* system requirements, including required onboard analog-to-digital converter (ADC), serial driver support, and built-in memory capacity, small size, and ease of integration. In addition, students' previous experience with *Rabbit Semiconductor's Dynamic C Integrated Development Environment* allowed for more rapid software development and prototyping in this legacy environment.

The CDH (and TTC) hardware was integrated on a 10 × 15 cm printed circuit board (PCB), as shown in Fig. 3.11. The circuit board layouts for all of the *Hokulua*

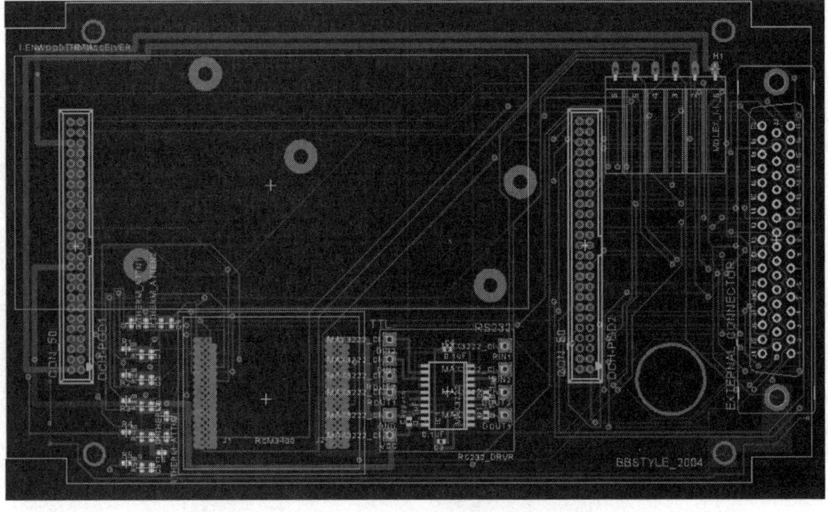

Fig. 3.11 CAD layout of CDH/TTC circuit board.

PCBs were designed in EAGLE software (data available online at http://www.cadsoftusa.com) and manufactured by PCBexpress (data available online at http://www.pcbexpress.com).

2. *Tracking, Telemetry, and Control*

The hardware configuration for the *TTC* subsystem was changed considerably between the first phase of the UH Small Satellite Program (*Mea Huaka`i*) and the second phase (*Hokulua*). For *Mea Huaka`i*, the TTC subsystem consisted of three primary hardware components: 1) a terminal node controller (a modified version of John Hansen's KISS TNC [4,5]), which encodes the serial data from the on-board processor, 2) a modified Yaesu VX-1R transceiver, to broadcast the encoded data, and 3) a double monopole antenna for an efficient radiation pattern. The TTC subsystem for *Hokulua* used the Kenwood TH-D7A (shown in Fig. 3.12), a COTS radio with an integrated TNC and transceiver in a single unit (data available online at http://www.kenwood.com). In addition to the combined TNC/transceiver design, this radio was selected because of its small size, low power consumption, and its flight heritage on other small satellites. Although some modifications were required to integrate the TH-D7A radio into the satellite, the decision to use a COTS component considerably simplified the development of this TTC subsystem.

3. *Power Generation & Distribution*

One important aspect of the *PGD* subsystem is power management. The description of *Mea Huaka`i*'s PGD subsystem is limited to the subsystem

Fig. 3.12 Modified Kenwood TH-D7A radio.

Fig. 3.13 Thermo Oriel solar simulator for solar-cell characterization (AM 1.5).

requirements and a brief discussion of the basic block diagram. However, to develop a reliable power system, it is also important to prepare power generation and consumption budgets based on the characteristics of the system's electronics.

The single-junction Spectrolab solar cells used on *Hokulua* were characterized by UH students at the Hawaii Natural Energy Institute (HNEI) solar-cell laboratory using a Thermo Oriel solar simulator (shown in Fig. 3.13). These measurements were performed to compare the actual performance of individual cells to theoretical values, as well as to select the highest quality solar cells based on the test results.

Based on the actual power-generating characteristics of the solar cells, and the power consumption of various onboard electronic components, a power budget was created. Shown in Table 3.2, this budget illustrates the worst-case scenario for the satellite system.

During a 90-minute orbit (in 350–425 km, 28–55 deg inclination LEO), the photovoltaic solar cells are capable of generating 21.6 Wh of energy during the nearly 53 minutes of sun-on time. However, the high continuous bus power consumption and payload (experiment) power consumption throughout the 90-minute orbit, including 37 minutes of solar eclipse, produces a depth-of-discharge of 27.7% per orbit. This means that within four orbits (assuming the system starts fully charged), the battery will be completely discharged. As this worst-case power budget shows, *Hokulua* requires a power-management scheme, either software or hardware-based, to ensure continuous and reliable power production and distribution throughout its mission lifetime.

4. Mechanical Structure and Analysis

The mechanical structure for *Hokulua* was developed to conform to the 1.5U ($10 \times 10 \times 15$ cm) cubesat standard size. This standard (1U – 3U) specifies the

Table 3.2 *Hokulua*, Worst-Case Power Budget

PV cells	7	W
Voltage boost	90	%
Reg. dc/dc converter	90	%
Sun-on time	0.8766	h
Sun-off time	0.615	h
Cycles	16	orbits/day
Battery energy	21.6	Wh
D.O.D.	30	%
Experiment time	0.25	h
Continuous power weight	95.7	%
Continuous power max.	3.543248	W
Useful cont. power max.	3.188923	W
Experiment power max.	15.19814	W
Useful exp. power max.	13.67833	W
Actual DOD	273.67886	%

amount of space filled by a satellite within the 10 × 10 × 30-cm holding volume of a PPOD, a CalPoly-designed picosatellite storage and deployment device (data available online at http://cubesat.atl.calpoly.edu/pages/documents/launch-providers.php).

A 1.5U structure was necessary to fit the large battery pack used to satisfy the high power consumption of the scientific payload. Because there was no commercially available 1.5U structure, the mechanical structure (shown in Fig. 3.14) was custom-designed by UH mechanical engineering student Lance Yoneshige. This anodized aluminum structure was designed with mounting brackets for PCBs

Fig. 3.14 *Hokulua's* custom-designed mechanical structure.

Fig. 3.15 Part of the *Hokulua* development team, demonstrating the satellites' retrodirective antenna technology.

and an integrated battery box, as well as a skeleton shell for minimized mass while maintaining structural integrity.

5. System Integration and Testing

The *SIT* team for *Hokulua* included a few key and dedicated members from the core satellite design team. The SIT team consisted of project engineer and PGD leader (Blaine Murakami), as well as subsystem leaders for RAT (Justin Roque), TTC (Thomas Mizuno), MSA (Lance Yoneshige) and CDH (Colin Yamaguchi), shown from left to right in Fig. 3.15 (missing: Colin Yamaguchi).

This small five-member team was well suited for leading the system integration and testing effort. Because all members were leaders for their corresponding subsystems, each had a detailed, working knowledge of a single aspect of the satellite. Although hardware and software integration between the various subsystems was quite complex, the close working environment provided an opportunity for all integration and test concerns to be addressed and resolved immediately.

To show the full functionality of the integrated satellite and payload system, a demonstration board was developed to output the satellite's telemetry data. Shown in Fig. 3.16, this board served as a single unit test bench for the *Hokulua* system.

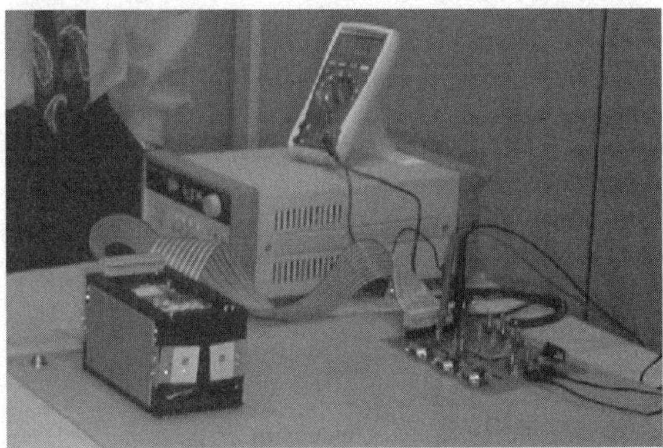

Fig. 3.16 Integrated system testing configuration for *Hokulua*.

Fig. 3.17 Integrated *Hokulua* system with retrodirective antenna array.

B. Scientific Mission and Payload

The primary scientific mission for this second phase of the UH Small Satellite Program was the development and demonstration of quadruple subharmonic phase-conjugating retrodirective antenna array technology for nanosatellites (Fig. 3.17) [6–8].

Although interest in retrodirective antenna array technologies has increased in the last decade, because of the rapid growth of wireless communication technologies, UH was the first to demonstrate wireless communication through retrodirective crosslinks on small satellites. This technology demonstration was the first step in the realization of a distributed, ad hoc network of small satellites, which is discussed in the following section.

IV. *Ho`okele (Way Finder)*, 2005–2006

The most critical element in the aftermath of a terrorist attack, natural disaster, or other large-scale emergency is for response teams to establish rapid, accurate, and consistent situational assessment. Terrestrial-based sensing and communication systems are infeasible in these situations, as they are often the target or are compromised in these attacks or disasters.

A more attractive solution is a small-satellite surveillance and tracking network that is autonomous, reconfigurable, redundant, and readily deployable. As mentioned earlier, their small size and weight makes launching a network of these satellites more cost efficient than launching a single large satellite. One of the goals of the UH's Small-Satellite Program is to create a distributed network of small satellites. This sparked the third-generation project, *Ho`okele*, which demonstrated a proof-of-concept small satellite that could conceivably operate as a node in a crisis-management or disaster-monitoring network. This project was fitting, following the tragic events taking place in Southeast Asia (December 2004 tsunami) and the southern United States (Hurricane Katrina, August 2005).

From this proof-of-concept goal, the mission of this phase of the program was to provide low-Earth orbit satellite images for use in crisis management and disaster mitigation, and demonstrate intersatellite communication in an expandable satellite network. This small-satellite project was developed through a collaborative effort between UH and Novasol, a local engineering firm. The UH development team was responsible for the development of the image acquisition and tagging payload, which included an onboard charge-coupled device (CCD) network camera, a global positioning system (GPS), and an inertial measurement unit (IMU). The UH team was also responsible for designing an engineering model of the satellite bus capable of supporting the imaging payload. To achieve the primary mission goal, an engineering model of the satellite bus was developed to support a geo-referenced imaging payload. Although the engineering model would not be launched into orbit, testing procedures on the ground simulated the basic functionality of the satellite bus and payload.

A. Project Organization

Bus development tasks for this project were delegated to the main subsystem teams similar to the previous phases: *power generation and distribution* (PGD); *tracking, telemetry, and control* (TTC); *command and data handling* (CDH); and *mechanical structure and analysis* (MSA). The organizational chart for this phase is shown in Fig. 3.18.

As the organizational chart in Fig. 3.18 shows (based on the number of members per subsystem), the design team for *Ho`okele* was considerably smaller than the first two phases of the program. Although development of the first two satellites was successful, 70-member and 40-member teams (*Mea Huaka`i* and

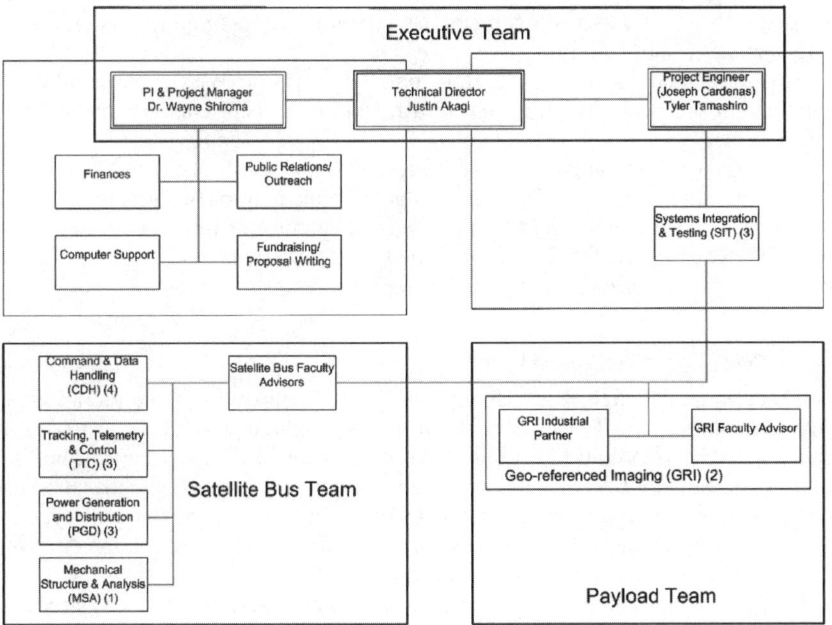

Fig. 3.18 ***Ho`okele* organizational chart.**

Hokulua, correspondingly) were too large for a simple small-satellite project. For this third phase, a nine-member team, consisting of eight undergraduate students, one graduate student, and a single faculty advisor developed and tested the entire satellite system in a nine-month timeframe. This new organization proved highly successful because all of the students were responsible for a specific part of the satellite system, as well as the final integration and testing effort (similar to the *Hokulua* SIT team).

B. System Bus Development

The *Ho`okele* design team followed the basic approach for developing the satellite bus, as explained in previous sections. However, new technologies for bus components provided the team with an opportunity to upgrade a number of hardware and software designs.

1. Command and Data Handling

The *CDH* subsystem was redesigned considerably from the previous two phases. The microcontrollers used for *Mea Huaka`i* and *Hokulua* were the Rabbit Semiconductor RCM2000 and RCM3400 modules. Although this technology was initially considered for its legacy value and reasonable 16 MB of onboard flash storage, the imaging payload requirements posed a problem. Because of the large amount of data each image would consume, combined with the limitations of the Rabbit microprocessor architecture, storage of an uncompressed 2 megabyte (MB) image to flash memory using the FAT file system would require ~20 seconds. However, based on camera specifications, a maximum transfer rate of ~15 seconds would be required.

This integration issue with the imaging payload, along with other serial communication protocol issue, resulted in a major component change for the CDH subsystem.

The legacy microcontroller was replaced with a Linux-based, ACME Systems Fox board (shown in Fig. 3.19; data available online at http://www.acmesystems.it). The board came with an integrated Ethernet connector for simple integration with the imaging payload. In addition, the Linux OS provided an excellent development platform for rapid software development, and easy software porting.

2. Tracking, Telemetry, and Command

There were also significant changes in the *TTC* subsystem from the previous two generations of UH satellites. Preliminary research provided a list of four specific radios that could be used at the core of the TTC subsystem. The first radio—the Kenwood TH-D7A—was the most commonly used on cubesats. This radio operates in the UHF band in the 430-MHz amateur radio frequency range, making frequency allocation and coordination fairly simple through the AMSAT (Amateur Radio Satellite Division) service.

Despite its familiarity among cubesat developers, the frequency and modulation used by this radio limit the maximum data rate to 9,600 bps. The payload images to be transferred from satellite to ground station would be ~0.5 MB

Fig. 3.19 Linux-based Fox board microcontroller, integrated onto satellite printed circuit board.

(compressed) or 2 MB (uncompressed). Based on a communication link budget analysis, it would require seven minutes to transfer one compressed image at the 9,600-bps maximum theoretical rate. Latency, distortion, and normal losses would reduce the maximum data throughput across the wireless link, increasing the transmission time another few minutes. This would have presented a problem because the communication window for a LEO is approximately 8–10 minutes, meaning that only one image or less could be transferred per pass.

The next two radios considered for the *TTC* subsystem operated in the 900-MHz frequency range. The MaxStream 9XTEND-PKG-R and Microhard MHX-900 radios featured an output power up to 1 W, and data throughputs greater than the Kenwood radio. The MaxStream radio featured a maximum data rate of 115.2 kbps; the Microhard radio provided a maximum rate of 230.4 kbps. These radios would both provide sufficient data rates for intersatellite image transfer.

Microhard also manufactures a similar radio operating in the 2.4-GHz frequency range, the MHX-2400 (shown in Fig. 3.20), with a 115.2-kbps maximum data rate (data available online at http://www.microhardcorp.com). This radio provided a number of the same features as its lower frequency counterpart, including built-in networking capability. Both the MHX-900 and MHX-2400 allow for each radio node to have a unique address as well as a network address for each group of nodes, which would be critical in the development of a wireless network of satellites.

The 430/144-MHz Kenwood radio was not selected because of its limited data throughput, which was a critical requirement based on mission specifications. The two 900-MHz radios were not selected because of frequency allocation issues.

The MHX-2400 was selected based on data bandwidth and frequency allocation concerns, as well as its reliability with built-in cyclic redundancy check (CRC) error detection and automatic retransmission upon error detection. This radio proved to be an excellent choice for small satellites, with its high power output capabilities (up to 1 W), spread-spectrum frequency-hopping communication

Fig. 3.20 Microhard MHX-2400 S-band (2.4 GHz) transceiver.

protocol (useful for wireless networks), and original equipment manufacturer (OEM) package for simple system integration.

3. Power Generation and Distribution

The *PGD* subsystem was redesigned to use power generated with Spectrolab 28.3% ultra-efficiency triple-junction GaAs solar cells, an improvement in efficiency over single-junction solar cells. The 14 solar-cell configuration, each cell measuring 6.9 × 4.0 cm, would provide 1.5 W of continuous power for the satellite during the satellite's sun-on period in orbit.

The solar cells charged a 4200-mAh Varta Microbattery lithium-ion battery pack (shown in Fig. 3.21) for use during solar eclipse. The battery pack included built-in overcharging and overcurrent protection for safe operation. In addition, a

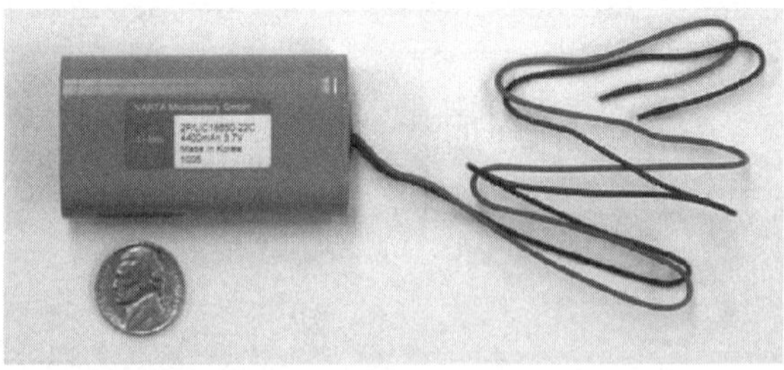

Fig. 3.21 Varta Microbattery 3.7 V, 4200-mAh lithium-ion battery pack (2P/LIC18650-22C) (data available online at http://www.varta-microbattery.com).

Dallas/Maxim DS2760 battery monitor was used to measure battery telemetry for continuous battery characterization and power system management.

C. Scientific Mission and Payload

1. Geo-Referenced Imaging

The imaging payload consisted of an Axis 206M network camera, a Garmin GPS unit, and a MEMSense nano-inertial measurement unit (nIMU) for geo-referenced imaging (data available online at http://www.axis.com, http://www.garmin.com, and http://www.memsense.com). The Axis camera (shown in Fig. 3.22) is capable of producing high-resolution motion JPEG images up to 1280 × 1024 pixels, using a high-quality progressive scan CMOS image sensor. This camera also interfaced directly to the CDH subsystem's ACME Fox board using Ethernet for ideal software control and image storage.

The Garmin global positioning system (GPS) unit was integrated into the payload system to tag the GPS location of each captured image. Although the GPS unit could not be utilized for the space environment (operable under an elevation of 18 km and a velocity of 515 m/s), this payload provided the students with an opportunity to learn the basics of developing a low-cost, geo-referenced imaging system.

The MEMSense nIMU (shown in Fig. 3.23) was integrated in the payload to provide satellite orbital telemetry data, including the attitude and rotational velocity, based on the gyro, accelerometer, and magnetometer data.

The growth of the technologies implemented in the satellite bus between these first three satellite projects not only coincided with advances in the scientific community, but also with the experience and knowledge gained within the UH Small-Satellite Program. Although there are a number of technology and architecture design changes between the design of each system, continued success at each phase of the program allowed us to move forward, improving the basic system platform for the next generation of small satellites.

Fig. 3.22 Axis 206M network camera.

Fig. 3.23 MEMSense nIMU.

V. *Ho`okia`i (Watchman)*, 2005–2007

UH's latest small-satellite, *Ho`okia`i*, is focused on integrating an active microthruster propulsion system and improved retrodirective array technology into the current satellite bus architecture. This satellite bus incorporates many technologies developed in previous satellite generations such as software-level housekeeping, attitude determination, and COTS-based S-band wireless communication. Because a major advantage of small satellites is their potential to operate as a distributed network, wireless networking for an ad hoc satellite constellation is also under development.

The overall goal of this fourth-phase project is to provide enabling technologies for an autonomous, flexible, dynamically reconfigurable, redundant and readily deployable constellation of nanosatellites.

A. Project Organization

Similar to previous phases of the program, satellite bus development for this project is divided into the basic CDH, TTC, PGD, and MSA subsystems. However, a new subsystem, *ground station facilities* (GS), discussed in Chapter 16, was created to develop a terrestrial-based satellite command station. In addition to the satellite bus team, two payload teams consisting of graduate students (and one undergraduate) worked on the *retrodirective antenna array* and the *microthruster propulsion* systems.

Students and mentors were divided into teams based on their engineering discipline; however, the small size of the design team provides an opportunity for students to work closely with each other throughout the development process. Like the previous phase of the project, *Ho`okele*, this project was being carried out by a small cross-disciplinary team, consisting of only eleven graduate and undergraduate students, advised by two faculty members (Fig. 3.24). As the executive team learned in earlier phases, a large organization is not necessarily as effective as a small design team. The large organization provides a number of advantages, such as more hands and heads and less work per person. However, the smaller design team has a better learning experience in all aspects of satellite development and considerable professional growth by leading the development of an entire satellite subsystem.

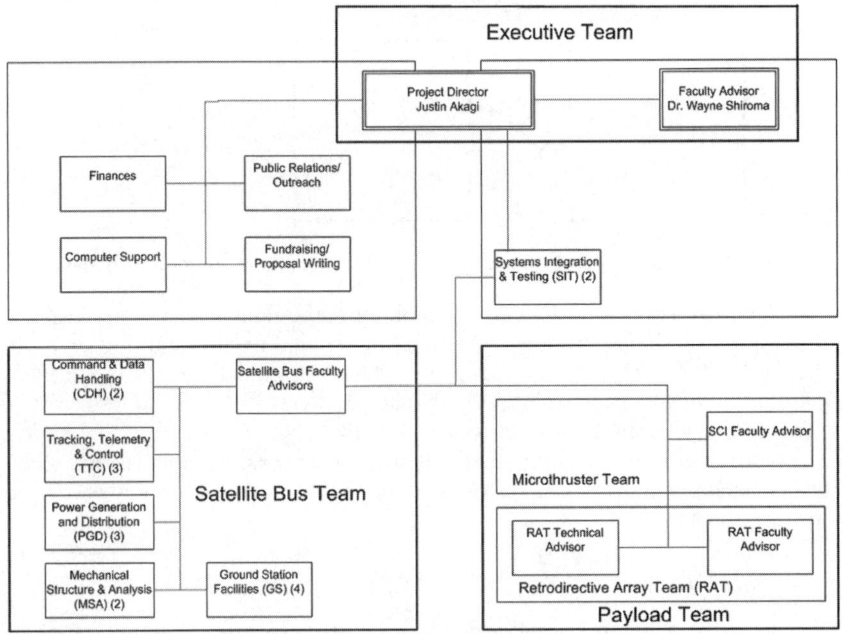

Fig. 3.24 ***Ho`okia`i*** **organizational chart.**

B. Scientific Mission and Payload

1. Phase-Detection Phase-Shifting Retrodirective Antenna Array

One of the primary scientific missions for this phase of the UH Small-Satellite Program is the development of improved retrodirective antenna array technology for small satellites. A *quadruple subharmonic phase-conjugating retrodirective antenna array* was developed and demonstrated on *Hokulua* in the second phase of the program. However, the phase-detection, phase-shifting architecture currently being researched by UH provides an advantage of plug-and-play capability for simple integration with COTS transceivers, although the goal of this payload is to demonstrate the functionality of a plug-and-play retrodirective antenna array for secure data communication in a wireless network. Further details regarding this technology and implementation can be found in Chapter 17.

2. Microthruster-Based Attitude Control

One of the design concerns faced by small-satellite developers is addressing the limitations of attitude control caused by strict mass and volume limitations. UH is currently working on a novel type of microthruster technology that is characterized by its stability, lightweight, low toxicity, and simple controls. This technology, illustrated in Fig. 3.25, involves the use of low-impulse thrusters fueled by sublimating naphthalene, that may or may not be heated by a low-current electrical resistance. The thrusting jet of naphthalene gas is controlled by an equilibrium vapor pressure chamber and electrical current (temperature) input

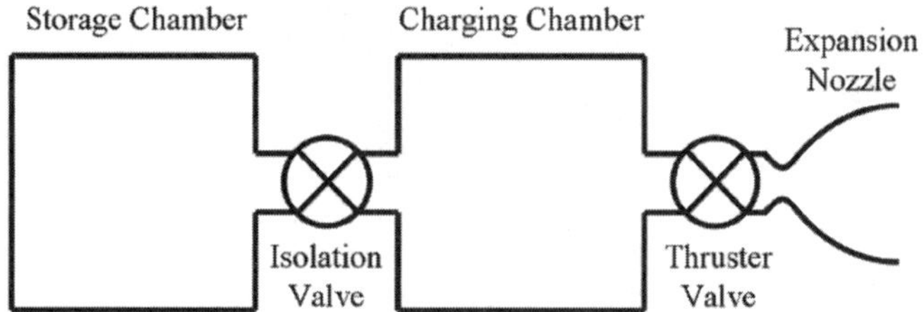

Fig. 3.25 Microthruster propulsion design using sublimating naphthalene.

alone. The key feature of the new technology is the lightweight, non-sloshing, ultra-low-power, long duration of the thrusting jet and its ability to fire multiple times or continuously. This makes it an ideal replacement for passive attitude control systems like hysteresis rods and magnets.

The goal of this microthruster system is to demonstrate its ability to continuously correct attitude of LEO small satellite to account for atmospheric drag. Further details regarding UH research on this microthruster technology can be found in Chapter 10.

3. *Wireless Networking for ad hoc Satellite Constellations*

To demonstrate the potential to create a distributed satellite network, a major advantage of small-satellite applications, UH is developing networking capabilities for small-satellite systems. This capability will create the potential to launch clusters of identical satellite nodes, increasing processing power for the constellation, network redundancy, and tolerance to single-point failure.

The primary satellite radio for *Ho`okia`i*, the Microhard, Inc., MHX-2400, is used as the communication unit for the wireless link. This spread-spectrum, frequency-hopping transceiver was selected based on its built-in wireless networking capability. The embedded Linux kernel in the onboard microprocessor performs the networking protocol, based on existing wireless networking topology.

The ring topology for this satellite network (Fig. 3.26) specifies a single active master in the network. The master node is responsible for communicating with all of the slaves (within direct communication range, as well as those requiring indirect communication using repeater nodes). Because the master is unable to communicate directly with all slaves in the network, each slave also acts as a repeater, transparently relaying data to slaves out of direct contact.

The node-to-node hopping (between master and slave, through intermediate repeaters) is based on primary and secondary hopping patterns, set in the radios' configurations. Using this method, all data received on a repeater's primary hopping pattern will be broadcasted on the repeater's secondary hopping pattern for data not addressed to it. Although this technique of data transmission allows all nodes within the network to communicate with the master, data transmission is unidirectional around the ring (from primary to secondary). To prevent a zero-master scenario, if the master unit fails, the network has the ability to reconfigure itself and designate a new slave to become the master unit.

Time Division Multiple Access (TDMA) is used to divide communication time between master and slaves (Fig. 3.27). When TDMA is enabled on the master, it

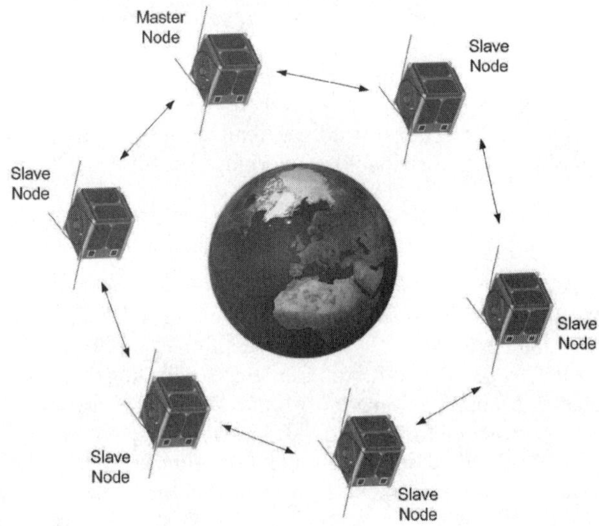

Fig. 3.26 Satellite network communicating using ring topology.

polls each of the slaves in the network, allocating a fixed amount of time (~30 to 500 ms) to each slave to receive/transmit as much packets as possible before moving onto the next slave. This cycle is split equally between all nodes to ensure a regular communication schedule. Because slaves cannot directly communicate with other slaves, the master is also be responsible for interslave communication routing. This method of routing must be encoded in the protocol framing to be used between the various nodes.

VI. Chapter Summary

During the first few generations of satellites developed by the UH Small-Satellite Program, many UH undergraduate students gained technical know-how

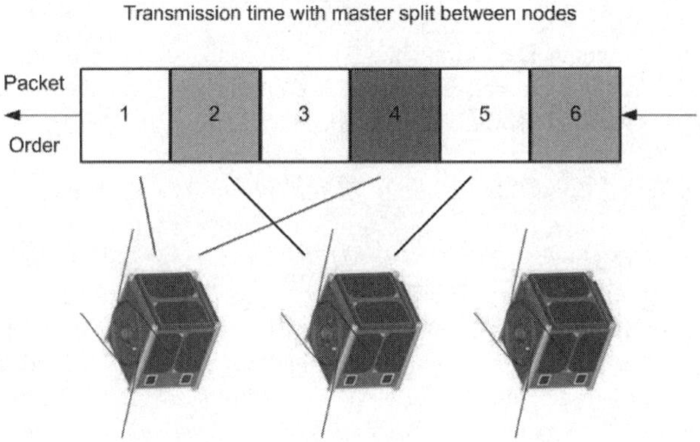

Fig. 3.27 TDMA: transmission time with master split equally between nodes.

in satellite-systems engineering as well as professional experience through their involvement in a real-world engineering project. Even though the program began with the design and manufacture of a simple satellite bus using commercial-off-the-shelf components, the lessons learned in each subsequent phase have continued to lead the program toward the realization of a launch-ready small-satellite bus platform with distributed networking capabilities.

Acknowledgments

We would like to thank all the previous and current members of the UH Small-Satellite Program, without whom this program could not have been successful. We extend huge thanks to Aaron Ohta, Michael Tamamoto, Byron Wolfe, Blaine Murakami, Scott Kawakami, and Joseph Cardenas for leading their teams to success as undergraduates. Many, many *mahalos* ("thank you" in Hawaiian) to everyone who stayed on the project until the end, even after they graduated! We would also like to thank Jason Akagi, Toy Lim, Dr. Carlos Coimbra, Dr. Tep Dobry, and all the other advisors and industry mentors who have helped guide this program.

References

[1] Shiroma, W. A., Ohta, A. T., and Tamamoto, M. A., "The University of Hawaii CubeSat: A Multidisciplinary Undergraduate Engineering Project," 33rd ASEE/IEEE Frontiers in Education Conference, AIAA Paper S3A-7-11, Nov. 2003.

[2] Piket-May, M. J., and Avery, J., "Service Learning First Year Design Retention Results," 31st ASEE/IEEE Frontiers in Education Conference, AIAA Paper F3C-19, Oct. 2001.

[3] Murakami, B. T., "The Design, Fabrication and Testing of a Nanosatellite System: A How to Approach," Capstone Design Project Rept., Univ. of Hawaii, Honolulu, May 2005.

[4] Hansen, J. A., "PIC-et Radio: How to Send AX.25 UI Frames Using Inexpensive PIC Microprocessors," *17th ARRL and TAPR Digital Communications Conference*, ARRL, Newington, CT, 1998, p. 29.

[5] Hansen, J. A., "PIC-et Radio II: How to Send AX.25 UI Frames Using Inexpensive PIC Microprocessors," *19th ARRL and TAPR Digital Communications Conference*, ARRL, Newington, CT, 1998.

[6] Sung, S. S., Roque, J. D., Murakami, B. T., Shiroma, G. S., Miyamoto, R. Y., and Shiroma, W. A., "Retrodirective Antenna Technology for CubeSat Networks," *IEEE Topical Conference on Wireless Communication Technology*, IEEE, New York, Oct. 2003, pp. 220, 221.

[7] Murakami, B. T., Roque, J. D., Sung, S. S., Shiroma, G. S., Miyamoto, R. Y., and Shiroma, W. A., "A Quadruple Subharmonic Phase-Conjugating Array for Secure Picosatellite Crosslinks," *2004 IEEE MTT-S International Microwave Symposium Digest*, Vol. 3, June 2004, pp. 1687–1690.

[8] Roque, J. D., Sung, S. S., Murakami, B. T., Shiroma, G. S., Miyamoto, R. Y., and Shiroma, W. A., "A Coupled-Antenna Interrogator/Receiver for Retrodirective Cross-links in a Distributed Nanosatellite Network," *2005 IEEE/ACES International Conference on Wireless Communications and Applied Computational Electromagnetics*, IEEE, Piscataway, NJ, April 2005, pp. 610–613.

II. Missions

Chapter 4

Survey of Atmospheric and Other Research Projects Employing Small University Satellites

Purvesh Thakker* and Gary Swenson†
University of Illinois at Urbana-Champaign, Urbana, Illinois

I. Introduction

PICOSATELLITES have been applied to many research problems. This section provides an overview of those research projects. Many of them relate to atmospheric research, but there is also research activity in other areas.

II. Atmospheric Activity

Although most university satellite programs have focused more on their educational missions, a few have included instruments that demonstrate the potential scientific return of university satellites. Investigators employ remote sensing techniques, which detect phenomena from a distance, and in situ techniques, which directly measure the phenomena of interest, to make atmospheric measurements. The University of Illinois provides the primary example of remote sensing instruments for university satellites. Taylor University, Montana University, and a number of other universities have also employed university satellites to perform atmospheric science missions. The STEDI/UNEX are listed after the other university satellites because they occupy a space between university satellites and industry satellites. As shown in this list, the University of Illinois has developed four remote sensing instruments and has proposed two additional remote sensing instruments:

1. ION-1 Oxygen airglow photometer
2. ION-2 Neutral hydrogen photometer
3. TEST 630-nm CCD imager
4. TEST Hertzberg photometer

---

*Program Manager, ION Cubesat, Department of Electrical and Computer Engineering.
†Professor, Department of Electrical and Computer Engineering.

5. Spectra-photometer (proposed)
6. RF beacon (proposed)

The completed ION-1 satellite, a 2-kg cubesat, had a photometer that was to perform the first global assessment of wave perturbations of the O_2 atmospheric band in the Earth's mesosphere. It was to measure waves of oxygen airglow perturbations carried by wind (762 nm, nadir, 89 km). The currently in development ION-2 project, another 2-kg cubesat, has a two-channel photometer for measuring the brightness of airglow emission by neutral hydrogen (H) atoms in the Earth's exosphere (656.3 nm, zenith, >500 km). The partially developed TEST [1] satellite, a 30-kg nanosat jointly developed by Taylor University and the University of Illinois, contains a number of instruments including two optical instruments developed by the University of Illinois. The first optical instrument from Illinois, a 2.5-kg, 630-nm charge-coupled device (CCD) imager, studies irregularities in ionospheric electron densities caused by acoustic gravity waves (AGWs). The second optical instrument, a two-channel 2.5-kg photometer, studies the intensity of the O_2 Hertzberg bands in the 260–290-nm spectral region (nadir, 95 km). Illinois has also proposed two instruments that have not yet been developed. The 6-kg limb spectra-photometer would monitor spectral emissions including auroral emissions vs altitude across all latitudes. The 3.4-kg RF beacon uses differential Doppler shift on multifrequency beacon signals to measure total electron content (TEC) along propagation links.

As discussed earlier, the TEST satellite also contains a number of in situ and remote sensing instruments from Taylor University, which study thunderstorm effects in space:

- TEST Langmuir plasma probe
- TEST electric field booms
- TEST very low-frequency (VLF) receiver
- TEST solid-state detector spectrometer (SSD)
- TEST transient photometer
- TU Sat-1 Langmuir plasma probe
- TU Sat-1 tether
- TU Sat-1 Nitol tether

The Langmuir plasma probe has a low energy (0 to 6 eV) plasma probe circuit for measuring the thermal plasma density and temperature of the satellite environment. Two electric field booms provide differential voltages (0.005 V to 1 V/m DC, 3-kHz ac) giving E-field strength information in the horizontal and vertical directions. The VLF receiver consists of a loop antenna that detects electromagnetic information in the 100- to 30-kHz range, such as produced during thunderstorm charge buildup and lightning discharge. The solid-state detector spectrometer (SSD) measures electrons ($E = 10$ keV to 1 MeV) and ions ($E = 80$ keV to 2 MeV). The transient photometer (viewing in the nadir direction) measures light intensity at 391.4 nm (nitrogen line) to examine peak light emission of lightning and the aurora. The three-axis magnetometer measures the direction of the magnetic field (40 μGauss to 2 Gauss). In addition to TEST, Taylor University (TU) has developed TU Sat-1 [2], a 2-kg cubesat with some instruments. The Langmuir plasma probe on TU Sat-1 measures plasma density and temperature for assisting in the study of the ionosphere including auroral regions, equatorial fountains, plasma waves, and other plasma phenomenon. Long-term plans include deployment of a constellation of satellites with these probes to make multipoint measurements in similar areas simultaneously. TU Sat-1 also includes a tether, which is used both for attitude control and as an electric field data collector. Finally, TU Sat-1 measures magnetospheric VLF radio waves in the kilohertz range, which are impossible to measure at the Earth's surface, using a 30-m electrically conductive Nitol tether (10 Hz to 1 MHz).

Montana State University has developed some instruments (MEROPE Geiger counter and Maia solid-state silicon detectors) including MEROPE [3], which contains a reproduction of the scientific payload aboard the United State's first satellite, Explorer-1 (1 February 1958). MEROPE was to carry a single Geiger counter into a 600-km, sun-synchronous polar orbit to measure the corpuscular radiation that characterizes the Van Allen radiation belts. MEROPE also provides an example of how satellite instrumentation and electronics have become smaller. The Explorer-1 was 14 kg, whereas MEROPE is a 1-kg cubesat. In addition to MEROPE, Montana has developed Maia [4], a 25-kg nanosat. Maia characterizes variations in the energetic charged particle environment in the topside ionosphere by measuring precipitating electrons (with energies >1 keV) and ions (with energies >50 keV). Ionization produced in the ionosphere by precipitating energetic particles contributes the largest single unpredictable component of ionospheric density. There is a need for small satellite operational monitoring of these phenomena as they are sometimes influenced by violent and unpredictable solar eruptions. Maia also focuses on the development and testing of solid-state silicon detectors in a new lower energy domain (down to a few keV) that has no prior space usage.

III. Other Research Areas

Some other universities have also demonstrated the utility of cubesats for scientific research of various types:

- Stanford—QuakeSat extremely low-frequency (ELF) detector
- Stanford—bio-nano satellite explorer yeast cell experiment
- University of Washington—plasma impedance probe/dc probe
- University of Washington—GPS scintillation measurements
- Italy—AtmoCube magnetometer and dosimeter
- University of Kansas—KUTESat dosimeters
- Hankuk—Hausat-2 animal tracking system
- Hankuk—Hausat-2 electric plasma probe

Stanford University's QuakeSat [5], a 3-kg cubesat, detects ELF magnetic field signals worldwide, which correlate with earthquakes in both the weeks leading up to an earthquake and the months following an earthquake. This inexpensive experimental satellite was built and flown to help determine the design parameters and values needed to build a larger research satellite. Bio-Nano Satellite-1 [6], a 10-kg satellite from Stanford University and NASA Ames Research Center, studies the growth rate of yeast cells in microgravity.

A University of Washington Space Design class [7] partially designed and developed two cubesat mission architectures for ionospheric measurements based on two instrument packages. The first architecture consists of a combined plasma impedance probe (PIP)/dc probe system on two satellites separated by a tether. The second consists of two separate cubesats that make GPS scintillation measurements. The scientific objective of both missions is to take distributed measurements within the ionospheric plasma to aid the understanding of ionospheric density structures and to contribute to the creation of accurate models.

The goal of AtmoCube [8], a 1-kg cubesat from Italy, is to build a precise map of the Earth's magnetic field and radiation. The satellite uses a magnetometer, dosimeter-radiometer, and a global positioning system to develop these maps. The University of Kansas KUTESat program consists of two cubesats and a nanosat. The cubesats contain a miniature dosimeter, while the nanosat contains an array of miniature dosimeters. Hausat-2 [9], a 25-kg nanosat developed by Hankuk Aviation University, studies the scope of activities and ecology of animals using animal tracking system (ATS). It also collects space information data from an electric plasma probe (EPP).

The STEDI/UNEX satellites include SNOE (STEDI 1, 115 kg), TERRIERS (STEDI 2, 125 kg), CATSAT (STEDI 3, 168 kg), CHIPSat (UNEX 1, 60 kg), and IMEX (UNEX 2, cancelled). SNOE contains an ultraviolet spectrometer for measuring nitric-oxide altitude profiles, a two-channel auroral photometer for measuring auroral emissions beneath spacecraft, and a five-channel solar soft X-ray photometer. TERRIERS contains a gas ionization solar spectal monitor (GISSMO), a single-element imaging spectrograph (SEIS), and photometers. CATSAT contains a soft x-ray spectrometer (SXR), a hard x-ray spectrometer (HXR), and a directional gamma x-ray spectrometer (DGS). CHIPSat contains a cosmic hot interstellar plasma spectrometer (CHIPS). Finally, IMEX studies the response of the Earth's Van Allen radiation belts to variations in the solar wind.

References

[1] Voss, D. L., Kirchoff, A., Hagerman, D. P., Zapf, J. J., Hibbs, J., Dailey, J., White, A., Voss, H. D., Maple, M., and Kamalabadi, F., "TEST—A Modular Scientific Nanosatellite," *Space 2004 Conference and Exhibit*, AIAA Paper 2004-6121, Sept. 2004.

[2] Holmes, W. C., Bryson, J., Gerig, B., Oehrig, J., Rodriguez, J., Schea, J., Schutt, N., Voss, D., Voss, J., Whittington, D., Bennett, A., Fennig, C., Brandle, S., Dailey, J., and Voss, H. D., "TU Sat 1—A Novel Communications and Scientific Satellite," *16th Annual AIAA/USU Conference on Small Satellites*, Paper SSC02-I-1, Aug. 2002.

[3] Obland, M., Hunyadi, G., Jepsen, S., Larsen, B., Klumpar, D. M., Kankelborg, C., and Hiscock, W. A., "The Montana State University NASA Space Grant Explorer-1 Science Reflight Commemorative Mission," *15th Annual AIAA/USU Conference on Small Satellites*, Paper SSC01-III-2, Aug. 2001.

[4] Larsen, B., Klumpar, D. M., Obland, M., and Hiscock, W.,"The Montana Nanosatellite for Science, Engineering, and Technology for the AFRL/NASA University Nanosat Program," *17th Annual AIAA/USU Conference on Small Satellites*, Paper SSC03-IX-4, Aug. 2003.

[5] Flagg, S., Bleier, T., Dunson, C., Doering, J., DeMartini, L., Clarke, P., Franklin, L., Seelback, J., Flagg, J., Klenk, M., Safradin, V., Cutler, J., Lorenz, A., and Tapio, E., "Using Nanosats as a Proof of Concept for Space Science Missions: QuakeSat as an

Operational Example," *18th Annual AIAA/USU Conference on Small Satellites*, Paper SSC04-IX-4, Aug. 2004.

[6] Twiggs, B., and Kuroki, S., "BioExplorer Bus—Low Cost Approach [Satellite Design]," *IEEE Aerospace Conference Proceedings*, Vol. 1, 2002, pp. 1-427–1-434.

[7] Waydo, S., Henry, D., and Campbell, M., "CubeSat Design for LEO-Based Earth Science Missions," *IEEE Aerospace Conference Proceedings*, Vol. 1, 2002, pp. 1-435–1-445.

[8] Gregorio, A., Bernardi, T., Carrato, S., Kostadinov, I., Messerotti, M., and Stalio, R., "AtmoCube: Observation of the Earth Atmosphere from the Space to Study 'Space Weather' Effects," *Proceedings of Recent Advances in Space Technologies*, Aeronautics and Space Technologies Inst., Istanbul, 2003, pp. 188–193.

[9] Young Keun, C., Byoung Young, M., Ki-Lyong, H., Soo-Jung, K., and Suk-Jin, K., "Development of the HAUSAT-2 Nanosatellite for Low-Cost Technology Demonstration," *Recent Advances in Space Technologies, 2005. RAST 2005*, Aeronautics and Space Technologies Inst., Istanbul, 2005, pp. 173–179.

Chapter 5

ION-1 and -2 Cubesat Optical Remote Sensing Instruments

Purvesh Thakker,* Gary Swenson,† and Lara Waldrop‡
University of Illinois at Urbana-Champaign, Urbana, Illinois

I. Introduction

FROM 1999 to 2006, over 80 organizations registered intentions to develop cubesat-class satellites (1 to 3 kg), and more than two dozen have launched. This work illustrates the utility of this class of satellites for performing optical remote sensing missions. Cubesat deployers [1] typically hold three $10 \times 10 \times 10$-cm cubesats, which conform to a simple interface defined in the cubesat specification document (data available online at http://cubesat.atl.calpoly.edu/pages/documents/developers.php). These cubes can be combined to form larger double or triple cubesats that measure up to $10 \times 10 \times 33$ cm. The deployers are carried to orbit using unutilized space on existing launches. Thanks to this growing secondary payload launch infrastructure, universities and other organizations have begun using these spacecraft for education [2], scientific measurement [3–6], and space technology testing missions [4,7].

A. ION-1 and -2 Projects

This work describes the optical remote sensing payloads onboard ION-1 and -2, which are both double cubesats measuring $10 \times 10 \times 21.5$ cm. ION-1 (see Fig. 5.1) was completed in April 2005, but the satellite was lost when its rocket crashed back to Earth in July 2006. Nevertheless, there are many important results from the ION-1 project that are being published. The most important lesson is the demonstration that these satellites can support scientific remote sensing payloads. This capability allows new measurement opportunities for atmospheric scientists.

---

*Program Manager, ION Cubesat, Department of Electrical and Computer Engineering.

†Professor, Department of Electrical and Computer Engineering.

‡Research Scientist, Department of Electrical and Computer Engineering.

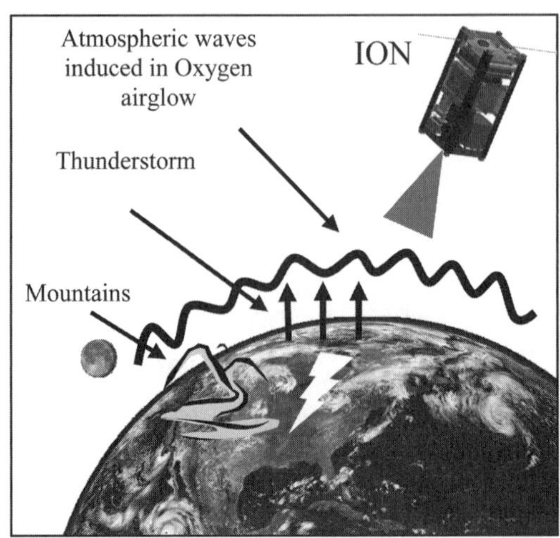

Fig. 5.1 ION-1 oxygen measurement.

ION-2 is currently under development and does not yet have a specified launch. Its design is split into two parts: the satellite bus and the optical payload. Duplicates of the satellite bus can be built to support a duplicate of ION-1's payload or future payloads.

B. Paper Overview

This paper first discusses the capabilities and limitations of the cubesat satellite platform for optical remote sensing missions. The following two sections describe the ION-1 and -2 missions in detail. Because the payloads of both satellites are composed of optical photometers, Section IV describes a photometer design procedure. Section V describes the application of the photometer design procedure to each spacecraft, while Section VI summarizes the conclusions from this work.

II. Conducting Science with Cubesats

A. Advantages and Limitations

The unique capabilities of cubesats offer a number of advantages over larger spacecraft. First, they can be developed in as few as one to two years by small teams of less than ten people. This keeps their cost low and makes the programs much more risk tolerant thanks to the smaller investment. The tolerance for risk is particularly evident in a university environment, where students earn valuable hands-on engineering experience even if the satellite does not function properly [2,8]. As illustrated by the ION-1 and -2 missions described in this work, cubesats also offer an affordable space-based sensor platform enabling potentially vast scientific return [3,4]. In addition, cubesats can serve as important preliminary steps toward the realization of larger spacecraft missions. They can do this by

Table 5.1 System Limitations

Size	10 × 10 × 33 cm
Lens diameter	8 cm
Sensor diameter	8 cm
Focal length	20 cm
Power	3 W (triple cube, 27.5% eff solar cells, four panels deployed to fixed positions)
Communications	1 MB/day (9600 baud), 135 kB/day (1200 baud)

testing new technologies, such as sensors [4], or by performing preliminary measurements to verify anticipated signal and noise levels [9]. Finally, cubesats are well suited to multispacecraft measurements including multipoint and multi-angle measurements.

The primary restriction on cubesat mission design is the small spacecraft size, measuring at most 10 × 10 × 33 cm. The scientific payload must fit in the designated space together with all support systems. The small size limits the surface area of any lenses used in remote sensing, for example, which in turn limits the light gathering power and therefore the sensitivity of the instrument. Similarly, cubesats face strict power limitations. Even the largest cubesats using state-of-the-art solar cells and extended solar panels cannot support more than a few watts of power. For communications, cubesats typically use commercially available hardware from the ham radio community. Faster rates might be possible, but currently data rates generally do not exceed 9600 baud resulting in a maximum download per day of about 1MB, for a single ground station (Table 5.1).

One challenge that most cubesat programs face is securing adequate resources. Because no formal sources of funding currently exist, cubesats are typically developed in a university environment. Although this infrastructure offers modest expertise, equipment, and time commitment from a particular student, university cubesat programs have flourished owing to their significant educational benefits and potential scientific merits.

B. Science Payloads

The ION-1 and -2 cubesat missions have objectives that provide simple, unique, and global measurements of geophysical interest. The general scope of sensor technology for low-light-level emissions in the UV-visible spectra include classical photomultipliers, photodiodes (including avalanche photodiodes, or APDs), and charged-coupled-device (CCD) arrays. Photon count-limited photomultipliers and APDs can be packaged together into small volumes with associated electronic amplifiers and power supplies. As demonstrated by ION-2, a two-channel photometer can be packaged, with optics, into less than one cube, while simple electronics can count and store measurements.

CCD sensors are more difficult to include on these small satellites because of the limited downlink available. To make the downlink budget manageable, these small satellites can use image compression, undersampling, or selective download of certain pixels. For example, it is reasonable to consider using an imaging spectrometer that samples only the spectral regions of interest. As another example, the TEST nanosatellite incorporated a 50-mm aperture CCD into two cubes in order to conduct tomographic imaging of thermospheric airglow from two orbital locations [10]. With this approach, small amounts of imaging information can be routinely downloaded.

As a final note, photometry and imaging experiments should always have a measurement of the background. Where one is attempting to measure either a point source or image a monochromatic emission, a second channel should be dedicated to measure the backgrounds.

III. ION-1 and -2 Missions

A. ION-1 Mission Overview

ION-1's design targets the first global assessment of wave perturbations of the O_2 atmospheric band in the Earth's mesosphere (Fig. 5.1). This is accomplished by performing photometric measurements of molecular oxygen airglow emissions. The measurements should show brightness perturbations in the airglow showing atmospheric waves that are generated by various effects. These effects include convection from shear convective instabilities, including thunderstorms, and orographic forcing, such as winds over mountains. The targeted perturbations are expected to vary in brightness by 0.5 to 50% of the layer mean brightness.

These perturbations propagate through the atmosphere with 15-km to >2500-km horizontal wavelengths. By studying these waves, atmospheric scientists learn how energy transfers across large spatial regions, contributing to knowledge of upper atmospheric dynamics. Liu and Swenson [11] have modeled the response of the layer brightness perturbations to atmospheric waves propagating through the layer. These small-scale atmospheric gravity waves (AGWs) are a major source of momentum and energy in this upper-altitude region of the atmosphere, where little observational morphology currently exists [12].

ION-1's photometer consists of a simple remote sensing instrument designed to observe the Earth's chemiluminescent emission from the O_2 atmospheric band (0,0) at 762 nm. ION-1 is equipped with a filter in front of its photometer to isolate the 762-nm emission. The bright airglow emission layer has peak brightness at about a 94-km altitude, and it has a vertical thickness of about 8 km. This emission has a mean, nadir brightness of about 6 kR. It is also self-absorbed in the lower atmosphere, and so a downward-viewing instrument at this wavelength will not be contaminated by city lights, for example. The self-absorption also prevents measurement of the signal by ground-based sensors, which in turn view the much weaker (0-1) band. Satellite detection offers the major benefit of global coverage. ION's polar inclination (98-deg) orbit gives it the opportunity to gather these data around the Earth's polar regions, which offer pole-to-pole coverage of particular interest to atmospheric scientists.

The mission places an attitude stabilization requirement on the system within the Earth's reference frame. The attitude knowledge and stabilization objective of pointing within ±5 deg of nadir (downwards) ensures that the instrument

Table 5.2 ION-1 Parameters

Science Parameters	
Source	O_2 atmospheric band (0,0)
Thickness	8 km
Target altitude	94 km
Intensity	6000 Rayleighs (mean), 0.5% to 50% variation
BG intensity	20 R/nm
Wave size	15 km to >2500 km
Wavelength	762 nm
Integration time	1 s
Signal-to-noise target	50
Satellite Parameters	
Satellite altitude	500 km
Inclination	98 deg
Attitude	<5 deg off nadir
Speed	7.6 km/s

observes the targeted waves correctly. If the satellite does not achieve its attitude objective, rough attitude stabilization of ±20 deg of nadir would still provide useful data. In fact, even without attitude stabilization useful data could be obtained by modeling the spin period of the spacecraft and time correlating the data to the phase of the spin cycle. Then synchronized observations could be mapped to the airglow layer. Higher attitude stabilization accuracy allows the location of the data on the Earth to be mapped and cross analyzed with measurements from other instruments. The instrument has a target signal-to-noise ratio of 50. Because the satellite moves at 7.6 km/s through the atmosphere, the integration time must be no larger than about 1 s to sample the wave component in the orbital plane of 15 km or longer. The data of interest are globally sampled brightness perturbations that show wave amplitudes and variance (see Table 5.2).

B. ION-2 Mission Overview

ION-2's mission objective is to measure the brightness of airglow emission by neutral hydrogen (H) atoms in the Earth's exosphere, an extremely tenuous atmospheric region located above the exobase (~500 km) and extending out to tens of Earth radii. Although many of the H atoms in the exosphere are gravitationally bound to the Earth, an important fraction ultimately escapes from the Earth's atmosphere and is lost to space. This escape is enhanced during periods of high geomagnetic activity because efficient charge-exchange reactions between exospheric H atoms and magnetospheric storm-time H^+ ions further energize the neutral particles [13]. Knowledge of the exospheric H distribution therefore is important for understanding both the long-term evolution of the terrestrial atmosphere as well as the response and recovery of the atmosphere and magnetosphere following geomagnetic storms.

Remote sensing of airglow emission from exospheric H atoms is one of few means to characterize exospheric structure and dynamics. The ION-2 cubesat mission will measure the brightness of H Balmer-alpha (Hα) emission at 656.3 nm, which is excited by resonant fluorescence of Lyman β photons originating either directly from the sun or from successive resonant scatterings by other H atoms in the exosphere. As demonstrated by He et al. [14], measurements of Hα emission brightness can be inverted using a model of Lyman β radiative transfer to yield estimates of exospheric hydrogen density [H]. These estimates can be used to study periodic abundance variations, arising from solar or seasonal cycles, to characterize the exospheric response to sporadic geomagnetic activity, and to assess secular trends, which are potentially of anthropogenic origin.

ION-2 employs two photometers to measure the Hα column emission brightness, typically having nighttime values of ~5–15 R, but varying by up to a factor of two over a few days [15]. The first photometer directly measures exospheric Hα emission intensity, in counts, at 656.3 nm, while the second photometer measures the spectral background at a nearby wavelength (690.0 nm), expected to have a nighttime intensity of about 7–20 R or more if potential extraterrestrial sources of Hα are in the field of view (Table 5.3). Hα emission brightness is linearly proportional to the difference between the measured signal and background intensities, and the instrument has a target signal-to-noise ratio of 50. To avoid the difficult task of removing strong solar continuum background from the relatively weak Hα signal, observations will be conducted when the spacecraft is within the Earth's solar shadow.

Like ION-1, ION-2 is expected to have a circular orbit near 500 km and the same ±5-deg pointing accuracy requirement. Unlike ION-1, ION-2's photometers point upwards (zenith) towards the exospheric source of Hα emission. This viewing geometry avoids two significant complications of ground-based remote sensing of the Hα emission. It avoids contamination by OH emission at 656.9 nm and 655.4 nm from the lower atmosphere below the spacecraft orbit, and it minimizes signal attenuation through the atmosphere. Because timescales of H variation are much longer than the waves targeted by the ION-1 mission, the integration time is correspondingly longer at 6 s.

Table 5.3 ION-2 Parameters

Science Parameters	
Source	H alpha
Intensity	5 R to 15 R
BG intensity	7 to 20 R/nm
Target altitude	>500 km
Wavelength	656.3 nm (main channel) 690.0 nm (background channel)
Integration time	6 s
Signal-to-noise target	50
Satellite Parameters	
Satellite altitude	500 km (goal)
Inclination	90 deg (goal)
Attitude	<5 deg off zenith
Speed	7.5 km/s

Table 5.4 Photometer Design Parameter Units

r_1, r_2, f	m
A_1, A_2	m^2
FOV	deg
LGP	m^2 str
PhotonRate, DarkCountRate	photons/s
Signal	photons
Intensity	Rayleighs
IP	s
Signal	photons
Temperature	K
σ	Sqrt photons
SNR	Sqrt photons
SampleSize	kB
SampleTime	s
DataRate	kB/day

IV. Photometer Design Procedure

Each of the three photometers on ION-1 and -2 is designed in a similar manner using Eqs. (5.1–5.11). In practice, a spreadsheet is used to calculate different possible values until a viable design is found. In this chapter, only the final design values are shown. This section describes the photometer design procedure, and then Section V shows the final results for ION-1 and -2. Table 5.4 shows the units for all variables in Eqs. (5.1–5.11).

A. Signal to Noise

The optical design consists of an optical filter in front of a lens that focuses light onto a sensor (see Fig. 5.2). The lens is precisely one focal length f away from the sensor, and the field of view (FOV) of the instrument is determined by the size of the sensor r_2 and the focal length of the lens as follows:

$$\text{FOV} = 2 * \arctan\left(\frac{r_2}{f}\right) \tag{5.1}$$

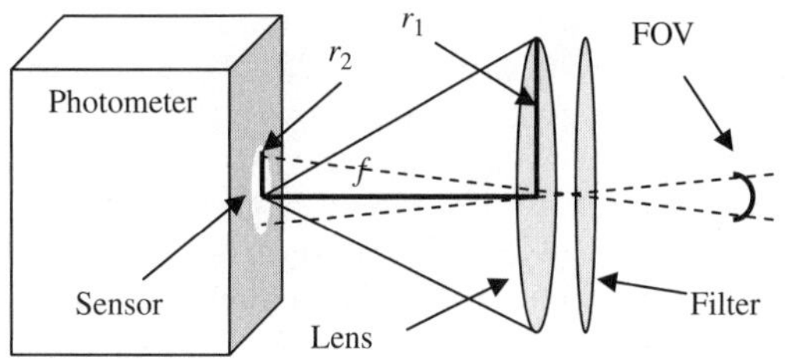

Fig. 5.2 Photometer, lens, and filter.

The light gathering power (LGP) of the instrument is determined by the area of the sensor (or stop in front of the sensor) A_2, the area of the lens A_1, and the focal length of the lens as follows:

$$\text{LGP} = A_1\Omega = \frac{A_1 * A_2}{f^2} = \frac{\pi r_1^2 * \pi r_2^2}{f^2} \tag{5.2}$$

The rate at which photons enter the system is the product of the light gathering power and the source intensity as shown in Eq. (5.3), where *intensity* is in units of Rayleighs, or 10^{10} photons/m^s.

$$\text{Photon Rate} = \text{Intensity} * \frac{10^{10}}{4\pi} * \text{LGP} \tag{5.3}$$

As photons land on the photometer, they are optically amplified through several stages within the photometer. The resulting signal is then further amplified electrically into pulses. The rising edges of these pulses can be counted over an integration time to obtain the *signal*. Photon counts are reduced by the system quantum efficiency QE, which is the product of the individual filter, lens, and photomultiplier tube quantum efficiencies. The measured signal, in units of photon counts, is the rate of photon detection integrated over a period I, and reduced by the system quantum efficiency QE, as follows:

$$\text{Signal} = \text{Photon Rate} * \text{IP} * \text{QE} \tag{5.4}$$

In addition to pulses triggered by the reception of photons, heat triggers pulses creating dark counts. The dark count test results from ION-1 are shown in Fig. 5.3, and the linearized version of this figure is shown in Fig. 5.4. Once the temperature dependence of this undesired dark count is characterized, it can be removed with the help of onboard temperature measurements using Eq. (5.5). The linear constant A corresponds to the slope of the line in Fig. 5.4, while C corresponds to the vertical intercept of that line. *Temp* corresponds to the temperature of the photometer in Kelvin.

$$ln(\text{DarkCountRate}) = \text{A} * \text{Temp} + C \tag{5.5}$$

Although the dark count can be removed, the additional noise introduced by the dark counts σ_{DC} cannot be removed. As discussed earlier, ION-2 also has a second background channel measurement that is subtracted from the main channel to find the signal. The background signal *Background* is calculated in a manner identical to signal. Just like the dark count noise, the noise from the background channel $\sigma_{\text{Background}}$ and its dark count noise $\sigma_{\text{DC(BG)}}$ cannot be removed from the signal. The signal noise σ_{Signal} makes up the third component of the total system noise σ_{Total}. The signal noise [Eq. (5.6)], background noise [Eq. (5.7)], and two dark count

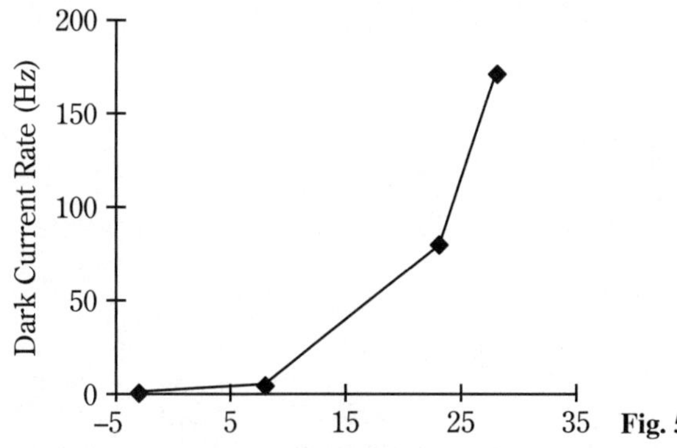

Fig. 5.3 Dark current.

noises [Eq. (5.8)] can be combined to find the total noise [Eq. (5.9)] and then the signal-to-noise ratio (SNR) [Eq. (5.10)].

$$\sigma_{\text{signal}} = \sqrt{\text{Signal}} \tag{5.6}$$

$$\sigma_{\text{Background}} = \sqrt{\text{Background}} \tag{5.7}$$

$$\sigma_{\text{DC}} = \sigma_{\text{DC(BG)}} = \sqrt{\text{DarkCountRate} * \text{Integration Time}} \tag{5.8}$$

$$\sigma_{\text{Total}} = \sqrt{\sigma^2_{\text{Signal}} + \sigma^2_{\text{Background}} + \sigma^2_{\text{DC}} + \sigma^2_{\text{DC(BG)}}} \tag{5.9}$$

$$\text{SNR} = \frac{\text{Signal}}{\sigma_{\text{Total}}} \tag{5.10}$$

The minimum signal intensity that can be viewed with the photometer is determined by the minimum signal-to-noise ratio goal. The maximum measurable signal intensity is determined by the photometer's maximum pulse rate and the system's counter. The data rate is a function of the sample size and the time between samples (Sample Time). The product of these two values is multiplied by the operating duty cycle of the instrument. For ION-1 and -2, this duty cycle is

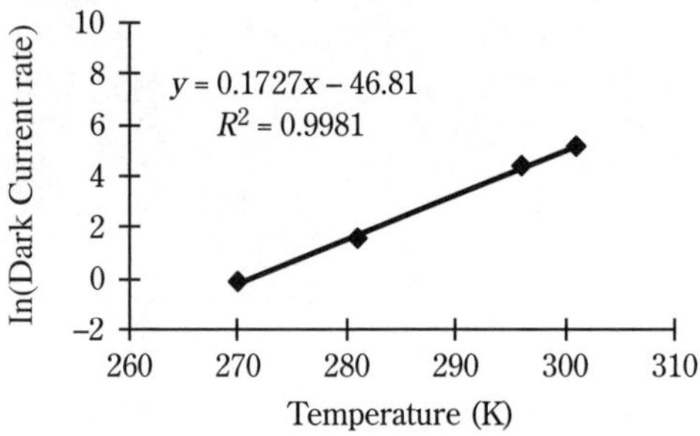

Fig. 5.4 Linearized dark current.

approximately 30% because the satellites are in eclipse for roughly 1/3 of their orbit. The amount of data generated per day in kB is as follows:

$$\text{Data Rate} = \frac{\text{Sample Size}}{\text{Sample Time}} * \frac{3600\ \text{s}}{1\ \text{hr}} * \frac{24\ \text{hrs}}{1\ \text{day}} * 30\% \tag{5.11}$$

B. Optical Measurement Limitations

As discussed in Section II, the optical measurement limits of the photometer are determined by the size of the instrument. The light-gathering power equation (5.2) shows that the instrument's performance is limited by the surface area of the lens and the sensor. With maximum lens and sensor diameters of 8 cm each, the light-gathering power is limited to 2.02×10^{-2} m^2 str.

Photometers can burn out if exposed to too much light, and so the photometer must only be scheduled to operate at night. In addition, a backup hardware shutdown circuit can be implemented with photodiodes to automatically cut power to the photometer when the diode detects too much ambient light. The shutdown circuit compares an exposed photodiode with a covered photodiode eliminating systematic effects. The exposed photodiode also uses a lens to give a field of view slightly larger than the photometer triggering the shutdown before the photometer is exposed to the higher light level. The circuit can be designed such that the processor can override the automatic shutdown in case the threshold is not set properly.

V. ION-1 and -2 Photometer Designs

A. ION-1 Photometer

ION-1's photometer and shutdown diodes are circled in Fig. 5.5. The lens and filter housing are attached directly under the photometer. The other system elements are included here to provide a sizing reference frame, but are not discussed. Note that the photometer occupies a small amount of this double cubesat's space.

Table 5.5 contains some additional photometer design parameters required for the calculations described in the preceding section. The resulting value of the equations is then shown in the photometer calculations portion of Table 5.5. Note that the signal-to-noise ratio is not affected by temperature because of the strength of the signal. In this case, the interest is in the signal variation. Because the temperature varies slowly, dark count removal is not as critical either.

The photometer consumes 88 mW with about a 30% duty cycle, which fits comfortably into ION's 1.1-W power budget. ION-1 has a 1200-baud communication system, which can support the 57 kB/day that the photometer generates (see Table 5.1). ION-1 is also designed with an attitude control system that should meet the ±5-deg attitude requirement.

B. ION-2 Photometer

Table 5.6 contains parameters and photometer calculations for each of ION-2's photometers. The exact hardware layout for ION-2 has not yet been determined,

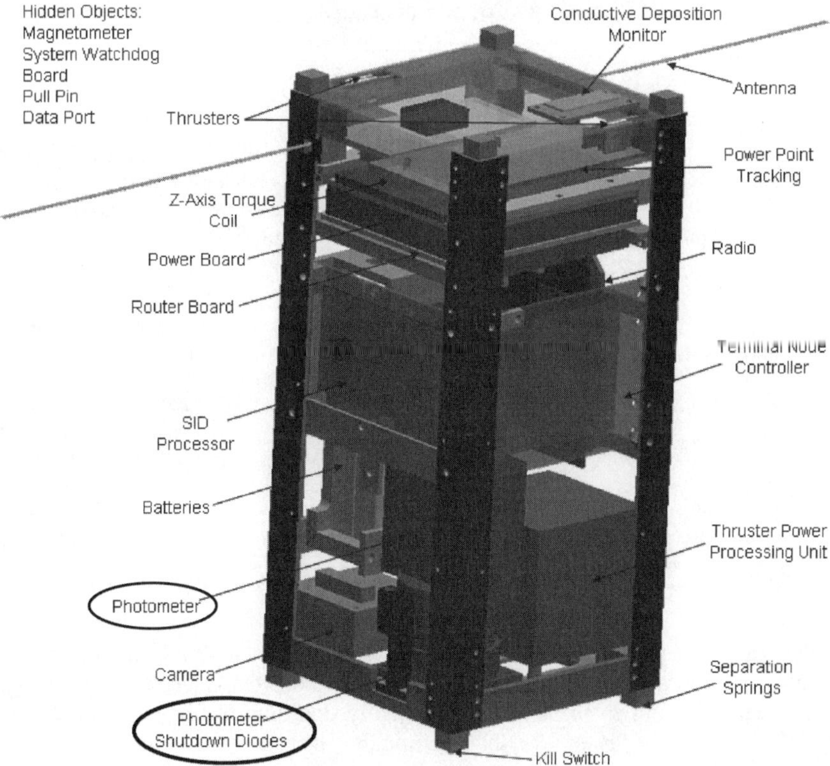

Fig 5.5. ION-1 photometer location.

Table 5.5 ION-1 Photometer Design

	Input Parameter Values
r_1	1.5×10^{-2} m
r_2	7.5×10^{-4} m
i	2.2×10^{-2} m
QE	filter (0.42), lens (0.9), photometer (0.78%)
Filter width	6 nm
	Calculation Results
FOV	4.0° (M)
LGP	2.7×10^{-6}
Photon rate	1.3×10^{7}
Signal	38329
σ_{Signal}	196
$\sigma_{Bsckground}$	28
$\sigma_{DarkCurrent}$	0 @ –20°C, 9 @ 23°C
S/N	194 @ –20°C, 193 @ 23°C
Data rate	52 kB/day

Table 5.6 ION-2 Photometer Design

	Input Parameter Values
r_1	1.27×10^{-2} m (M[a]), 1.27×10^{-2} m (B[b])
r_2	4.75×10^{-3} m (M), 4.75×10^{-3} m (B)
f	2.41×10^{-2} m (M), 1.42×10^{-2} m (B)
QE	both filters (0.6), both lenses (0.95), photometer (1.5%) (M), (0.86%) (B)
Filter width	1.3 nm (M), 10 nm (BG)
	Calculation Results
FOV	22.3 deg (M), 37.0 deg (B)
LGP	6.18×10^{-5} (M), 1.78×10^{-4} (B)
Photon rate	2.5×10^{5} (M), 2.8×10^{7} (B)
Signal	1.3×10^{4} (M), 8.3×10^{5} (B)
σ_{Signal}	112 (M)
$\sigma_{\text{Background in Main Photometer}}$	256 (BG in main)
$\sigma_{\text{DarkCurrent}}$	1 @ −20°C (M&B), 21 @ 23°C (M&B)
S/N	45 @ −20°C, 45 @ 23°C
Data rate	17 kB/day

[a]M = Main channel.
[b]B = Background channel.

although one of the two cubes is reserved primarily for ION-2's dual photometer. ION-2 should have much more power available than ION-1 because panels are being deployed out to fixed positions roughly doubling the power available to 2 W. ION-2 has a faster 9600-baud communication system, which more than meets the 20-kB/day requirement (see Table 5.1). ION-2 also has an upgraded version of ION-1's attitude control system and should be able to meet the same requirements.

VI. Conclusion

ION-1 and -2 clearly demonstrate that scientists can utilize cubesats to perform optical remote sensing missions. Cubesats have limits, but offer tremendous advantages for the missions that they can perform. The ION examples should help expand the perceptions of what these satellites can do allowing them to be deployed for a greater array of missions in the future.

References

[1] Nason, I., Puig-Suari, J., and Twiggs, R., "Development of a Family of Picosatellite Deployers Based on the CubeSat Standard," *IEEE Aerospace Conference Proceedings*, Vol. 1, IEEE, Piscataway, NJ, 2002, pp. 1-457–1-464.

[2] Lan, W., Brown, J., Toorian, A., Coelho, R., Brooks L., Suari, J. P., and Twiggs, R., "CubeSat Development in Education and into Industry," *AIAA Space*, AIAA Paper 2006-7296, 2006.

[3] Holmes, W. C., Bryson, J., Gerig, B., Oehrig J., Rodriguez, J., Schea, J., Schutt, N., Voss, D., Voss, J., Whittington, D., Bennett, A., Fennig, C., Brandle, S., Dailey, J., and

Voss, H. D., "TU Sat 1—A Novel Communications and Scientific Satellite," *16th Annual AIAA/USU Conference on Small Satellites*, Paper SSC02-I-1, Aug. 2002.

[4] Rysanek, F., and Hartmann, J. W., "MicroVacuum Arc Thruster Design for a CubeSat Class Satellite," *16th Annual AIAA/USU Conference on Small Satellites*, Paper SSC02-I-2, Aug. 2002.

[5] Gregorio, A., Bernardi, T., Carrato, S., Kostadinov, I., Messerotti, M., and Stalio, R., "AtmoCube: Observation of the Earth Atmosphere from the Space to Study 'Space Weather' Effects," *Proceedings of Recent Advances in Space Technologies*, Aeronautics and Space Technologies Inst., Istanbul, 2003, pp. 188–193.

[6] Obland, M., Hunyadi, G., Jepsen, S., Larsen, B., Klumpar, D. M., Kankelborg, C., and Hiscock, W. A., "The Montana State University NASA Space Grant Explorer-1 Science Reflight Commemorative Mission," *15th Annual AIAA/USU Conference on Small Satellites*, Paper SSC01-III-2, Aug. 2001.

[7] Simburger, E. J., Liu, S., Halpine, J., Hinkley, D., Srour, J. R., Rumsey, D., and Yoo, H., "Pico Satellite Solar Cell Testbed (PSSC Testbed)," *Conference Record of the 2006 IEEE 4th World Conference on Photovoltaic Energy Conversion*, 2006, IEEE, Piscataway, NJ, pp. 1961–1963.

[8] Almindc, L., Bisgaard, M., Bhanderi, D., and Nielsen, J. D., "Experience and Methodology Gained from 4 Years of Student Satellite Projects," *Proceedings of 2nd International Conference on Recent Advances in Space Technologies*, Aeronautics and Space Technologies Inst., Istanbul, 2005, pp. 94–99.

[9] Flagg, S., Bleier, T., Dunson, C., Doering, J., DeMartini, L., Clarke, P., Franklin, L., Seelback, J., Flagg, J., Klenk, M., Safradin, V., Cutler, J., Lorenz, A., and Tapio, E., "Using Nanosats as a Proof of Concept for Space Science Missions: QuakeSat as an Operational Example," *18th Annual AIAA/USU Conference on Small Satellites*, Paper SSC04-IX-4, 2004.

[10] Voss, D. L., Kirchoff, A., Hagerman, D. P., Zapf, J. J., Hibbs, J., Dailey, J., White, A., Voss, H. D., Maple, M., and Kamalabadi, F., "TEST—A Modular Scientific Nanosatellite," *Space 2004 Conference and Exhibit*, AIAA Paper 2004-6121, 2004.

[11] Liu, A. Z., and Swenson, G. R., "A Modeling Study of O_2 and OH Airglow Perturbations Induced by Atmospheric Gravity Waves," *Journal of Geophysical Research*, Vol. 108(D4), 2003, pp. ACH11.1–ACH11.9, doi:10.1029/2002JD002474.

[12] Swenson, G. R., Mende, S. B., and Llewellyn, E. J., "Imaging Observations of Lower Thermospheric O(1S) and O2 Airglow Emissions from STS-9: Implications of Height Variations," *Journal of Geophysical Research*, Vol. 94, No. A2, 1989, p. 1417.

[13] Tinsley, B., "Effects of Charge Exchange Involving H and H+ in the Upper Atmosphere," *Planetary and Space Science*, Vol. 26, No. 9, 1978, pp. 847–853.

[14] He, X., Kerr, R., Bishop, J., and Tepley, C., "Determining Exospheric Hydrogen Density by Reconciliation of Ha Measurements with Radiative Transfer Theory," *Journal of Geophysical Research*, Vol. 98, No. A12, 1993, pp. 21,611–21,626.

[15] Kerr, R., Garcia, R., He, X., Noto, J., Lancaster, R., Tepley, C., Gonzalez, S., Friedman, J., Doe, R., Lappen, M., and McCormack, B., "Periodic Variations of Geocoronal Balmer-Alpha Brightness due to Solar-Driven Exospheric Abundance Variations," *Journal of Geophysical Research*, Vol. 106, No. A12, 2001, pp. 28,797–28,817.

Chapter 6

Cubesats for GPS Scintillation Science

Bryan Doyle,* Akshay Patel,† Terence Brauneis,‡ Lucy Cohan,§ Jin-Woo Lee,¶ and Mark Campbell**
Cornell University, Ithaca, New York

I. Introduction

THE Ionospheric Scintillation Experiment cubesat (ICE Cube) satellite project consisted of two 10-cm cube cubesat satellites known as ICE Cube 1 and 2. The project's mission was to use these two satellites with global positioning system (GPS) receivers to collect both spatial and temporal data to help characterize ionospheric scintillations.

The ionosphere is a region of the atmosphere that is ionized by solar radiation. It is located between approximately 50 and 500 km above the surface of the Earth. Of the most interest to the ICE Cube project is the F layer, which is located in the upper part of the ionosphere. This region is characterized by high fluctuations in the electron density induced by turbulence in ionized gases. These fluctuations, termed scintillations, occur with the most frequency shortly after sunset near the equatorial region. Therefore, the more specific area of scientific interest in studying scintillations is what is known as the equatorial anomaly, which lies within ±20 deg of the magnetic equator and is a very dynamic region where most scintillations occur. These scintillations and plasma density irregularities can interfere with communication, GPS, and other signals that propagate through the ionosphere [1]. Not only can the scintillations disrupt signals and reduce accuracy, but they can also cause complete loss of lock on GPS satellites, which can be dangerous for systems relying on GPS navigation [2,3].

*Undergraduate Student, ECE B.S. 2006.
†Undergraduate Student, M&AE B.S. 2003, M. Eng. 2004.
‡Undergraduate Student, M&AE B.S. 2003, M. Eng. 2004.
§Undergraduate Student, M&AE B.S. 2005.
¶Research Associate.
**Associate Professor.

When the ICE Cube project began, there had been other research that had characterized the ionosphere; however, all had been ground based. Also, because the ionospheric models are stochastic in nature, any additional data that can validate and update the models are considered useful [4]. Therefore, the ICE Cube project was in a unique position to provide space-based temporal and spatial measurements of the scintillations. These data could then ultimately be used to decrease the effect of the scintillations on radio frequency signals propagating through the ionosphere.

The scintillation characterization was to be accomplished using a GPS receiver capable of high sampling rate measurements. The receiver would acquire GPS satellites and then store and transmit the signal-to-noise ratios (SNR) and navigation solutions. When a GPS signal passes through an area of the atmosphere that is dense with scintillations, the resulting SNR measurements (temporal) and the SNR measurements distributed across the two satellites (spatial) could be used to develop highly sophisticated models of the phenomena.

The Cubesat program provided a unique opportunity for a low-cost satellite and launch in order to study the ionospheric science. To comply with the launch mechanism, the Cubesat program dictated that each satellite was a 10-cm-sided cube with a mass of no more than 1 kg. Each Cornell University cubesat satellite was a self-supporting bus with the following basic subsystems: attitude determination and control (ADCS), command and data handling (CDH), communications, power, thermal and science. The science was a GPS receiver that was modified from a receiver used at Cornell for other experiments. The ADCS system used a gravity-gradient and a torque-coil-based stabilization scheme to keep the communications antenna pointed nadir and the GPS antenna pointed zenith.

The satellites were originally scheduled to launch in 2002, but various delays continually pushed back the launch until 2006. Finally, on 26 July 2006, a Dnepr launch rocket carrying two ICE Cube satellites and many cubesats from other institutions had a failure, resulting in an unsuccessful launch. The first-stage booster of the Dnepr rocket shut down prematurely before the second stage could kick in to send the satellites into space. The rocket and its multiple payloads crashed just more than 100 miles from the underground silo where the mission was launched at Baikonur Cosmodrome in the former Soviet republic of Kazakhstan. The following chapter describes the ICE Cube design, paying attention to some of the key lessons learned.

II. GPS Subsystem Overview

As discussed in the introduction, the ICE Cube mission requires the GPS subsystem to provide SNR data while traveling through the ionospheric "sweet spot" illustrated in Fig. 6.1. This required the subsystem to acquire GPS satellites and produce a navigation solution while traveling at 7 km/s at an altitude of 400–600 km. Further, because of the quick traversal of the sweet spot and the dynamic characteristics of the scintillations, the GPS system required that the navigation solution converge in less than 5 min, and the SNR data be sampled for at least one GPS satellite at a sampling rate greater than 50 Hz. The nature of the cubesat size imposed further constraints on the GPS subsystem design, which translated into a mass constraint of 80 g and low power usage (1.6 W). This power requirement

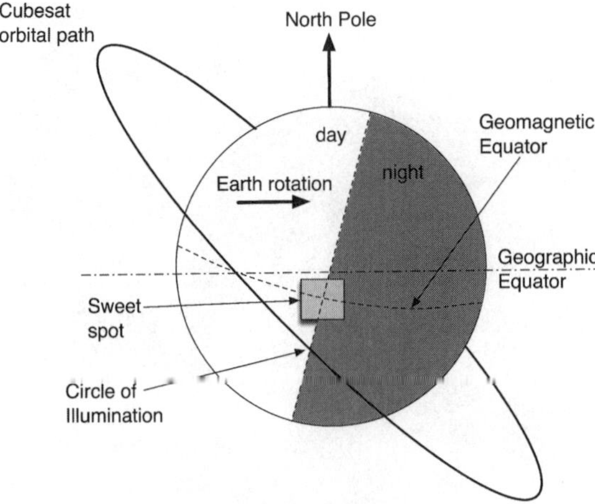

Fig. 6.1 Ionospheric sweet spot.

was still too large to keep GPS powered during the whole orbit. Therefore, in addition to the requirement of acquiring a navigation solution quickly, it was also necessary to cycle power to the GPS board, using a propagator to facilitate a warm start even when the GPS board was initialized from a powered-down state. Additionally, a warm start from power down also required the almanacs in memory to be maintained during power loss. A broad overview of the GPS subsystem and its interactions can be found in Fig. 6.2.

A. GPS Receiver

Because the requirements of the GPS subsystem implied direct modifications to the receiver software, the design solutions available to the ICE Cube project were limited. Fortunately, the project utilized an inexpensive solution that the GPS lab at Cornell University had worked with and updated previously. Specifically, the Orion GPS board, Fig. 6.3, was a COTS GPS receiver manufactured by Mitel

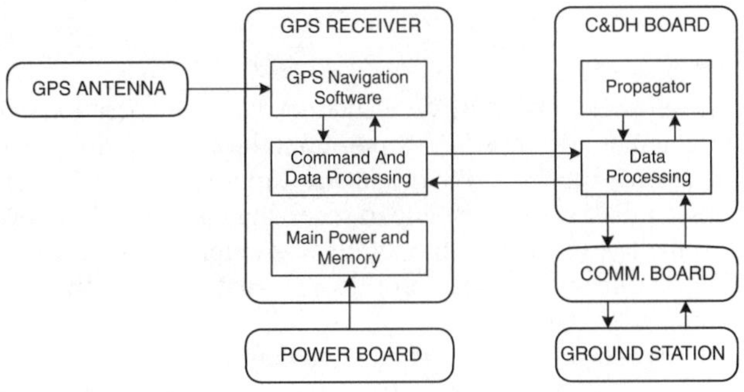

Fig. 6.2 GPS subsystem functional diagram.

Fig. 6.3 GPS board with RF shielding; an additional box of tin was placed around the entire board.

Semiconductor, GPS Group. At the time of this writing, Mitel Semiconductor is Zarlink Semiconductor, and the board is no longer being manufactured.

When the GPS board was powered down during operation, two methods were used to retain the almanac in memory. The primary retention method was a constant supply of 5 V to the GPS memory even when the GPS receiver was considered to be off. The secondary method was the use of a high-capacity "supercap" (a 0.22 F capacitor the size of a small watch battery) on the GPS board. This retained memory contents for about 4 h after power loss. This memory retention, combined with a propagator, provided warm start capability on power-up.

The satellites were designed for an altitude of 400–700 km and speeds of over 7 km/s. Because the GPS receivers were to be used at an altitude greater than 18.2 km and faster than 1.8 km/h, they fell under ITAR and the receivers, software, and any modifications required tight control. Further, export restrictions required waivers prior to shipping the flight units to an international launch site. Because this was anticipated by Cornell and the launch providers, the long lead times and paperwork had minimal effect on the project, but planning was necessary.

B. GPS Antenna

The primary requirements of the GPS antenna, from an operational perspective, were that the antenna accept a 5-V input, contain a 12–26 dB preamp, and be capable of acquiring the L1 GPS frequency (approximately 1.5 GHz). It was desirable to find a GPS antenna capable of receiving GPS signals at low elevations with respect to the top of the satellite, but it was determined that antennas of the same form factor had very similar profiles and that the sensitivity and preamp values were more important.

Given the fixed receiver choice, the antenna was ultimately selected based on a size constraint because most of the GPS antennas researched were either larger than the entire cubesat or required protrusions that did not meet cubesat program requirements. The initial antenna trade resulted in the selection of the SMK-4 antenna by Synergy Systems, LLC, with a 34×25.3-mm footprint and 24-dB gain, shown in Fig. 6.4a. The top wall panel of the satellite functioned as the patch antenna's ground plane, and the antenna's small footprint allowed for more solar-cell coverage on the top face of the satellite. A series of interference issues developed, however, and these antennas were thought to play an integral part; a complete discussion of the issues and solutions is given next. Therefore, the patch antenna decision was revisited, and a helical antenna similar to those found on today's cell phones was tested and seriously considered. Although it was determined that the helical antenna did not function properly inside of the structure, and thus was not a potential design solution, it was determined that the antenna in general was not causing the interference problems. Primarily because of the size and packaging constraints in the small cubesat, the flight unit design ultimately used a patch style antenna, the P1MAL antenna by Group Allis, Fig. 6.4b, which had the best shielding characteristics and was recommended by the Cornell GPS group. This antenna was slightly larger than the antenna from Synergy Systems, but it still provided the same amount of solar-cell coverage once a small modification to the top panel was made. This modification had already been applied to the flight units and was ultimately the reason why the P1MAL antenna was used over the SMK-4 for flight.

The patch antennas came enclosed in a plastic casing with magnetic mounts. Because of outgassing, plastic is not a good candidate for space use, and the magnet would cause bias and uncertain signals from the satellite's magnetometer. As a result, the casing was removed, and all integration and testing was performed using the exposed antennas. Based on testing and signal levels, the casing removal did not have a significant effect on the tuning of the antenna. For integration, the provided coaxial cable was replaced with one of shorter length, and the antennas were mounted to the top panel of the satellites with fasteners (see Fig. 6.9).

C. GPS Testing

Aside from traditional subsystem-level testing, two main functional tests were performed. The receiver was tested using 1) a simulation of the GPS constellation in a lab environment and 2) the actual GPS constellation in an open field. Simulation testing provided the project with the ability to simulate the satellite moving at speeds and altitudes equivalent to those found in orbit, whereas field testing provided integrated, end-to-end validation of the GPS hardware using actual GPS satellites outside of the lab environment. The use of a GS1010 Voyager GPS Simulator by Welnavigate, Inc., shown in Fig. 6.5, was provided by Cornell's GPS group. More advanced simulators are now available, but the Voyager met

a) 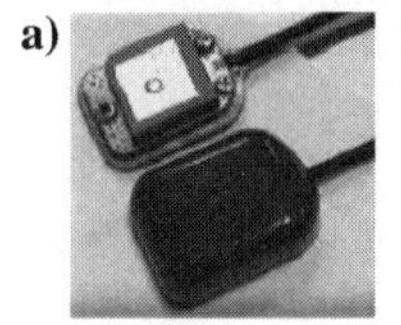

b) 

Fig. 6.4 Modulate, amplify, transmittion occurs. you'ives the stability of your satellite. Make a solid, shielded power design and yGPS Patch Antennas.

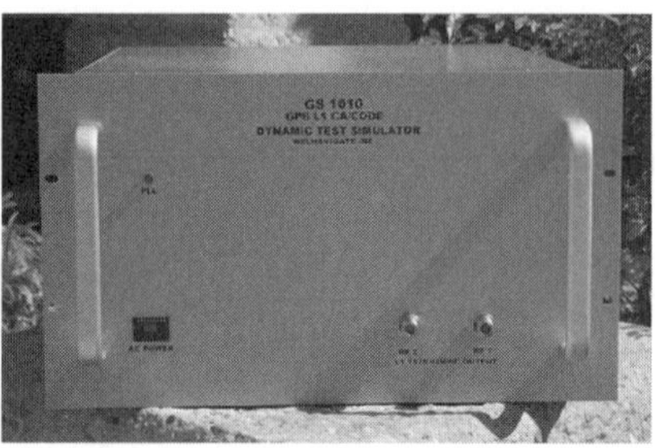

Fig. 6.5 GPS GS1010 Voyager Simulator.

ICE Cube's testing requirements. The simulator provided GPS signal validation for a dynamic space vehicle in motion, whereby a typical satellite traversal path could be specified, and the simulator would generate signals that mimic those measured by the antenna, providing the receiver with incoming simulated GPS signals. Additionally, power level calibration was used to provide signal levels that are observed in the space. The simulator also allowed for the specification of space-vehicle parameters, including the Earth's drag and gravity effects, into testing. The reproducible nature of the simulator scenarios was invaluable to the project as it helped solidify the software while outdoor tests experienced some serious hardware interference and integration issues.

D. GPS Interference Issues

One of the most critical problems encountered in the ICE Cube project, which caused very large delays, was interference in the GPS system. The main symptom of the interference was the inability of the GPS receiver to achieve lock when the antenna was mounted to, or was in close proximity to, the ICE Cube structure. Although the problems appeared to be straightforward to debug, it was a challenge. Each component of the full spacecraft system (structural components, power board, communications board, etc.) was suspected of this interference at one point during the diagnostics. Meticulous testing was performed to troubleshoot the problem; the antenna design was even completely overhauled using an internal helical antenna to see if the fundamental use of a patch antenna was causing the issue. This systematic testing led to the conclusion that the culprit was actually the GPS receiver itself! It turned out that the GPS receiver emitted noise at the L1 frequency from both its digital and RF components, and the very small proximity between the antenna and receiver created an interference feedback loop. Once this was discovered, the symptoms, problems, and possible solutions finally made sense.

Many potential solutions were attempted in order to reduce or eliminate the interference problems, but given the *required* close proximity of the antenna to the GPS board because of the small satellite size, it was determined that shielding was the best solution. Materials that did not work to eliminate the interference included aluminized Mylar sheets, copper shielding mesh, and copper tape. Ultimately, at the suggestion of the antenna manufacturer, tin boxes worked significantly better than any other potential solution in terms of enabling the integrated system to track GPS satellites and develop navigation solutions. Although metals with higher conductivity could have worked even better, the malleable property of tin helped with the integration process, where fitting the tin box around the GPS board and inside the satellite was a large packaging problem. This tin shielding was placed around the GPS RF components and was soldered to the ground plane in the board to separate the RF circuitry from the digital circuitry, Fig. 6.3. The board was further shielded from the antenna and other electronics by encasing the entire board in tin as well. Because the interference was the L1 signal, any gaps in the tin had to be smaller than a fraction of a wavelength. Based on discussions with the board manufacturer and outdoor testing, a 0.5-mm thickness was selected, and no gaps were larger than 5 mm (approximately 1/38 of the wavelength of the L1 frequency). Although tin is not a space-friendly material because of metallic whiskers forming in a vacuum environment and the resulting potential shorts, the whiskers were mitigated by wrapping the tin in Kapton tape. Although this was not an ideal solution, it allowed the satellite to function properly. In addition, while tin is high in density, there was enough mass budget to accommodate the tin cover for the GPS board. This change used the last of the mass budget contingency, bringing the final mass of the ICE Cube satellites to approximately 1 kg each, the limit of the cubesat mass launch requirement.

E. Lessons Learned: GPS

Harmful interference was not considered in the initial ICE Cube electronics design. Given the close proximity of electronics in a cubesat, all boards and RF components should be shielded from the start to prevent unwanted and unanticipated electrical interactions. When systematically diagnosing the GPS interference, inconsistencies were found. These were falsely attributed to an unstable board in the setup but were actually caused by the false assumption that the GPS board and receiver worked together without issue at close proximity. If this assumption had been properly tested, the issue would have been found and solved sooner. In general, small patch antennas worked well on a cubesat, but they did not achieve the same performance as an antenna mounted on a car roof because of the smaller ground plane on the satellite. The helical antenna design should be revisited. The helical antenna was functional while mounted inside the satellite without the GPS receiver, and so with further integrated testing it might prove to be another design choice. This antenna type would allow for an additional cell on the top panel, eliminating the flap mechanism, but it comes at the expense of internal volume. Finally, GPS can be a sensitive area because of ITAR, so that if GPS is to be used, ensure that the restrictions are accounted for in your budget, personnel, and schedule.

III. Attitude Determination and Control System Overview

The attitude determination and control subsystem (ADCS) was responsible for two-axis stabilization, pitch and roll, of the satellite to facilitate GPS operation as well as ensure communication with the ground. The pointing requirement for both antennas in their respective directions was ±30 deg. The goal of the ADCS was to keep the satellite within these pointing requirements while meeting mass and power constraints.

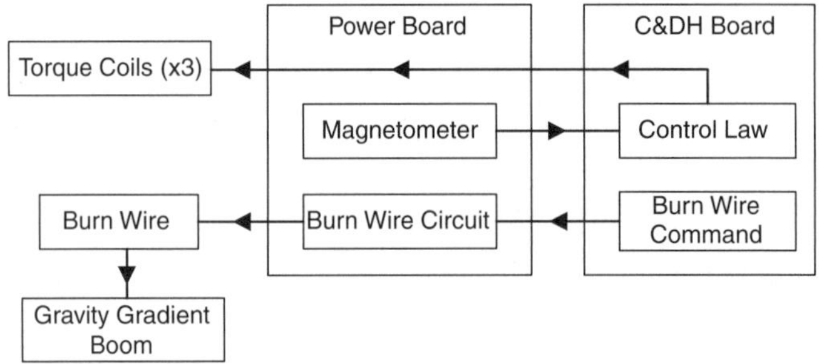

Fig. 6.6 Functional block diagram of ADCS.

The ADCS consisted of both passive and active control elements. The active control system utilized three orthogonal torque coils and a magnetometer. Based on the magnetometer measurements, the control law algorithm determined that torque coils to activate in order to reorient the satellite. Commands were then sent to the power board to activate the appropriate torque coils. The coils then produced a magnetic moment that would turn the satellite as desired. The coils also actively damped out any librations in the roll and pitch axes. The passive control element consisted of a deployable gravity-gradient boom that stabilized the spacecraft. An ADCS overview is depicted in Fig. 6.6.

A. Passive Control Background

The ADCS design went through several major modifications concerning the passive control element. One option involved gravity-gradient stabilization, which provides passive stabilization for two axes. However, gravity-gradient booms are not effective at very low altitudes because of high atmospheric drag. For LEO, the primary nongravitational force on a satellite is aerodynamic drag. Above 800 km altitude, solar pressure begins to dominate over the drag force, and by 900 km the atmosphere is so sparse that the drag force can nearly be neglected. Early in the ICE Cube program, the range of Cubesat launch orbits was assumed to be circular orbits of various inclinations and altitudes between 450 and 500 km. This assumption was based on previous design documents as well as interactions with the cubesat launch provider.

The initial design was a tether gravity-gradient system in order to validate the concept in space. After much consideration and some preliminary design work, however, the tether system was ruled out because of its high complexity and low reliability. It was discovered that a tethered mass presented a much more complicated control problem than a mass connected to a boom. The next design involved an aerodynamic stabilization system known as the Aerofin design.

B. Aerofin Design

The Aerofin design, see Fig. 6.7, in conjunction with torque coils, provided a more functional and reliable attitude determination system at altitudes between 450 and 500 km. The Aerofins essentially took advantage of the atmospheric drag

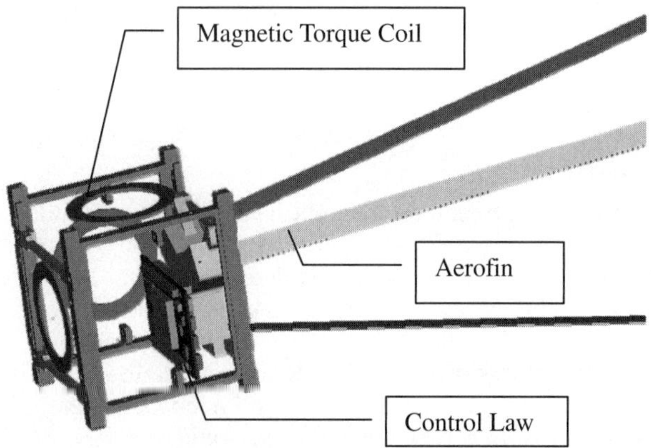

Fig. 6.7 Aerofin system design.

at lower altitudes to stabilize the spacecraft. The fins provided a stable equilibrium about which the torque coils acted such that the pitch and yaw directions would not revert to undesirable attitudes when the local magnetic field lines were aligned with those axes. The roll direction was uncontrolled passively, and so the coils were to be used to stabilize this axis and meet the pointing requirements. The design of the fins was dictated by the following factors: aerodynamic stiffness, the ballistic coefficient of the satellite, its effect on lifetime, and the ability to stow the fins in the satellite during launch. The aerodynamic stiffness affected the robustness of the passive system to disturbances. A system with a high stiffness coefficient will have less oscillation at steady state and reject disturbances more effectively [5].

The aerodynamic stiffness is primarily a function of the size of the fins and the angle between the deployed fins and the negative velocity vector. A maximum stiffness is achieved at 45 deg. However, smaller angles provide larger ballistic coefficients and longer lifetimes. Thus, the general rule of thumb in designing the fins was to make them as long and thin as possible and then tune the angle to optimize between the controller stiffness and the ballistic coefficient. The final angle chosen was 12.5 deg, and each deployed Aerofin was 1 m in length, 0.0127 m wide, and roughly 0.0001 m thick. Similar to many other cubesat antennas and gravity-gradient boom designs, the Aerofins were made from Stanley Tape measures. The paint was removed from the fins using Zip-Strip Trigger Spray paint and finish remover and steel wool.

The rolled configuration of the fins caused considerable difficulty when fitting them into the satellite. Original designs suffered from crimping and yielding over time, and this was undesirable for aerodynamic stabilization. Given that yielding should be avoided at all costs, a 1-in. outside diameter design was chosen based on test results from a fin that was rolled to the same dimension for several months with no visible deformation. The fins were also to be painted with Aeroglaze Z306 black paint to prevent oxidation and rust of the steel. The black color was chosen to minimize solar pressure on the fins. The rolled fins were tied together using a nylon fishing line, which, when connected to a simple burn circuit, provided a simple yet reliable deployment mechanism. Packaging of the Aerofin concept is similar to what was done for the gravity-gradient housing, which is described in more detail in the structures section.

C. Gravity Gradient

Midway through the project, the cubesat launch integrators selected a launch, and therefore they were given orbital parameters; the altitude was now 650–700 km. After a long series of simulations, it was determined that the Aerofin concept would not work reliably above 550 km because of the lack of atmospheric drag. Because much of the ICE Cube satellite had been designed and constructed, a systems-level decision was made, with time and packaging being big factors, to move to a passive gravity-gradient stabilization system using similar materials.

A gravity-gradient system, which in this case would provide control of the pitch and roll axes, utilized two connected masses, modeled as point masses. Each of these masses has its own center of mass, as does the satellite system as a whole. Unless the satellite is oriented perfectly tangent to the Earth, one of the two masses will be slightly closer to the Earth than the other. This difference might only be a meter, as compared to the satellite's orbit of 700 km. However, this small difference is enough to create a small torque, which will cause the satellite to orient itself perpendicular to the Earth.

Although the gravity gradient does produce a torque that causes the satellite to orient itself perpendicular to the Earth, there is also a small torque caused by the aerodynamic drag forces. Because these two torques cause opposing rotations, and assuming a damping torque exists as a result of the torque coils, the satellite will slowly decrease its oscillation angle until the angle is reached at which the torques are equal and opposite to each other. Therefore, with gravity-gradient control and torque coil damping, the satellite will eventually settle into an equilibrium position. For the constants associated with the ICE Cube orbit altitude, this equilibrium was calculated to be approximately 6 deg. A second concern of the gravity-gradient design is the bistable nature of the system. After deployment, the system could stabilize about the nadir (with the communications antenna pointing towards Earth) or antinadir direction (with the GPS antenna pointing towards Earth). This depends on the angle and the oscillation rate of the satellite when the boom is released (Fig. 6.8). To counteract this problem, the ICE Cube satellites (Fig. 6.9) were designed with a "flip sequence" mode, which used the torque coils and a novel "pumping action" similar to a child on a swing set in order to flip the satellite and allow it to resettle in the desired orientation.

The gravity-gradient boom design had two main components. The first component was a boom made of a 1-m-long piece of 0.0127-m (1/2-in.)-wide Mylar-coated spring steel from a Stanley tape measure. The thickness was roughly 0.0001 m. This material was chosen because it maintains rigidity once unraveled (even after storage for two months) and allows the boom to be coiled into a compact size. Lessons learned from the Aerofin design were incorporated into the gravity-gradient design. The boom was secured to the central bracket using two M2 fasteners to ensure that it maintained its orientation with respect to the satellite. All edges of the boom had to be parallel to the vertical edges of the satellite to ensure maximum gravity-gradient length and no undesired torques. The second

Fig. 6.8 Deployed boom (boom not to scale).

component was a cylindrical tip mass made of tungsten. This tip mass was 0.01905 m diam, was 0.01549 m thick, and had a mass of approximately 71 g. The boom was inserted into a 0.000305-m slice in the tip mass and was secured by a #6-32 set screw.

Ultimately, the design came full circle as the gravity-gradient boom was used in the flight units. In-depth research, analysis, and system-level evaluations were made during each of the design iterations. This allowed the team to fully understand the implications of the altitude change, and, as a result, a more reliable and

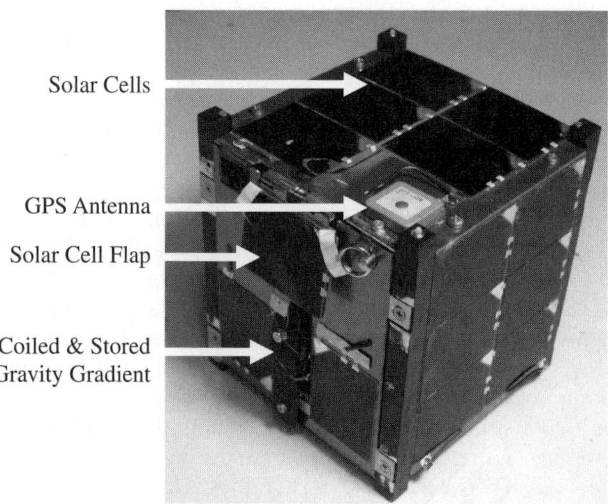

Fig. 6.9 Fully assembled ICE Cube.

technically effective ADCS design for the mission was quickly implemented into the design.

D. Lessons Learned: ADCS

Passive stabilization using a gravity-gradient tether or a tape measure gravity-gradient boom both provide the required torques once deployment has been made. However, the use of the tether requires more parts and requires a deployment scheme that is very difficult to test on the ground. If the tether is let out too quickly, the system could oscillate and become unstable. If the tether is let out too little, not enough gravity-gradient torque will be generated. For passive control, aerodynamic stabilization will work better at lower altitudes while a gravity-gradient boom will work better at higher altitudes. The spring steel from tape measures can undergo undesirable crimping and yielding over time if not packaged correctly in the satellite. Design documentation helped to retain research and analysis from previous design iterations allowing students to revisit design concepts when mission requirements changed. Finally, although both the gravity-gradient and Aerofin solutions appeared to work well in hardware and software, developing a deployable solution that could easily and quickly be retested should be a priority. The use of the nylon fishing line was appropriate, but the time to repackage after deployment was very high because the fishing line had to be tied to the burn circuit on the inside of the satellite, which required a partial satellite disassembly.

IV. Structures and Mechanisms Overview

The structural subsystem had requirements on mass, packaging of components, and the ability to withstand the structural loads (static, vibration, acoustic, and shock). The small size was actually a benefit when it came to the design for loads. The main structural frame of each satellite was constructed of 7075-T6 aluminum. All non-load-bearing components were manufactured using 6061-T6 aluminum. The structure was required to house four integrated circuit boards, three lithium-ion rechargeable batteries, both passive and active attitude control systems, a string of temperature sensors, a data port for an external interface, and a flight pin and redundant kill switches for power inhibition during mechanism integration and deployment, respectively. All components were to be held together with countless screws, nuts, wires, and connectors. Additionally, mounted to the exterior of the satellite were six wall panels that held a total of 26 solar cells (one of which was mounted on a spring-loaded flap that flipped up after deployment to add an additional cell pointing zenith), an antenna for communication with the ground, a GPS antenna to collect science data, and separation springs used to facilitate deployment. It might sound ambitious to house these components within the size and weight requirements; however, with the proper amount of planning and preparation it was quite possible.

ICE Cube's core structure consisted of two aluminum wall frames connected with four crossbars using countersunk screws; a central bracket ran across the middle of the satellite between the two wall frames. All circuit boards were fastened to this bracket using rods that screwed into the wall frames through the bracket and were secured on the opposite side with nuts. The batteries were housed within an

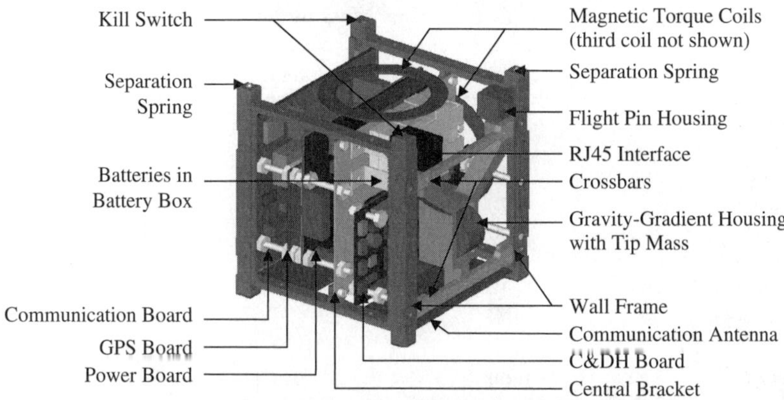

Fig. 6.10 ICE Cube structure and main internal components.

aluminum box that was also fastened to the central bracket with four screws and nuts. All metal components were first designed using IDEAS three-dimensional modeling software. These models were then converted to two-dimensional engineering drawings and DXF files for use with CNC (computer numerical control) manufacturing. To prevent cold-welding on contact surfaces, both wall frames were hard anodized. The anodized layer allows the material to be dyed any color, and so black was chosen for its obvious thermal benefits. Figure 6.10 shows a CAD drawing of the full ICE Cube satellite, with each of the internal components and structure viewable and annotated.

A. Mass Budget

One of the primary objectives for the program was to maintain a realistic mass budget for each subsystem. With a strict mass limit of 1 kg, it was important to consider the mass of each component prior to the final fabrication. Each subsystem had a mass allocation to work within, which made it easier to ensure that the final assembled product was within the mass limitation of 1 kg. Of course, it was not easy to estimate the mass of unspecified components, and so the mass budget was revisited and reallocated as necessary to meet the individual needs of the subsystems. See the final mass budget in Table 6.1.

B. Gravity Gradient

As discussed in the preceding section, the gravity-gradient design was quite unique. A tape measure with a cylindrical tungsten mass secured to the tip was attached to the central bracket and was then fed through an opening in the bottom of the satellite. In the nondeployed state, the tape measure was rolled around the tip mass, which was then wrapped around the corner of the structure and stored in the housing connected to the adjacent side. The coiled tape measure was secured with nylon fishing line that could be melted upon the activation of a burn wire release mechanism which allowed the system to deploy the gravity gradient at the

Table 6.1 Final Mass Budget

Subsystem	Allocated, g	Actual, g	Deviation, %
ADCS	140	130.30	−6.93
C&DH	40	34.50	−13.75
COMM	80	73.12	−8.60
GPS	70	64.12	−8.40
Power	200	179.30	−10.35
Structure	280	261.70	−6.54
Thermal	40	25.72	−35.70
Fasteners	100	73.65	−26.35
Contingency	50	155.00	+310.00
TOTAL	**1000**	**997.41**	**−0.03**

appropriate and specified time. Once the burn wire mechanism was released, the tip mass was then ejected from the satellite causing the tape measure to unravel. In orbit, the gravity-gradient torque and active magnetic torque coils would then orient the satellite properly.

The gravity-gradient deployment mechanism design had many complications. The main consideration was to ensure that the gravity gradient would not jam in the housing upon deployment. The first three attempts at a housing design had flaws that could have potentially led to a mission failure, while the fourth and final design was robust and reliable. Figure 6.11a shows the first design that stored the coil in a rectangular, flat-bottomed cutout that would pinch the coil and prevent it from unraveling. After testing the first design, it was realized that once released, the coil would start to unravel and get stuck in the housing. To correct for this, the body of the housing was elongated, and the contact surface was contoured, as shown in the second design in Fig. 6.11b. The logic behind this design was that the contour would allow the coil to unravel naturally. However, this once again failed because it was still possible for the coil to get pinched at the very top of the housing. Another problem with this design was that because the inside of the housing was so much bigger than the coil itself, the coil would naturally unravel while it was stowed (before deployment while under the tension of the nylon fishing line), making it very difficult to secure. The third design, shown in Fig. 6.11c, shortened the housing to be only slightly larger than the coil itself; this prevented the coil from expanding in the housing prior to release. Additionally, the depth of the housing was shortened so that the top of the walls ended below the centerline of the coil to prevent pinching. This design ensured successful deployment; however, a gap was created between the wall panel of the satellite and the housing because the depth of the housing was decreased. Also, it was not an option to make this housing flush with the wall panel as this would violate the maximum overhang allowed in the cubesat launch specifications. The gap could also potentially cause thermal issues and with sensitive electronics inside the satellite this was not an option. The problems with this design were corrected in a fourth and final version that incorporated a shield that connects to the wall panel; see Fig. 6.11d. By spacing this shield far enough away from the coil contact surface, it did not interfere with the deployment.

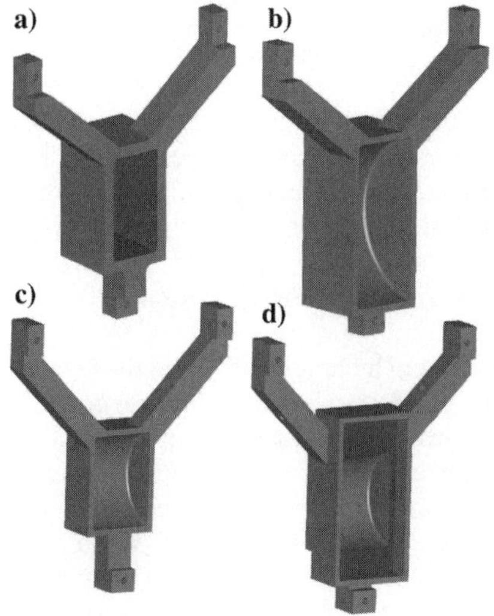

Fig. 6.11 Gravity-gradient design options.

C. Inhibits

The flight pin and kill switches were critical features because the satellites had to remain powered off until deployment. ICE Cube utilized ultrasubminiature snap action switches manufactured by Cherry Corporation for this job. These switches were activated when they were in the depressed state. The switches performed identically to kill all power to the satellite, but had independent functions, and so different models were required. The flight pin action switch, secured with epoxy, was placed into a slot in the flight pin aluminum housing and utilized a lever actuator. The lever actuator was toggled by inserting a metal dowel with a key ring on one end into a hole that ran through the middle of the housing. With this dowel inserted, the ICE Cube satellites could be handled in an off state outside of the P-POD deployment mechanism. A kill switch was integrated into each of the two wall frames for redundant power inhibition during the flight. Two small holes were drilled through each kill switch, and they were secured directly to the wall frames with screws. These switches employed a button actuator that was toggled by an aluminum plunger that passed through the top of the wall frame. Once the satellites were secured in the launch mechanism, the kill switches were depressed by either a neighboring satellite or the launch mechanism cover. At this point, the power to the satellite was killed, and the flight pin could be safely removed without powering the system. Once the satellites were ejected from the launch mechanism, the kill switches release, and the satellite powers on. It was found that the switches should not take the full load of the separation force of the launch mechanism. Instead, they should only be partially but solidly depressed, to mitigate switch failure.

D. Lessons Learned: Structures and Mechanisms

Most of the structural lessons learned came up during the prototype integration and assembly phase. While four of the exterior wall panels were secured using a combination of countersunk screws and epoxy, the ICE Cube design called for two of the panels to be secured using only epoxy; neither method was perfect. If the screws were not perfectly flush with, or below, the exterior surface, they had to be sanded down. This caused damage to the structure and made disassembly more difficult. Securing two of the sides with only epoxy made it very difficult to disassemble a fully assembled satellite without causing damage. Although some solvents would break the bond between the epoxy and aluminum, this was by no means a good practice. A better solution would be to use a thicker material for the wall panels, allowing for more room to countersink the screws and then do this for all six sides.

The biggest lesson learned had to do with wiring techniques. During the design, the space required for connectors was not well known. All data and power connections inside ICE Cube were made using standard wires and connectors. With all of the electrical components in this satellite, many connections were needed, and any free space inside the walls was quickly consumed. This led to squeezing connections where there was barely enough clearance and the connections sometimes broke during the assembly process. A better solution would be to use stackable boards or an electronics backplane, thereby eliminating the need for wires between boards, which would free up much needed space. Also, where wiring was necessary, it was critical to optimize the length of the wires, being careful not to make them longer than required in order to facilitate connection and assembly. It is important to critically examine the wire lengths before cutting as sometimes it will be necessary to feed the wire around or through other components. Finally, the burn circuit should have been designed such that retying the nylon line was possible outside of the satellite, making testing easier, faster, and repeatable. This would enable more deployment tests of the mechanisms, thus increasing reliability.

V. Power Subsystem Overview

The electrical power subsystem was required to provide the satellite with sufficient power to operate and successfully complete the science mission. It was determined early on that a rechargeable energy source was required for a six-month mission requiring an average of 1.2 W over each 90-min orbit. As such, the ICE Cube power subsystem was comprised of solar arrays, a battery pack, inhibits, and an electronics board. The electronics board harnessed the incoming solar power, controlled the charging of the batteries, regulated voltages, and distributed power to the rest of the subsystems. The power board also housed the ADCS torque coil circuitry, the magnetometer, and the burn-wire actuation circuitry. The battery pack was used to provide the satellite with power during times of eclipse. An overview of the space vehicle's electrical connections, including the components of the power subsystem is provided in Fig. 6.12.

A. Electronics Board

The electronics board contained several functional circuits in order to meet the requirements of the GPS science mission and the cubesat program. These are listed in Table 6.1. Largely, the power subsystem met the requirements, but there were some design paths that could have benefited greatly from earlier insight.

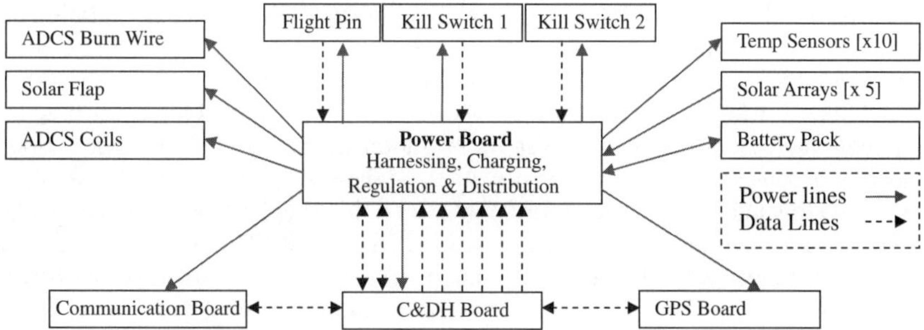

Fig. 6.12 Electrical connection overview of Cornell University's ICE Cube satellite.

Rather than describe the technical feats of the power subsystem, it is of greater benefit to go through general design decisions and the lessons learned during integration, testing and delivery.

ICE Cube, as designed, had an excessive number of wires connecting the C&DH board with the power board. Many of these lines were digital select lines that triggered the power board to turn on and off various components, such as the GPS board, directly. This was done using a simple circuit, shown in Fig. 6.13. A better design would have been to drive the same I/O circuitry from the power board microcontroller, described in Table 6.2, via a serial connection with the C&DH board, which was already in place. Although this would have required either a peripheral I/O chip or a microcontroller with more I/O pins, this benefit would have been fully appreciated during integration because of the large reduction of wires. The

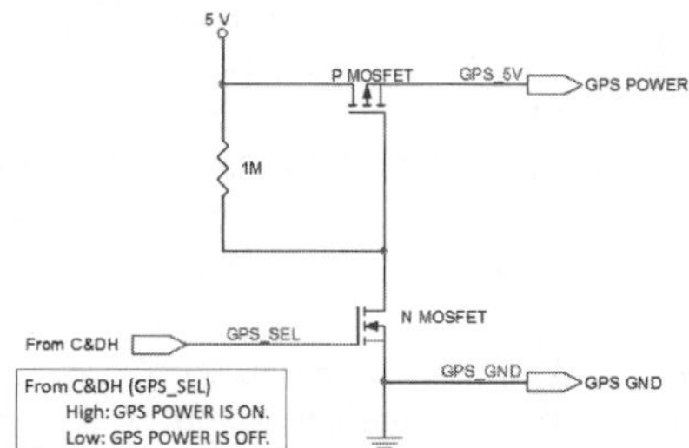

Fig. 6.13 I/O circuitry.

Table 6.2 Electronics Board Functions and their Purpose

Circuit	Function
Current sensors	Current readings on battery and solar cells
Voltage regulation	5-V, 3.3-V, and 1.5-V regulated buses
Power I/O selection	Switch on and off different subsystems
Torque coil drive H-bridge	Bidirectional drive for torque control
Magnetometer circuit	Home for ADCS sensor
Microcontroller	Thermal sense, I/V measure and charging
Flight pin/kill switch	Keeps satellite off during transport/launch

excessive I/O lines resulted in a more complex harness, which made integration an issue as discussed in the preceding section. Further, because many of these lines were coming from a COTS C&DH board, which initialized after receiving power from the power board, the power system's I/O select lines were momentarily in an unknown state as a result of power initialization transients. This was of serious concern, because the nature of the burn wires on the satellite are a one-shot deployment, and this interface provided for no intervention through software or otherwise.

For voltage regulation, ICE Cube used the Maxim MAX1649/1651 family of PFM buck regulators. The reference schematics were used verbatim, but a few design issues arose. The first was that one of the inductors used had an equivalent series resistance (ESR) that was too large. This resulted in an unacceptably large ripple on the regulated voltage outputs. The ripple voltage followed Ohm's law, where the change in voltage of the converter output equaled the equivalent series resistance times the change in inductor current. This ripple was large enough to cause the microcontroller and communication board to reset with each peak. Although all of the regulated lines had this ripple, these two components were the only ones functionally affected. Once this was corrected with a lower ESR inductor, bypass capacitors and better grounding, the ripple voltage on the regulated lines was reduced from over 1 V to an acceptable level of 200 mV_{pp} for the microcontroller. This voltage ripple fix worked well for the microcontroller, but did not reduce the ripple enough to provide the communication board with a reference voltage that was clean enough for RF circuitry. To address this, a regulator was placed on the communication board directly. This design mistake would not have happened if proper power electronics simulations were done, rather than largely trusting the reference application designs and using black boxes for the rest of the subsystems and power components. For future designs, a fully COTS regulator solution should be considered, and if custom boards are being made, the regulation is best to occur on the individual boards as this allows for more complete subsystem testing, reduces EMI occurring at the harness level, and will facilitate integration and system testing efforts. Further, all efforts should be made to shield power components, especially the switching regulators and their supporting components and to have a ground and battery bus voltage plane within the board.

The lack of proper power simulations also created problems with the kill switch inhibit and I/O select circuit designs as their discharge transient currents were, at times, large enough to cause short failures in the internal capacitors of the FETs. Ultimately, this meant that turning off the satellite could cause the satellite to never turn back on.

B. Solar Arrays

Given the small surface area and the stringent power requirements of the ICE Cube satellites, high-efficiency triple junction CICs were selected. A trade study was performed (see Table 6.3), but ultimately, 4 × 3-cm cells from Emcore were chosen. The cells ordered were quoted as having an average efficiency of 27.5%. Although best efforts were made to pair up cells such that they averaged over 27%, the cover glass and other factors degrade efficiencies. Ultimately, once assembled, the solar strings measured 26.7% on average. A total of 26 cells were used on each satellite, with cells being strung in pairs to create an operating

Table 6.3 Solar-Cell Trade Summary

Supplier	Dimensions	Efficiency, %	Power, W	#/SV
Spectrolab	Prototype	19	0.44	N/A
Emcore	4 × 3 cm	27.5	0.92	26
Emcore	7 × 3 cm	27	0.94	15
Spectrolab	8 × 3 cm	26.5	0.83	13

voltage of approximately 4.6 V. The 26 cells were split into five strings, four of which had six cells each with the final string having only two cells due to the gravity-gradient footprint on one of the side panels. For the strings with six cells, each had three parallel sets of two cells. Because the top of the satellite is the most critical surface for power generation, and because the GPS antenna consumed part of the top surface area, a flap with a single solar cell and hinge was added to the design. The flap, which utilized a spring loaded hinge, nicely integrated into the deployment of the boom such that a nylon string was used to deploy both subsystems. Figure 6.9 shows the solar-cell flap in the design.

The ICE Cube satellite project was fortunate enough to have a Xeon lamp to test the assembled strings and accept the best ones for flight. Surprisingly, even though some of the strings reported proper voltages, they did not generate the expected current because their combined efficiencies were up to 43% lower than expected. Without this lamp, verification of the cells would have been inconclusive and relative at best, and some of the underperforming panels would most likely have been used for flight, causing the satellite to be power negative, resulting in failure.

Another issue in the solar-array design was that of the reverse bias protection on the solar strings. While the individual cells had reverse bias protection diodes, the whole strings did not. This resulted in reverse biasing the cells during a failure of the PWM on the microcontroller, which was used to boost the solar-cell voltage above the battery bus of 12.6 V to charge them. Reverse biasing cells can permanently degrade their performance. However, one positive aspect is that reverse biasing a cell does show off its defects as they will not illuminate the reddish hue that the rest of cell does.

C. Batteries

Because of the size constraints of the satellite, the power subsystem required a dense energy storage solution. A trade study was performed, and ultimately the lithium-ion battery chemistry was selected. Lithium polymer cells were available at the time, but their lifetimes were lower. The ICE Cube charging model called for a discharge depth limit of 30% and was allowed to reach 80% max capacity during charging. This resulted in an expectation of 3000 charging cycles from the battery, which met the mission lifetime requirements. Ultimately, the battery pack required three cells, but connecting lithium cells in series was not trivial as a special protection circuit was needed. This protection circuit connected to the outside leads of the battery pack and to the internal nodes between the battery cells as well.

Although the batteries chosen were reliable, there were initial problems finding a supplier who was willing to provide small quantities of battery cells. In addition, suppliers also indicated that safety testing and assembly were cost prohibitive for small orders. Although initial success with one battery supplier was found, a second order of batteries from the same supplier had cells that failed short fairly quickly. The provider tried to make good on the order, but did not understand what went wrong themselves and ultimately determined that they could not devote their resources to a small order. Fortunately, the delay in the project made many more consumer products with smaller Li-ion cells readily available. After a thorough search, it was concluded that the Canon Digital Elph camera batteries were of similar size, cheaper, more readily available, and even better performing cells than the originals. The battery packs were assembled in house with protection circuits provided by ICCUS.com. The final battery packs were packaged in Kapton tape, with resistive heating elements placed between the batteries to maintain temperature.

D. Lessons Learned: Power

Many lessons learned for the power system have been brought up within their preceding contexts. In summary, one should use communication protocols wherever possible in lieu of signal lines. Further, wire reduction should be made by using a PCB harness or stackable boards where feasible. Functional circuits should always be rigorously tested, and proper power electronics simulations should always be completed prior to building and integration. When subsystems or the power board itself are off, ensure that there is minimal power leakage as it will negatively affect the power budget. All solar strings should undergo rigorous acceptance tests after assembly. Under no circumstance should anything be assumed about solar-cell string performance based on individual cell test data. While Cornell strung the cells in house, this was a complicated process. Get quotes, and consider paying to have the solar-cell strings fabricated and mounted properly as the process could end up being cheaper due to higher yields. A star-point, reference ground should be in place prior to integrating power with the rest of the system. When components, especially the batteries, are connected to the system, the negative reference terminal should be connected first. When using Li-ion cells, protection circuits and active temperature control are extremely important. Finally, the stability of the power system drives the stability of the satellite. Invest in a solid, shielded power design to ultimately save time and reduce problems during the integration process.

VI. Communications

The communication subsystem was required to reliably deliver telemetry and science data from the satellite to the ground segment, and to send operation commands from the ground to the satellite. The communication subsystem was split across the ground segment and space segment. The ground segment included a satellite tracking system, communication hardware, and an antenna. The ground station kept track of the satellites and probed for them when they were expected

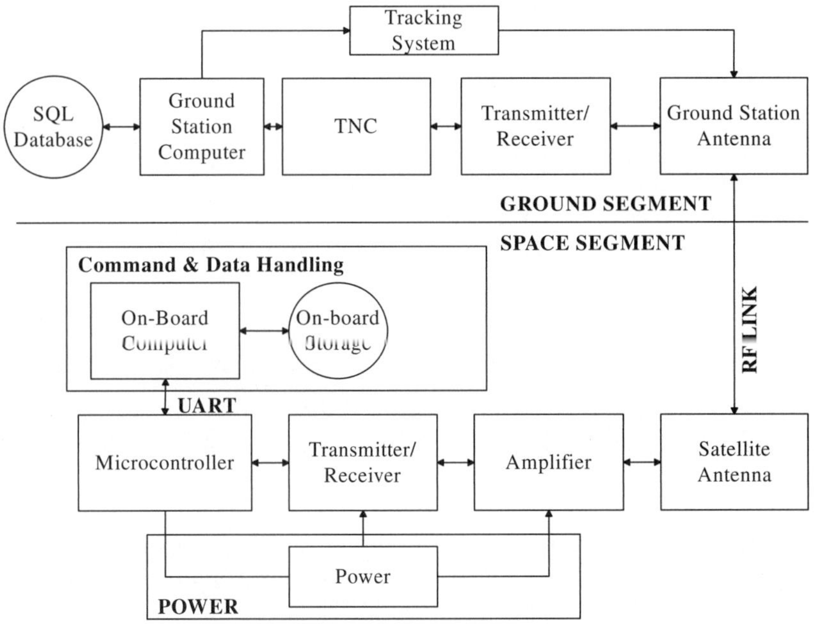

Fig. 6.14 Communication system functional diagram.

to be in range. Once the satellite responded to a ping, the ground was to initiate communication. The space segment contained an antenna, a radio and a terminal node controller (TNC). The space segment was required to modulate, demodulate, amplify, transmit and receive radio-frequency (RF) signals. Further, it was required to handle packets traveling in both directions.

The onboard software in the space segment controlled the tasks necessary for communicating with the ground station. A large effort was exerted in developing a reliable communication protocol. Considering its reliability and small overhead, AX.25 [6] was used as the base design, but because of several limitations in the communication system design, quite a few modifications were made. Figure 6.14 shows an overview of both the ground and space segments. The following sections provide an overview of the requirements, implementation, test results, and lessons learned.

The data download requirements were driven largely as a result of the large amount of science data to be collected very quickly over the sweet spot. Considering the amount of telemetry, science data and transmission overhead, the minimum data rate required was calculated to be 5900 bps. Including protocol overhead and further contingency, 9600 bps was chosen to be the target data rate. As shown in Table 6.4, the receiving sensitivity in the ground segment was required to be −99.547 and −106.019 dBm in the space segment. These numbers assumed 25 W of ground segment transmit power and 0.7 W of space segment transmit power. Given satellite size and mass constraints, the communication subsystem was required to be within $9 \times 9.5 \times 2.1$ cm^3 and be no more than 140 g.

Table 6.4 Initial Link Budget

	Ground, dBm	Space, dBm
Transmit power	43.979/25 W	28.451/0.7 W
Line loss from transmit to antenna	−1.0	−1.0
Transmit antenna gain	12.0	−1.0
Path loss	−147.998	−147.998
Atmospheric loss	−10.0	−10.0
Receiver antenna gain	12.0	−1.0
Receiver preamp gain	20.0	N/A
Edge of coverage loss (indirect aim)	N/A	−2.0
Receiving sensitivity	−99.547	−106.019

A. Space Segment Design

Various off-the-shelf radio systems were studied along with the requirements discussed in the preceding section. Unfortunately, no commercial radios satisfied all of the requirements. Ultimately, the size and the data rate requirements were the major offending parameters preventing the use of a commercial radio, and a custom communication board design was chosen.

The AT86RF211 transceiver chip from Atmel Corporation (data available online at http://www.atmel.com/) was selected for the communication board design because it provided simple control interfaces, a programmable operating frequency, 14-dBm high output power, and a sleep mode control. Because of limited power in the space segment, the satellite communication subsystem was designed to stay in sleep mode. In this mode the space segment communications board is only capable of detecting the ground station's ping command. Because of this design, the satellite did not have authority to initiate communication. Instead, the ground station was to keep track of the satellite and ping it when it was in range. If the satellite responded to the ping, then the ground station would initiate communication. The standard AX.25 protocol has no capability of this sleep mode control, and was designed to work with communication initiation in both nodes. Therefore, the AX.25 protocol was modified to handle this constraint. In addition to the AX.25 protocol implementation, the onboard software was responsible for the communication board hardware control, which included sleep/wakeup mode control, frequency changing and switching between receive and transmit mode.

The antenna had to fit in the satellite, transmit within the 70-cm radio band, and have a beam width of no less than 30 deg. Initially, a patch antenna was manufactured in house to fit in the satellite. The patch antenna was constructed on FR4 epoxy glass with a copper foil layer on the antenna side and an electroplated overlay on the ground-plane side. However, it turned out that the patch antenna was 19 dBm short from the required antenna gain. The next design, a dipole antenna, was designed with a length of $\lambda/4$ for the signal and $\lambda/8$ for the ground. This satisfied the design requirement of −1-dBm gain and had a 1.25-standing wave ratio (SWR) at the operating frequency 437 MHz.

A series of tests with the completed system indicated the space segment had 27.782-dBm (0.6-W) transmit power and −92-dBm receiving sensitivity. This result meant that the satellite segment was short by 0.1-W transmit power and required

14 dBm more sensitivity to meet the requirements shown in Table 6.4. This required compensation in the ground segment to complete the transmission link.

B. Ground Segment Design

Because one ground station was to track two ICE Cube satellites, one at 437.305 MHz and the other at 437.425 MHz, the ground station communication system was designed to be capable of changing the operating frequency and modulating the transmit frequency to compensate for the Doppler shift effect. Further, the ground station had to schedule when the downlink was to occur, parse the data, and send emails to both satellite operators and GPS researchers. The ground segment was very portable as it was in a case such that it could be brought to any remote location.

For the ground segment, initially, a high-gain helical antenna for 70-cm amateur radio band was designed. However, after a full system test of the ground station with the antenna and amplifier, it was realized that the ground segment transmit power could not compensate for the 14-dBm satellite segment deficiency. Further, the receive sensitivity was 7.5 dBm less than the required receiving signal strength. As a result, the ground segment antenna was to be replaced with a 6-m parabolic dish antenna facility located at Stanford University (data available online at http://ssoll.stanford.edu). This new antenna provided 10-dBm more gain, and an extra power amplifier was added to provide a total transmission power of 250 W. As a result, the link budget was completed with a 1.8-dBm margin in downlink and 6-dBm margin in uplink as shown in Table 6.5.

Ultimately, a custom communication board was made for both the satellite and the ground segments, and a dipole antenna was used on the satellite. As shown in Table 6.5, the developed communication system had enough capability to create a successful bidirectional data link. Table 6.6 shows additional final specifications of the space segment.

C. Lessons Learned: Communications

Although the tests of the communication systems were successful, a series of problems were observed during integration tests. One primary issue, as discussed

Table 6.5 Final Link Budget

	Ground, dBm	Space, dBm
Transmit power	53.979/250 W	27.782/0.6 W
Line loss from transmit to antenna	−1.0	−1.0
Transmit antenna gain	22.0	−1.0
Path loss	−147.998	−147.998
Atmospheric loss	−10.0	−10.0
Receiver antenna gain	22.0	−1.0
Receiver preamp gain	20.0	0
Edge of coverage loss (indirect aim)	0.0	−2.0
Receiving sensitivity	−92.0	−92.0
Received signal power	−90.216	−86.019

Table 6.6 Specifications of the Space Segment

Specification	Value
Transmit frequencies	437.305 Mhz 433.425 Mhz
Transmit rate	9600 bps
Transmit power	27.782 dBm
Receive sensitivity	–92 dBm
Mass w/shielding	150 g
Dimensions	$8.6 \times 9.1 \times 2.1$ cm^3
Operating voltage	3.3 V +/–20 mV
Operating power	1 W at transmit 150 mW at receive 40 mW at sleep

in the power section, was excessive ripples on the regulated voltage lines coming from the power board. These were only found once integration testing commenced. The initial design of the communication subsystem was to receive power from a centralized power subsystem, where all systems using a common voltage would share a regulator. The communication system was designed to accept up to 15-mV voltage ripples, but the actual voltage ripples after the integration was over 200 mV. Therefore, the space segment of the communication system was redesigned with a dedicated COTS power regulator onboard, which served to isolate the communication power line from all other subsystems.

The second problem found during integration was electromagnetic interference (EMI). The communication subsystem often received corrupted command packets, but only during a full integration test. Given the compact mechanical design, all of the satellite's electronics boards were assembled with only a centimeter of clearance in between them. Because of its proximity, the GPS system harmfully interfered with communications. As a last-minute partial solution, two grounded solder traces were added along the board edges, and 1.5-mm tin shield enclosures were soldered on the traces to create an enclosure, similar to the GPS board. This ultimately fixed the problem, but was not an elegant design.

Finally, the custom design based on the Atmel chip prevented a COTS radio to be used for the ground segment. This resulted in a ground segment design using the same board as the space segment. Testing a custom design from both ends proved to be difficult, so it is critical that a COTS communication solution be found, even at the expense of science data, unless one is extremely competent with RF design.

VII. Conclusion

This chapter presents the design solutions and lessons learned when building the ICE Cube satellites for a GPS scintillation science mission. Although many of the design solutions were well thought out and worked well, others did not, even with a recognition of the potential problems a priori. For example, time was spent to ensure the electronic boards would neatly line up and the structure itself was largely catered to facilitate integration. However, based on difficulties such as wiring, connectors, and the internal burn wire, the design could have been iterated further, and more design time and consideration should have been taken to ensure the satellite design had better integration feasibility, testability, disassembly, and electrical functionality.

The harness added complexity to the integration and testing of the satellite as wiring became a huge assembly issue. Tight clearances resulted in a reduced lifetime of harness connections. The harness should be considered during the paper design phase of the structure, and, as soon as a mockup is created and board positions are known, a mock harness with dummy elements and connectors should be made. This will give early insight to the mass and volume lost to the harness and demonstrates where assembly difficulties will arise.

The communication and power subsystems had design issues that made the system initially unstable. These issues were ultimately fixed, but it is very important to have the supporting electronics functioning properly on the satellite early in the integration process. Otherwise, test failures will only be as conclusive as the stability of the weakest component involved. Further, while it is difficult to find a COTS power subsystem that will meet the small cubesat design requirements, COTS communication boards are available. If a COTS solution is feasible, use it unless the board is being designed by an expert who has a large amount of experience with RF circuitry. The effort, complexity, cost, and risk are otherwise too high to use a custom board, even if using the COTS solution translates into a large reduction in mission critical elements such as science data.

Both the GPS and communication subsystems experienced interference during integrated tests. These issues were not always apparent when using the GPS simulator or the attenuated communication board to ground segment board direct connection. The interference for both subsystems was only found once the antennas were used. Also, the communication board interference appeared only when the satellites were more than a room away from one another. Therefore, integrated outdoor distance testing with antennas for any radio-frequency transmissions on a satellite is crucial. Do not assume that smaller tests will scale and provide the same results. Further, treat all electrical components in the satellite as antennas themselves and build conservative EMI shielding into the design. For example, inductors on switching regulators can cause harmful interference, especially when the boards are in such close quarters.

The satellite wall panels that used epoxy and the ADCS deployable components only provided one time disassembly and testing, respectively. This is not a good practice unless it is well known that the panel will never require disassembly. Clearly, every effort should be made to ensure that the design can be tested repeatedly without disassembly. Also, any disassembly should not be invasive, especially if testing is of a one-time nature where disassembly is required thereafter.

ICE Cube had two novel ADCS solutions, but they were constrained by altitude ranges. This proved to be limiting as the altitude of the orbit was changed halfway through the design. All design choices should maximize the range of altitude and orbit options as this will increase the number of launch opportunities and will minimize the risk associated with the design changing as a result of an orbit change.

COTS lithium-ion batteries worked well, but assembly into a proper battery pack was required because more than one cell was used. Ultimately, the battery pack was equivalent to those acquired from professional battery companies, but there can be significant safety risks associated with the handling of unprotected cells, and so this option should be carefully considered.

In a way, this chapter serves as a eulogy for the ICE Cube project. Although the project failed to provide GPS scintillation data due to a launch failure, the education provided to many Cornell students made the project a success. The project's lessons learned have been compiled here and are intended to serve those embarking on their own unique satellite mission.

References

[1] Kintner, P. M., Ledvina, B. M., and de Paula, E. R., "GPS and Ionospheric Scintillations," *Space Weather*, Vol. 5, Sept. 2007, pp. 1–23.

[2] Conker, R. S., El-Arini, M. B., Hegarty, C. J., and Hsiao, T. Y., " Modelling the Effects of Ionospheric Scintillation on GPS/SBAS Availibility," MP 00W0000179 Mitre Product, Mitre Corp., McLean, VA, Aug. 2000.

[3] Aarons, J., "Global Morphology of Ionospheric Scintillations," *Proceedings of IEEE*, Vol. 70, No. 4, April 1982, pp. 360–378.

[4] Secan, J. A., Bussey, R. M., and Fremouw, E. J., "High Latitude Upgrade to the Wideband Ionospheric Scintillation Model," *Radio Science*, Vol. 32, No. 4, 1997, pp. 1567–1574.

[5] Psiaki, M. L., "Nanosatellite Attitude Stabilization Using Passive Aerodynamics and Active Magnetic Torquing," *Journal of Guidance, Control, and Dynamics*, Vol. 27, No. 3, 2004, pp. 347–355.

[6] Tucson Amateur Packet Radio Corporation, "AX.25 Link Access Protocol for Amateur Packet Radio," Ver. 2.2, Tucson Amateur Packet Radio, Richardson, TX, July 1998.

Chapter 7

PowerSphere Development—An Example in Using Gossamer Technology on Picosatellites

E. J. Simburger*
The Aerospace Corporation, El Segundo, California

and

J. L. Lin† and S. E. Scarborough‡
ILC Dover, Frederica, Delaware

I. System Configuration Trade Study

A NUMBER of system configurations were explored in the initial phase of the PowerSphere development program [1–6]. A trade study was performed to evaluate six different system-level concepts: 1) Concept 1a—geodetic sphere with inflatable tube, 2) Concept 1b—geodetic sphere with center spring, 3) Concept 2a—gore sphere with inflatable torus, 4) Concept 2b—gore sphere with inflatable tube, 5) Concept 2c—gore sphere with center spring, and 6) Concept 2d—gore sphere without a center mechanism. To downselect the concept with the best potential for further development, a list of trade categories was generated from the requirement document. Table 7.1 shows the result of the system-level trade study.

From the results of the trade study, Concept 1a, namely, geodetic sphere with inflatable center deployment tube, was chosen for further development. This configuration is shown in Fig. 7.1.

II. Hinge Configuration Trade Study

Following the system-level trade study, PowerSphere solar-panel deployment hinge concepts were developed for the selected system concept. The hinges developed for the PowerSphere must be able to pack staging multiple layers and must

---

*Senior Engineer.
†Engineering Manager.
‡Senior Research and Development Engineer.

Table 7.1 PowerSphere System-Level Configuration Trade Study

a) Weights

Trade category	Weight factor	1a	1b	2a	2b	2c	2d
Packing	1	10	9	5	6	5	7
Controlled deployment	1	10	10	10	8	8	6
MASS	1	9.37	9.37	9.2675	9.2675	8.9475	8.9475
Number of panels with flex circuit integration beyond cell-hinge interface	1	8	8	2.5	2.5	2.5	2.5
Number of different sized panels	0.8	8	8	6	6	6	6
Capability to withstand thermal environment	0.9	8	8	10	9	9	8
Capability to withstand radiation environment	0.9	10	10	10	10	10	10
Structural capability	0.8	8	8	10	9	9	8
Design flexibility	0.6	7	7	8	7	7	6
Accuracy	0.7	8	8	10	8	8	7
Scalability	0.6	10	10	8	8	8	8
Complexity	0.8	8	8	8	7	7	6
Ground testability	0.6	10	10	10	10	10	10
Spacecraft compatibility	1	10	10	8	8	8	8
Cost	1	9.34	9.34	8.24	8.24	8.24	8.24
Total score		113.9	112.9	106.6	97.3	96.0	92.2

b) Design confirmation numbers

Design configuration number	Design configuration description
1a	Geodetic sphere with inflatable tube
1b	Geodetic sphere with center spring
2a	Gore sphere with inflatable torus
2b	Gore sphere with inflatable tube
2c	Gore sphere with center spring
2d	Gore sphere without a center mechanism

be able to unfold to the proper angle and maintain accuracy after deployment. Eight concepts were developed: 1) shape memory hinge, 2) UV-rigidizable composite "wave" spring hinge design, 3) UV-rigidizable "locking bridge" hinge design, 4) UV-rigidizable composite torsion spring hinge design, 5) UV-rigidizable "foldable truss" hinge design, 6) inflatable UV-rigidizable tubular hinge design, 7) inflation deployed/UV-rigidizable fabric hinge design, and 8) foam-rigidizable hinge design. The hinge configuration trade study is summarized in Table 7.2.

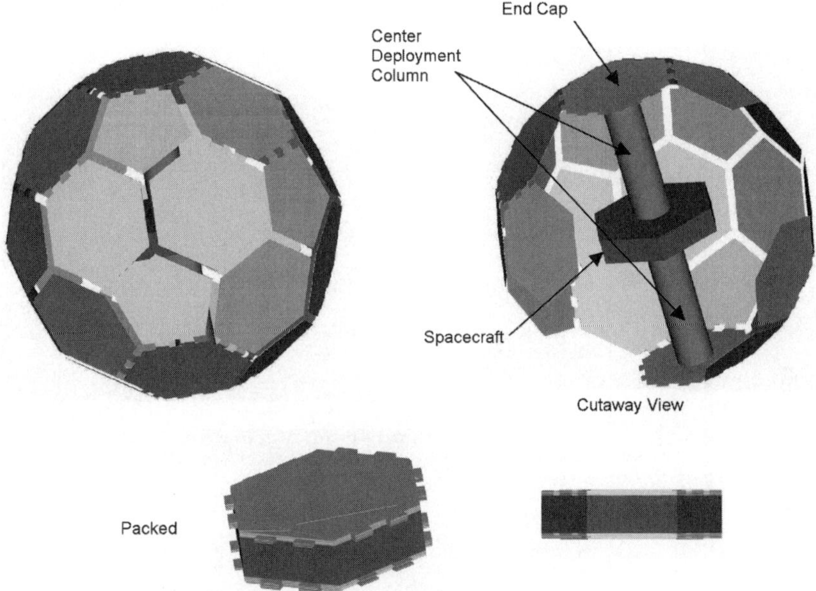

Fig. 7.1 Geodetic sphere with inflatable center deployment tube (Concept 1a).

Based on the results of the trade study, concept number 6, namely, inflatable UV-rigidizable tubular hinge design, is the preferred concept, with concept numbers 2 and 3 the next best options, respectively. To further develop the top concepts selected in this trade study, the following areas of development were recommended: 1) UV-rigidizable materials for PowerSphere structural application; 2) shape memory plastic (particularly Sub-Tg) for low-energy deployment; 3) ultra-low-mass inflation deployment system, particularly in the area of material sublimation; and 4) ultra-low-mass strain-energy deployment mechanism.

The top three hinge concepts, which have great potential for further developments, are presented in the following pages.

The inflatable UV-rigidizable tubular hinge (concept 6, Fig. 7.2) is constructed of an inflatable tube with UV-rigidizable wall. On the ground, under ambient laboratory condition, tie is triggered, the hinges will fully inflate, thus unfolding the panels to the proper shape. Once deployed, the hinges are exposed to natural UV, and they will rigidize. An alternative rigidizable material to the UV rigidizable is a sub-Tg thermoplastic material, which will rigidize at low temperature, but can be packed at room temperature. However, this might require the addition of a heater to each hinge location.

The UV-rigidizable composite "wave" spring hinge design (concept 2, Fig. 7.3) consists of two discrete components. The composite "wave" spring is a strain-energy-released deployment mechanism that allows the adjacent panels to be folded intimately or folded with a stack of several panels. The UV-rigidizable material, which is a prepreg of a fabric and a UV-curable resin with an embedded flex circuit, is the structural component of the hinge. To pack the array system, the

Table 7.2 PowerSphere Hinge Design Configuration Trade Study

a) Weights

		Hinge design configuration number[a]							
Trade category	Weight factor	1	2	3	4	5	6	7	8
Multiple deployment capability	1	5.3	9.7	9.7	10.0	9.7	7.3	7.7	1.0
Controlled deployment	1	8.3	3.3	4.0	3.7	3.3	5.7	4.7	7.3
Mass	1	9.0	7.0	5.7	3.3	3.7	8.3	5.0	4.3
Power required for deployment	1	2.3	9.3	9.3	9.3	9.3	9.3	9.3	3.7
Capability to withstand thermal	0.9	4.3	8.0	9.3	7.3	7.3	7.7	8.0	9.0
Capability to withstand radiation	0.9	6.3	7.3	7.3	7.3	7.3	7.7	7.3	8.7
Material availability/readiness	0.8	6.0	4.0	6.0	6.0	6.0	5.3	5.3	3.0
Structural capability per mass ratio	0.8	6.0	6.3	6.3	5.0	5.7	8.0	6.3	6.7
Design flexibility	0.7	7.7	6.3	6.7	4.7	4.7	7.3	6.0	6.7
Deployed accuracy	0.7	5.3	8.7	9.7	7.7	7.3	8.3	7.3	8.3
Storage life	0.9	8.3	7.3	7.3	7.3	7.3	6.7	6.7	6.3
Scalability	0.8	4.0	6.0	6.3	4.0	5.0	7.7	6.3	6.3
Complexity	0.6	7.7	7.3	7.7	5.0	3.3	8.0	4.7	5.7
Risk	0.6	3.3	4.0	5.0	4.3	3.0	5.7	3.3	3.0
Ground testability	0.8	7.0	8.7	9.3	9.3	9.3	7.3	7.7	2.0
Spacecraft compatibility	1	4.0	10.0	8.5	7.5	7.0	10.0	6.5	6.0
Compatibility with flex circuit	1	10.0	10.0	7.5	7.0	7.0	8.0	10.0	8.0
Cost	1	8.3	7.7	6.3	6.0	5.3	8.3	5.3	5.7
Total score		98.5	114.7	114.0	100.3	98.1	118.6	103.0	87.7
Ranking (overall)		6	2	3	5	7	1	4	8

[a]See Table 7.2b.

Table 7.2 Continued

b) Hinge design confirmation numbers

Design configuration number	Design configuration description
1	Shape memory hinge design
2	UV-rigidizable composite "wave" spring hinge design
3	UV-rigidizable "locking bridge" hinge concept
4	UV-rigidizable composite torsion spring hinge concept
5	UV-rigidizable "foldable truss" hinge concept
6	Inflatable UV-rigidizable tubular hinge concept
7	Inflation deployed/UV-rigidizable fabric hinge concept
8	Foam rigidizable hinge concept

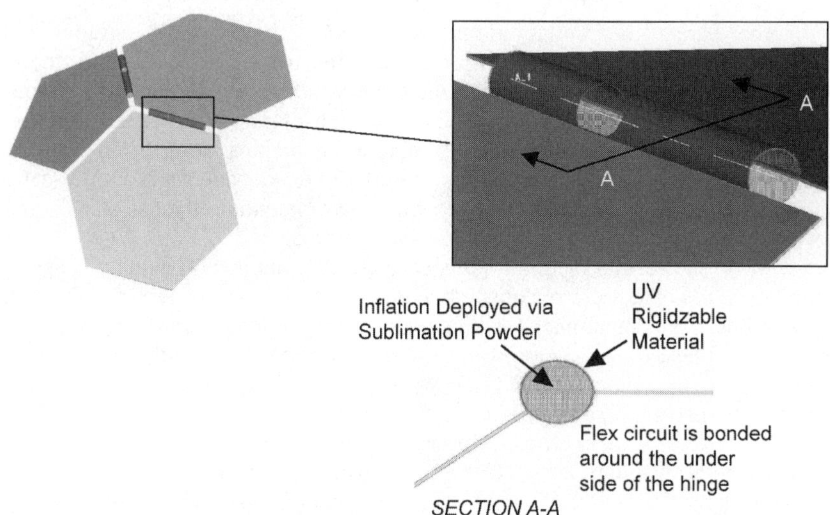

Fig. 7.2 Concept 6—Inflatable UV-rigidizable tubular hinge design.

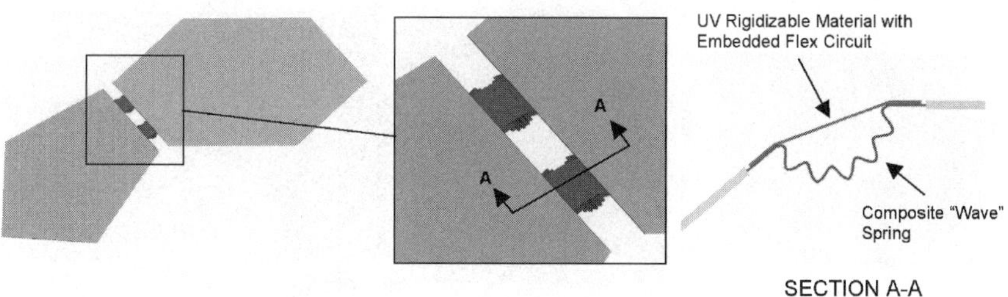

Fig. 7.3 Concept 2—UV-rigidizable composite "wave" spring hinge design.

panels are folded along the hinge in a predetermined sequence. (Note: The package must be secured by a deployment release tie because of the potential energy.) The deployment release tie is triggered to deploy the array system.

The UV-rigidizable "locking bridge" hinge design (concept 3, Fig. 7.4) consists of two independent components. The elastic band is a strain-energy-released deployment mechanism that stretches around the "locking bridge" when the adjacent array panels are packed. The UV-rigidizable "locking bridge," which consists of a prepreg of a fabric and a UV-curable resin, strips of mechanical stop, and an embedded flex circuit, is the structural component of the hinge. To pack the array system, the panels are folded along the hinge in a predetermined sequence. (Note: The package must be secured by a deployment release tie because of the stored potential energy.) The deployment release tie is triggered or cut to deploy the array system.

III. Center Deployment Column Trade Study

Another part of the configuration development is the concept generation and trade study of the center deployment column. In this part of the study, six different concepts were generated: 1) isogrid-inflatable rigidizable tube design, 2) foam-inflatable rigidizable tube design, 3) helical spring/isogrid-rigidizable tube design, 4) mechanical scissors spring design, 5) pure inflatable tube design, and 6) inflatable/shape memory isogrid tube design. Similar to other trade studies, trade categories were generated and defined to downselect potential concepts for further development. The summary of the trade study is presented in Table 7.3.

Based on the results of this trade study, concept number 4, namely, scissors spring design, is the preferred concept, with concept numbers 1 and 3 as the next best options. Concept number 4, which requires no material research but with size limitation, is best for applications in the small to intermediate PowerSphere systems (0.5- to 10-m-diam range). Concept number 3 is also best for applications in the small to intermediate PowerSphere systems. Concept number 1, with inflation deployed system, is most efficient for applications in the large PowerSphere systems (10- to 50-m-diam range). Concept number 5 is considered the highest risk option for long-duration missions. Concept numbers 2 and 6, with currently available material technology, required power to deploy and rigidize that might be

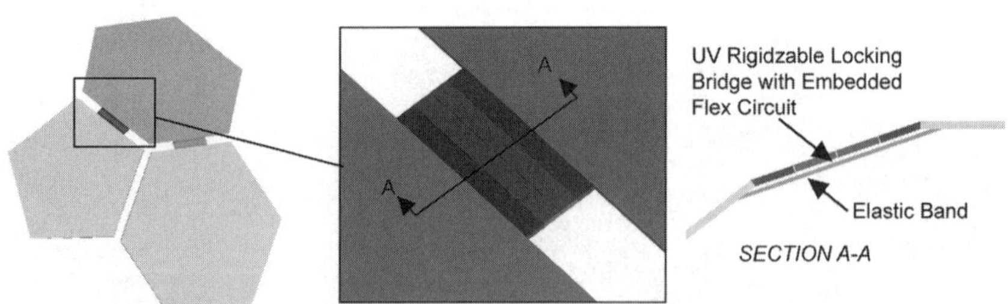

Fig. 7.4 Concept 3—UV-rigidizable "locking bridge" hinge design.

Table 7.3 PowerSphere Center Deployment Column Design Configuration Trade Study

a) Weights

Trade category	Weight factor	Deployment column design configuration[a]					
		1	2	3	4	5	6
Multiple deployment capability	1	4.3	2.3	8.3	10.0	7.3	6.3
Packing volume	1	8.0	7.3	7.0	5.3	9.3	8.0
Mass	1	8.3	7.0	8.3	6.3	9.3	8.0
Power required for deployment and rigidization	1	9.0	6.0	9.7	9.7	9.0	2.3
Capability to withstand thermal environment	0.9	8.7	8.3	8.7	9.3	5.0	5.0
Capability to withstand radiation environment	0.9	8.3	7.7	8.3	9.3	7.3	6.7
Structural capability	0.8	8.7	8.0	7.7	8.3	3.0	8.0
Design flexibility	0.8	9.0	5.0	7.0	6.0	5.0	7.3
Deployed accuracy	0.8	7.7	6.0	8.0	9.0	3.3	7.3
Storage life	0.8	6.3	6.3	6.3	10.0	9.3	7.0
Scalability	0.7	9.3	7.0	5.3	6.0	9.3	8.0
Complexity	0.6	6.7	6.0	6.7	5.7	9.0	5.7
Readiness of material	0.6	7.3	4.3	8.3	10.0	8.7	7.0
Risk	0.6	6.0	4.0	6.7	9.0	1.7	4.7
Ground testability	0.8	4.7	4.0	5.3	10.0	4.3	8.3
Spacecraft compatibility	1	7.3	6.0	8.7	8.3	7.3	6.7
Cost	1	7.3	7.0	7.3	5.7	9.0	6.7
Total score		107.2	87.0	108.8	115.8	100.6	94.9
Ranking (overall)		**3**		**2**	**1**		

[a]See Table 7.3b.

b) Design configuration numbers

Design configuration number	Design configuration description
1	Isogrid inflatable rigidizable tube design
2	Foam inflatable rigidizable tube design
3	Helical spring/isogrid rigidizable tube design
4	Mechanical scissors spring design
5	Pure inflatable tube design
6	Inflatable/shape memory isogrid tube design

constrained by the small spacecraft system. To further develop the top concepts selected in this trade study, the following areas of development are recommended:

- UV-rigidizable materials for PowerSphere structural application,
- Shape memory plastic (particularly Sub-Tg) for low-energy deployment,

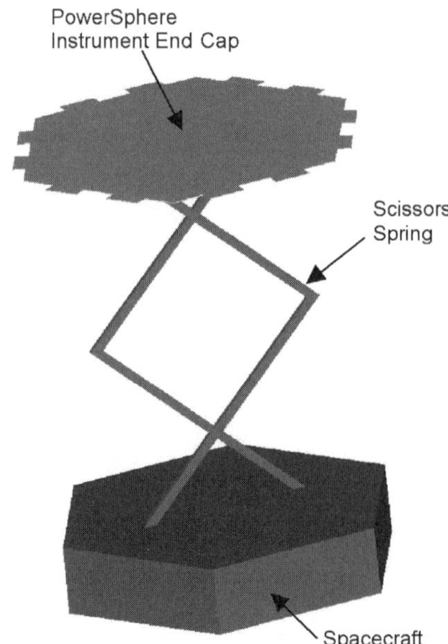

Fig. 7.5 Concept 4—Scissors spring design.

- Ultra-low-mass inflation deployment system, particularly in the area of material sublimation,
- And ultra-low-mass strain energy deployment mechanism.

The top three center deployment column concepts, which have great potential for further developments, are presented in the following pages.

The scissors spring design (concept 4, Fig. 7.5), which is based on a mechanical scissors lift concept, consists of composite linkages deployed and locked into place by a spring on the spacecraft. The ends of the linkages ride along slide tracks on the spacecraft and axis panels. Once the deployment release tie is triggered, the spring deploys the axis panels and locks into position.

The helical spring/rigidizable tube design (concept 3, Fig. 7.6) consists of a helical spring with a gridwork of UV-curable composite tows integrated into a

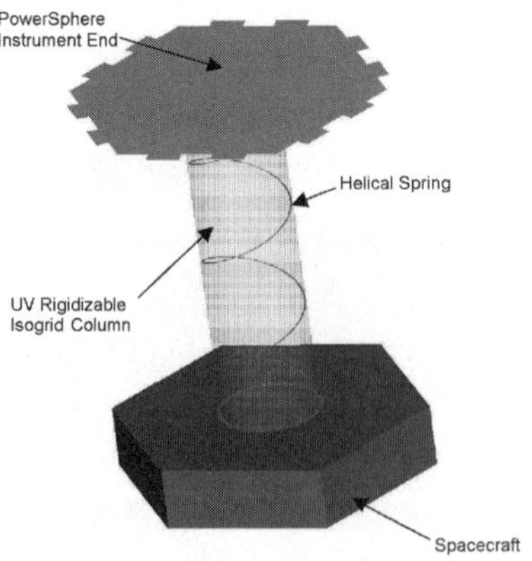

Fig. 7.6 Concept 3—Helical spring/rigidizable tube design.

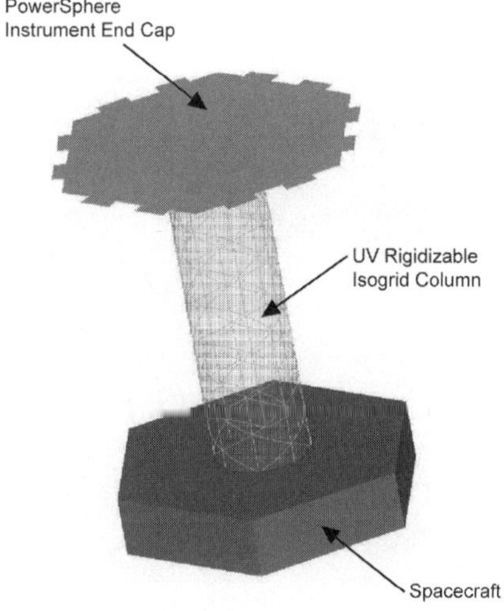

Fig. 7.7 Concept 1—Isogrid inflatable rigidizable tube design.

UV-transparent thin-film tube. The tube is packed by compressing the spring and by producing local z-folds in the tube. Once the deployment release tie is triggered, the spring deploys the tube. Then ultraviolet radiation cures the tube into a rigid supporting column.

The isogrid-inflatable rigidizable tube design (concept 1, Fig. 7.7) is a deployable tube consisting of a gridwork of UV-curable composite tows integrated into an airtight, UV-transparent, thin-film lay-up. It is z-folded for packing. Once the deployment release tie is triggered, the tube is inflated via sublimation of powder. Then the ultraviolet radiation cures the tube into a rigid supporting column. The subliming powder is needed only to inflate the tube and maintain pressure long enough for the tube to rigidize.

IV. PowerSphere Engineering Development Unit Design

The Engineering Development Unit (EDU) design of the PowerSphere solar array consists of one semispherical dome (0.6 m in diameter) connected to the centrally located spacecraft bus through an ultralightweight UV-cured isogrid

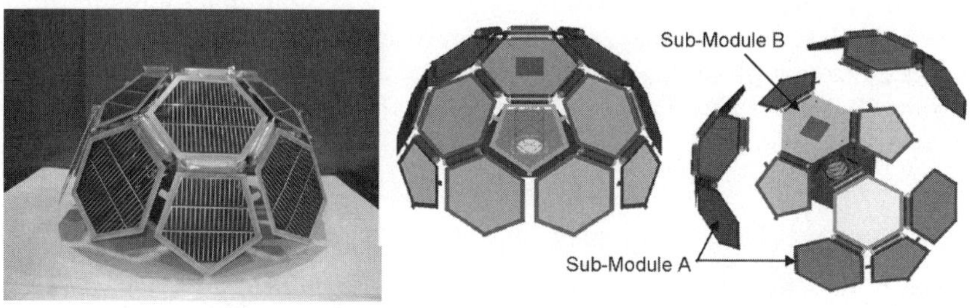

Fig. 7.8 PowerSphere engineering development unit.

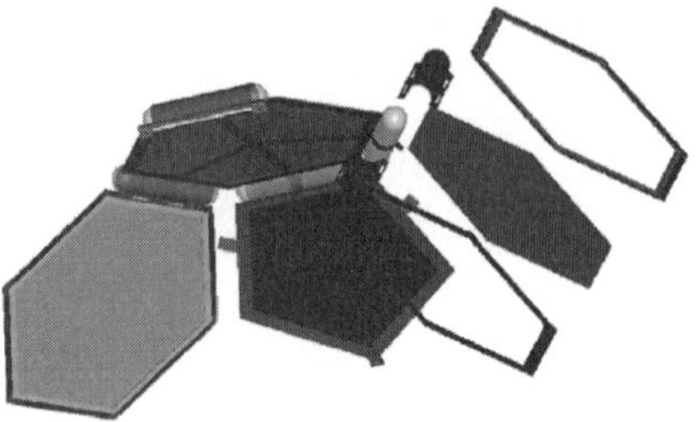

Fig. 7.9 PowerSphere Submodule A assembly.

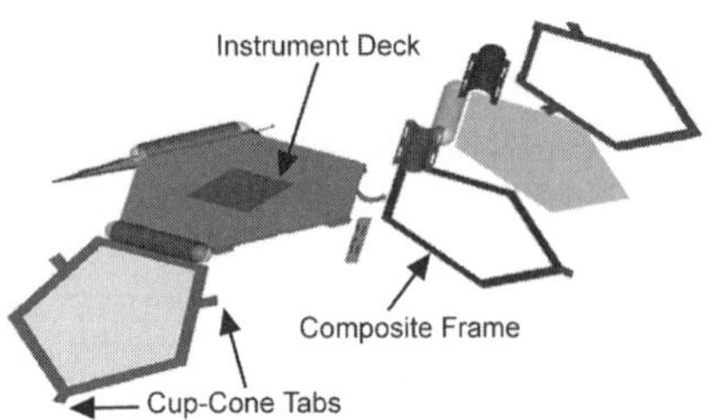

Fig. 7.10 PowerSphere Submodule B assembly.

composite boom (see Fig. 7.8). The isogrid composite boom, which is integrated with flexible wiring harnesses for power and signal transmission between the solar-array instrument deck and the spacecraft bus, is tightly folded into the spacecraft bus prior to deployment and rigidization. The semispherical EDU consists of two different subassemblies, namely, Submodule A (see Fig. 7.9) and Submodule B (see Fig. 7.10), which together form a geodetic spherical shape of hexagon and pentagon solar panels. A complete EDU semispherical dome requires three "Submodule A" assemblies and one "Submodule B" assembly.

Each Submodule A assembly consists of three hexagon and one pentagon solar panels. Solar panels are integrated with a flex-circuit blanket, which is on the backside of the solar panels, to form one subassembly component before integrating with mechanical hardware. Each Submodule A is equipped with a quick-disconnect feature to simplify the manufacturing process and to allow component replacement at the submodule level if a module fails during assembly or testing.

The Submodule B assembly consists of three pentagon solar panels that are fully integrated with the instrument deck. The mechanical interface between Submodules A and B is a bolted connection. The electrical interfaces between the two subassemblies are accomplished by using low-profile connectors manufactured by Molex. The instrument deck is designed and fabricated with patch antenna interfaces and launch restraint features. The composite frames for the solar panels of both submodules are equipped with cup-cone interface tabs for launch restraint tie down.

V. EDU Component Manufacturing and Assembly

Two EDU semispherical domes were produced during EDU production. One hemisphere went through packing and deployment trial, rigidization of structural

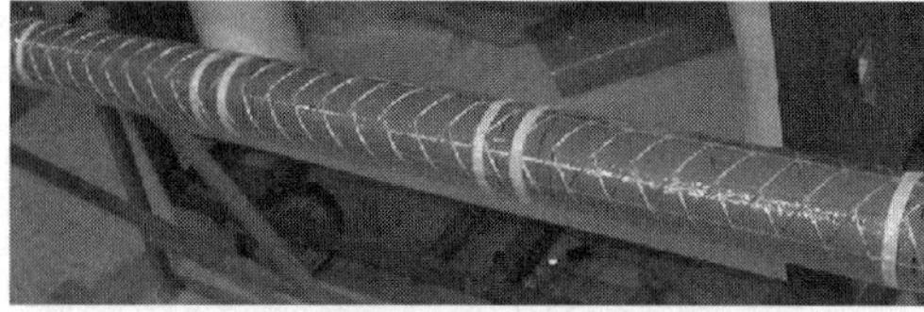

Fig. 7.11 Isogrid center column manufacturing.

components, and thermal cycling tests. The other hemisphere was packed and stored at the subassembly level at NASA Glenn for future testing.

One center column is used to connect and support each hemisphere onto the PowerSphere bus. Because of the anticipated low loading conditions, isogrid booms are used for the center columns. ILC fabricated three 0.3-m (12-in.) long, 76.2-mm (3-in.)-diam isogrid booms for the EDU (see Fig. 7.11). They were made from three rovings of 449-A S-glass impregnated with ATI-P600-2 UV curing epoxy resin (Adherent Technologies) and were encapsulated in 1-mil-thick Mylar Type LBT-2 film. Two of the isogrid booms were cured in sunlight while one was left uncured for the packed EDU hemisphere. The two cured isogrid booms were indexed and integrated with four flex-circuit strips for connecting the power and signal of the spacecraft to and from the solar array.

Thirty UV-rigidizable fiberglass hinges were fabricated for the hemisphere assembly (see Fig. 7.12). Each hinge has a bladder made from a thin-film material that is inflated to bring the corresponding solar panel to the correct angle. Once each hinge is inflated, the UV resin rigidizes via sunlight to permanently hold the shape. The dimensions on twelve of the hinges are designed for supporting the deployed pentagon frames at the correct angle, while the remaining 18 are designed for the hexagons. To control the deflection of each hexagon and pentagon solar cell on the PowerSphere, they are supported with G10 frames on all sides. The UV-rigidizable hinges are attached to these G10 frames. The flex-circuit, which transmits power from the individual solar cells to the spacecraft, runs through each hinge. The hemisphere was assembled in two sections: Submodule A and Submodule B components.

After the completion of both Submodule B's and all six Submodule A's, the EDU hemisphere was assembled (see Fig. 7.13). Assembly of the hemisphere was

Fig. 7.12 UV-rigidizable hinge manufacturing.

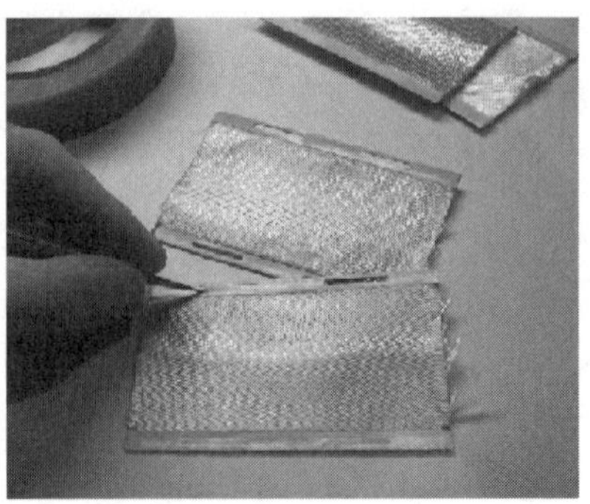

Fig. 7.13 Fully assembled EDU hemisphere components in flat layout.

completed by mechanically attaching the Submodule A's onto their corresponding locations on a Submodule B. After the submodules are secured, the flex-circuits from each submodule were plugged into their designated receptacles located on the flex-circuit bonded to the Submodule A.

After the final assembly of one EDU hemisphere, folding and packing trials were conducted to ensure that the hemisphere would pack as designed (see Figs. 7.14 and 7.15). Because of the tight fit between packed layers, tape was used to hold down the previously packed and aligned frame while the next frame was packed and aligned correctly over it. Alignment of the frames during packing involved lining up the cup-cone tabs on the G10 frames. The hexagonal frames were also required to line up with the edges of the Instrument deck.

However, as can be seen in Fig. 7.15, the thickness tolerances in the PowerSphere EDU solar panel made it difficult to fold the hemisphere to the designed thickness. The limiting component in the final fold is the length of the flex-circuit and the Mylar bottom hinge, which can be seen stretched taut at the right of Fig.7.15. This packing inaccuracy was the result of an underestimation of how much room the resin-impregnated fiberglass hinges would take up in the folded configuration. This problem will be addressed in the redesign by better thickness control of the solar panel during fabrication as well as changing the design to accommodate a longer Mylar bottom hinge.

Figures 7.16 and 7.17 show the z-folding and packed configuration of a center column into the spacecraft bus. The design requirement for the isogrid packing,

Fig. 7.14 Completed hemisphere assembly with folded submodules.

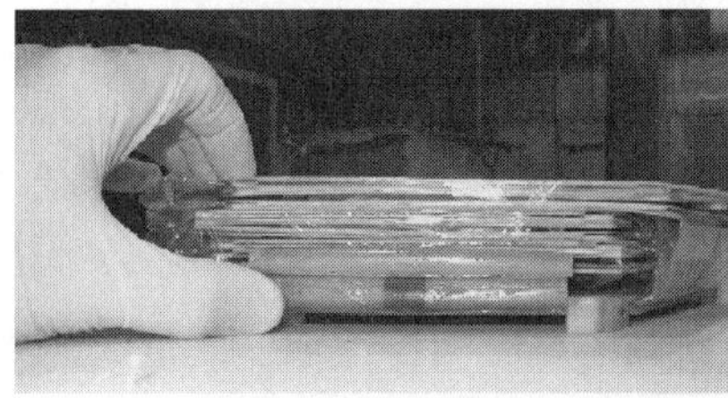

Fig. 7.15 Fully packed hemisphere assembly.

measured between the outside of the endcaps, was 47.5 mm (1.870 in.). The actual packed height of the manufactured isogrid and endcaps was approximately 33.5 mm (1.320 in.). Thus, the packing height requirement was met for the z-folded center column.

Plastic tubing and the bladders that are integrated into the hinges are used to facilitate the deployment of the EDU hemisphere. Because of the protrusions of the tubing, which were necessary to allow the full freedom of motion for the hemisphere to deploy, several difficulties were encountered during the deployment

Fig. 7.16 Z-folding of the center column into the spacecraft bus.

Fig. 7.17 Packed center column in the spacecraft bus.

Fig. 7.18 EDU hemisphere after deployment.

trials. The deployed EDU hemisphere is shown in Fig. 7.18. The submodule frames would frequently get caught on the tubing. However, the pressure in the hinges, which varied between 13.8 to 27.6 kPa (2 to 4 psi), under 1-*g* condition, was sufficient to open up the hemisphere from a packed state. The hemisphere successfully deployed, thus demonstrating the feasibility of the stowage concept. In Fig. 7.18, the effect of gravity on the EDU structure is evident because the hinges are unable to hold up the outlying frames to the correct angle. Given the "microgravity" environment of space, as well as the rotational motion of the PowerSphere, the hinges would be able to support the solar panels to the proper angles. A zero-gravity deployment experiment will be the natural next step to fully validate this stowage concept.

After the deployment trials, the EDU was cured in sunlight. Because of the effect of gravity, spacers based on the designed distance between the frames were secured between every pair of adjacent panels to help the EDU maintain its hemispherical shape while the hinge bladders were inflated and cured. These spacers can be seen in Figs. 7.19 and 7.20. The cure time for the EDU hinges is approximately one hour in sunlight, though it was cured longer to ensure that the UV reached every part of every hinge for a complete cure. After the EDU was rigidized, it was disassembled, packed, and shipped to NASA Glenn Research Center for thermal cycling test.

Fig. 7.19 Interior of hemisphere while being cured in sunlight.

Fig. 7.20 Curing of the EDU hemisphere in sunlight.

VI. Conclusion

The PowerSphere concept was envisioned in the late 1990s and has progressively gained maturity over the past several years and has reached the technology readiness level of 4 (in system level) at the end of the PowerSphere program in 2004. The fabrication and testing of the Engineering Development Unit further proved the feasibility of the PowerSphere design. Significant progress has been made through the systematic development of the PowerSphere solar-array system. What began as a solution to eliminate the problem of power choke in nano- and microsatellites has advanced the development of several technological areas that will directly benefit the development of ultralightweight structures in space.

The development of a multifunctional structure for PowerSphere pushed the developments of thin-film solar-cell process, integrated flex-circuit for thin-film solar-cell application, UV-rigidizable support structures and hinges, and electrostatic-discharge coating applicable for thin film and capable of folding.

The UV-rigidizable inflatable isogrid structure and hinge developments, accomplished by ILC Dover LP, gave space structural engineers a low-cost and low (to no)-power structural system capable of high compaction ratio. UV-rigidizable structures are particularly valuable when power for deployment is limited or not available.

The PowerSphere system and the associated technologies are ready for further refinements and flight demonstration.

References

[1] Prater, A., Simburger, E. J., Smith, D. A., Carian, P. J., and Matsumoto, J. H., "Power Management and Distribution Concept for Microsatellites and Nanosatellites," *Proceeding of 34th Intersociety Energy Conversion Engineering Conference*, SAE International, 1999.

[2] Gilmore, D. G., Simburger, E. J., Meshishnek, M. J., Scott, D. M., Smith, D. A., Prater, A., Matsumoto, J. H., and Wasz, M. L., "Thermal Design Aspects of the Power Sphere Concept," *Proceedings of Micro/Nano Technology for Space Application Conference*, 1999, pp. 451–458.

[3] Simburger, E. J., Matsumoto, J. H., Hinkley, D. A., Gilmore, D. G., Giants, T. W., and Ross, J., "Multifunctional Structures for the PowerSphere Concept," *42nd AIAA/*

ASME/ASCE/ASC Structures, Structural Dynamics, and Materials Conference and Exhibit, AIAA Paper 2001-1343.

[4] Simburger, E. J., Matsumoto, J. H., Lin, J. K., Knoll, C., Rawal, S., Perry, A. R., Barnett, D. M., Peterson, T. T., Kerslake, T. W., and Curtis, H., "Development of a Multifunctional Inflatable Structure for the PowerSphere Concept," AIAA Paper 2002-1707, April 2002.

[5] Simburger, E. J., Matsumoto, J. H., Giants, T. W., Tueling, M., Ross, J., Lin, J. K., Knoll, C., Rawal, S., Perry, A. R., Marshall, C., Barnett, D. M., Peterson, T. T., Kerslake, T. W., and Curtis, H. "Development of Flex Circuit Harness for the PowerSphere Concept," 29th Photovoltaics Specialist Conference, IEEE, Piscataway, NJ, May 2002, pp. 959–962.

[6] Simburger, E. J., Giants, T. W., Matsumoto, J. H., Garcia III, A., Liu, S., Lin, J. K., Scarborough, S.E., Rawal, S., Perry, A. R., Marshall, C. H., Peterson, T. T., Kerslake, T. W., and Curtis, H., "Engineering Development Model (EDM) Testing of the PowerSphere," AIAA Paper 2004-1570, April 2004.

Chapter 8

Multi-Application Survivable Tether (MAST) Experiment

Nestor Voronka,* Robert Hoyt,† and Tyrel Newton‡

Tethers Unlimited, Inc., Bothell, Washington

SPACE tethers have the potential to provide mission-enabling capabilities for a number of applications, including propellantless propulsion [1], formation flying [2], and momentum-exchange transportation [3–5]. A current significant impediment to the development of operational tether systems is the lack of data on the survivability of space tether structures in the micrometeoroid and orbital debris (M/OD), atomic oxygen (AO), and radiation environments in Earth orbit. The available data on the survival of tethers in the space environment are limited to a few inferred data points from the Tether Physics and Survivability (TiPS) Experiment, Small Expendable-tether Deployer System (SEDS) 1 and 2 experiments, and Tethered Satellite System (TSS) experiments, which have experienced tether lifetimes ranging from four days in the SEDS-2 experiment to nearly ten years for the TiPS tether. To address this issue, Tethers Unlimited, Inc. (TUI), in collaboration with the Stanford University Space Systems Development Laboratory (SSDL), developed the Multi-Application Survivable Tether (MAST) Experiment. The primary objective of the MAST experiment is to obtain in situ data on the degradation of tether structures by the micrometeoroid and orbital debris (M/OD) environment in low Earth orbit. The secondary objective is to obtain data on the dynamics of tethered spacecraft to provide a benchmark for tether simulation tool validation. To achieve these objectives affordably, this experiment takes advantage of the cubesat (data available online at http://www.cubesat.org) low-cost secondary payload launch opportunities. The MAST experiment consists of three picosatellites occupying a single P-POD and was designed to be launched through the University CubeSat program run by CalPoly.

---

*Vice President and Chief Technologist.

†CEO and Chief Scientist.

‡Lead Avionics Engineer.

When deployed in orbit, a space tether structure will experience a flux of impacts by micrometeorites and orbital debris. Because these M/OD particles impact with such high relative velocities, they can cut through tether lines that are several times larger than the characteristic size of the particles. This ratio of the size of the tether line an impactor will cut, to the characteristic size of the particle, is called the lethality coefficient K. This value of K is uncertain, but its average is commonly estimated to lie between 3 and 5. If the tether consists of a single line, a cut anywhere along the tether's length will sever the tether, and as a result the probability of survival of the single-line tether drops exponentially with time, as shown in Fig. 8.1. If the tether, however, is constructed with multiple lines, separated spatially, and periodically interconnected to provide redundant paths for carrying the tether tension, then the tether structure can potentially suffer many M/OD impacts that sever individual line segments while still providing a continuous path for supporting tensile loads. These redundant load-bearing paths can enable a multiline tether structure to survive the M/OD environment for long durations [6]. Figure 8.1 shows the predicted survival probability for the 1,000-meter-long tether structure planned to be deployed by the MAST experiment, calculated using M/OD fluxes predicted by the ORDEM2K Model and assuming a lethality coefficient of $K = 5$. The tether is expected to have a 99.75% probability of survival for the planned six-month operational period of the experiment, and a 96.5% probability of remaining intact until the system reenters the atmosphere due to atmospheric drag.

The MAST experiment is intended to verify the survivability of such multiline tether structures by obtaining in situ observations of the rate at which M/OD impacts cut segments of a tether deployed in low Earth orbit (LEO). The experiment is designed to do so by having one of the three picosatellites crawl slowly along the tether and take photographs of the tether. These photographs are to be transmitted to the ground and inspected, and the number of cuts of individual tether

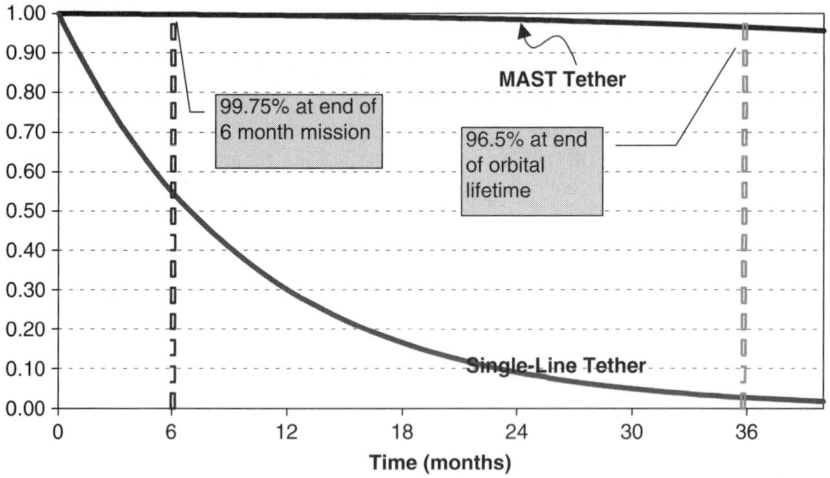

Fig. 8.1 Survivability of the MAST tether with respect to impacts by both M/OD and tracked objects, compared to a single-line tether of equal mass.

lines will be recorded over time. Figure 8.2 shows the expected observations of the cumulative number of cuts of individual lines in the tether over the operational duration of the mission—taking 14 days to go from one end to the other.

I. System Design

A functional block diagram of the MAST experiment configuration is shown in Fig. 8.3. The experiment consists of three picosatellites sized to fit together inside the 10 × 10 × 30-cm volume and 3-kg mass allocations of the CalPoly CubeSat PPOD deployer. All three satellites share identical avionics consisting of a custom-integrated command and data handling (C&DH) and power board, radio, and global positioning system (GPS) receiver unit. Each of the spacecraft also has a three-axis magnetometer and solar-panel current sensors to facilitate body-attitude estimation. All three picosatellites operate using identical software with their individual functions achieved through the loading of different command tables. This standardization of a common avionics bus and flight software was very beneficial in terms of controlling development and testing costs and schedule.

The tether deployer picosatellite, called "Ted," contains a tether spool with shroud, upon which is wound the 1,000-m long tether. In addition, the device that initiates satellite separation and therefore tethers deployment is located in Ted and is activated upon command from the C&DH processor, or when an internal watchdog timer expires. A camera is integrated into Ted, facing back along the tether and is configured to take slow frame video of the tether deployment process.

The tether inspection picosatellite, nicknamed "Gadget," contains a COTS VGA resolution (640 × 480) camera mounted internally with a view into a dark room enclosure. The tether line feeds through this darkroom, and pinch-rollers on either end of the enclosure can pull the tether through the satellite. Because of the power requirements to accomplish the crawling maneuver, Gadget has a "skirt" of external structure to support additional solar-cell area. This skirt fits over a portion

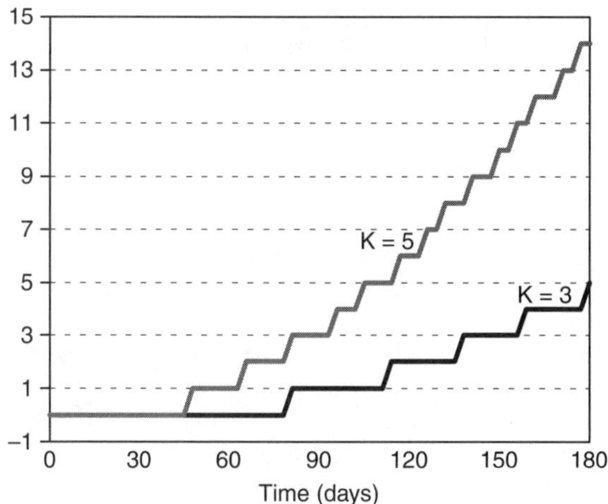

Fig. 8.2 Total cumulative number of cuts of individual lines in the tether.

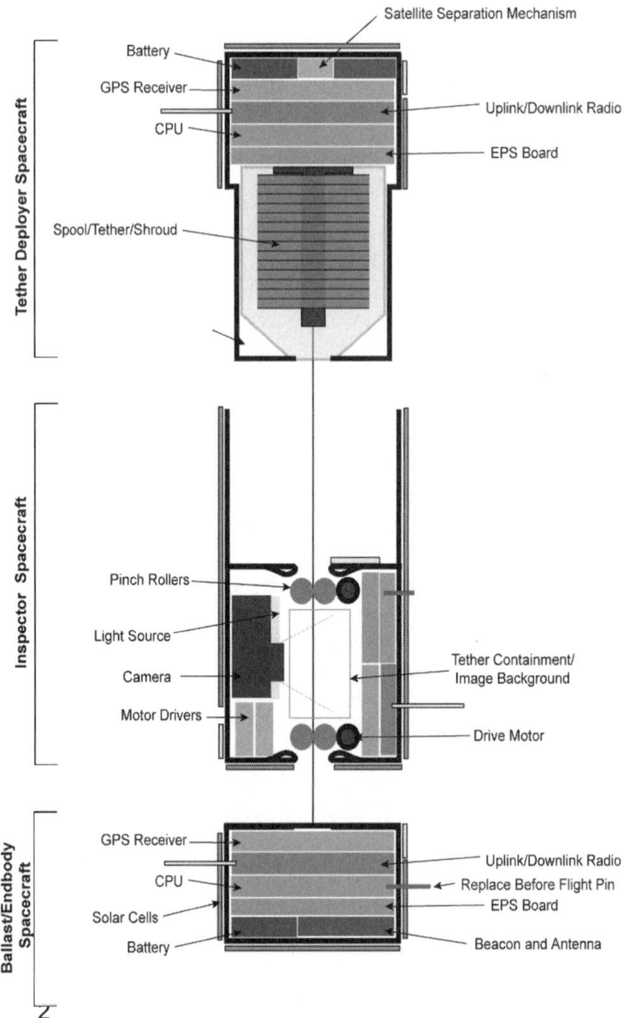

Fig. 8.3 Functional block diagram of the MAST experiment.

of the tether deployer picosatellite when the satellites are stacked for integration into the PPOD deployer. Additionally, a spring-activated pusher-plate slides within this skirt area to actuate the ejection of the tether deployer picosat.

The smallest picosatellite, nicknamed "Ralph," serves primarily as an endmass for the tether. Ralph is needed to provide a ballast end mass during deployed tether crawling operations and will also supply GPS position data to capture the dynamics of the tether system. As its sizing permitted the mounting of solar cells on only one of the spacecraft's faces, it will be operating mostly on batteries alone, and its operations will be limited by the lifetime of its batteries to one or two days at most.

Compared to other types of telemetry data, images are inherently very large. This makes them difficult to transmit to the ground because they require a very-high-bandwidth communications link—something not easily achieved on a picosatellites. The MAST mission in particular required the capture and transmission of approximately 1700 images per day in order to image the entire tether with sufficient temporal resolution. Although a number of techniques were used to minimize the size of these images (JPEG compression, controlled background and lighting, image resolution and size control), the average daily data rate required was still on the order of 1.5 MB.

Because of the small area available for body-mounted solar cells on picosatellites, the spacecraft's transmission capability is power limited. Although improvements in equivalent radiated power can be achieved with higher-gain antennas, this would impose spacecraft pointing requirements that are quite challenging to achieve on this scale. To enable the high-bandwidth communications required for this mission, we selected the highly integrated Microhard MHX-2400 2.4-GHz radio that has a maximum of 1 W of transmit power. The MHX-2400 is a frequency-hopping, spread-spectrum modem with an impressive list of features including forward error correction, fully asynchronous RS-232 interface, and a point-to-multipoint network architecture. The antenna used is a quarter-wave monopole strip antenna. To close the link with the picosat, a relatively large antenna dish is required, and the primary ground station to be used is an 18-m parabolic mesh dish with a 10-m solid surface to enable operation in the S-band frequency range. Laboratory testing of this radio prior to launch, along with measurements of interfering noise in the Palo Alto, California, area predicted a link margin of less than 10 dB.

The flight models of the three picosatellites are shown in Fig. 8.4. Figure 8.5 shows the three picosats stacked and bound together with the restraint line. Figure 8.6 shows the picosats integrated into the PPOD deployer.

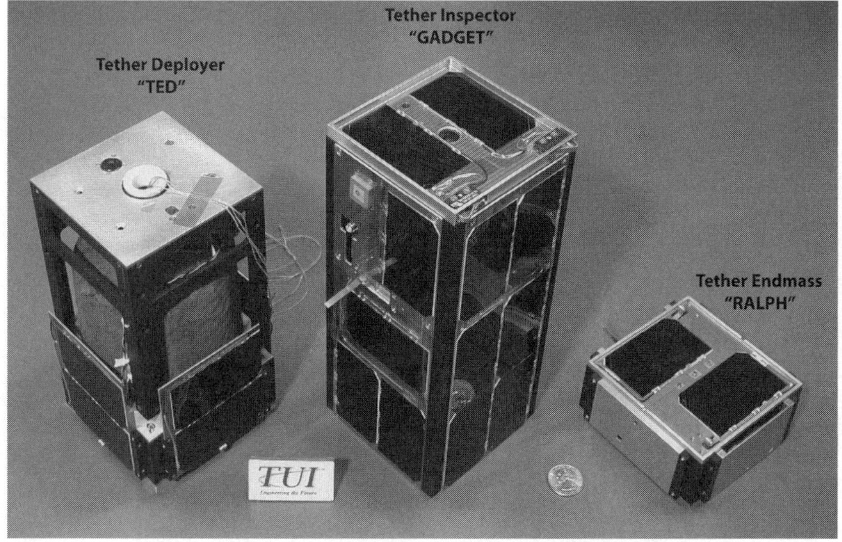

Fig. 8.4 MAST experiment picosatellite flight models.

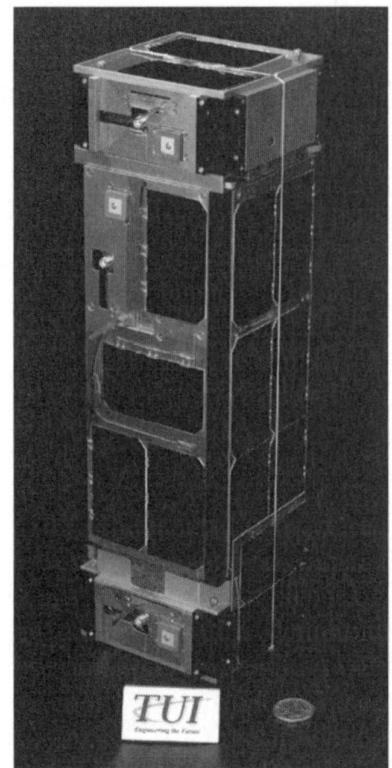

Fig. 8.5 MAST picosatellites stacked for integration into the PPOD deployer.

II. Preliminary Mission Results

A. Launch and Satellite Status

The MAST experiment was launched on 17 April 2007 aboard a Dnepr rocket launched out of Baikonur Cosmodrome, Kazakhstan into a 98.1-deg 782 × 647-km

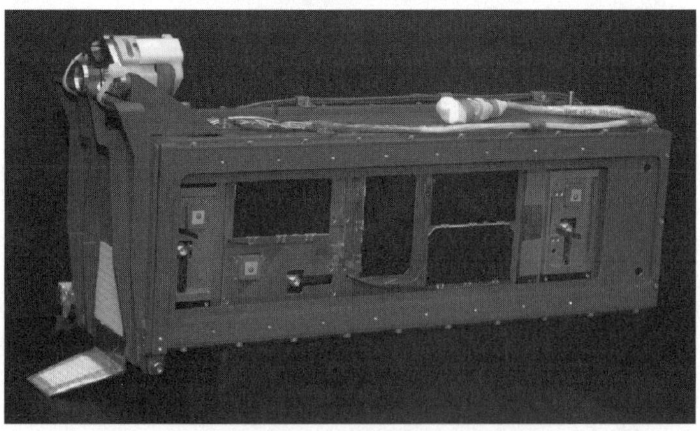

Fig. 8.6 MAST picosatellites integrated into the PPOD deployer.

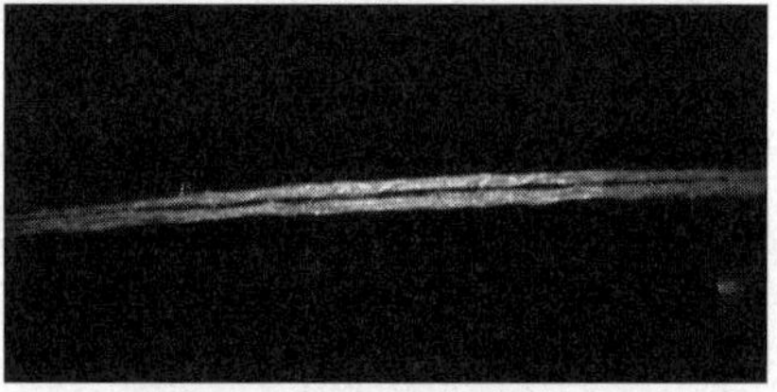

Fig. 8.7 Picture of the tether segment inside Gadget taken with the tether inspection camera 5 min after MAST was released from the PPOD.

orbit and successfully deployed from the PPOD. At the first opportunity late on 19 April, contact was established with the Gadget picosatellite. In subsequent contacts, an image taken by Gadget's inspection system 5 min after MAST was released from the PPOD was downloaded and is shown in Fig 8.7. As of this writing, Gadget is healthy, and the mission operations team is continuing to evaluate the state of the MAST experiment.

B. Communications Architecture Performance

Given the limited access time to the MAST picosatellites and the large amount of data they generate, the decision was made to implement a store-and-forward approach to data collection. This means that other than basic health updates, the majority of a communications session consists of the ground station requesting data and the satellite sending data. The size of the data package that we were able to successfully use is directly representative of the quality of the link. During the first communications pass, a miniscule package size of 64 bytes was used. During the early days of the mission when the satellites had turned down their radio's RF output power, the maximum package size we were able to use was only 256 bytes. After optimizing the satellite's operations and improving the ground software to dynamically adjust package size, at times the data package size peaked at 2 kB. From the first days of the mission to the time of this writing, we have seen the total data downloaded in a single ~10-min communications pass increase from a mere 10 kB to over 250 kB. Peak download rates during a single communications pass have been measured at approximately 15 kbps. Although much less than the prelaunch estimated theoretical of 30 kbps, the primary limitation on download rates is currently the ground software and the inherent latency between requesting a data package, receiving the result, then requesting the next data package. Nonetheless, the overall phenomenal performance of the MAST experiment's communications architecture has allowed us to download in excess of 2 MB of data in the first three weeks of operation.

References

[1] Forward, R. L., and Hoyt, R. P., "Space Tethers," *Scientific American*, Feb. 1999, pp. 86–87.

[2] Farley, R. E., and Quinn, D. A., "Tethered Formation Configurations: Meeting the Scientific Objectives of Large Aperture and Interferometric Science," AIAA Paper 2001-4770, Aug. 2001.

[3] Hoyt, R. P., and Uphoff, C. W., "Cislunar Tether Transport System," *Journal of Spacecraft*, Vol. 37, No. 2, 2000, pp. 177–186.
[4] Hoyt, R. P., "Design and Simulation of a Tether Boost Facility for LEO->GTO Payload Transport," AIAA Paper 2000-3866, July 2000.
[5] Sorensen, K., "Conceptual Design and Analysis of an MXER Tether Boost Station," AIAA Paper 2001-3915, Aug. 2001.
[6] Forward, R. L., and Hoyt, R. P., "Failsafe Multiline Hoytether Lifetimes," AIAA Paper 95-2890, July 1995.

III. Attitude Determination and Control Systems

Chapter 9

Attitude Determination, Control, and Related Operation of the Illinois Observing Nanosatellite

Andrew Pukniel*

University of Illinois at Urbana-Champaign, Urbana, Illinois

I. Introduction

THIS chapter describes both the hardware and software used for attitude determination and control of the ION satellite. The chapter is divided into four sections. The first section explains how the initial orientation of the satellite is obtained and propagated to any desired future time. The second section focuses on methodology used to optimally despin and orient ION into a nadir-facing attitude using magnetic torque actuation. The third section demonstrates the envisioned "long-duration feedback" operational plan that involves outsourcing the majority of the computational burden to the ground station. This method is especially applicable to systems with highly constrained power, mass, and computational capabilities as only sensor sampling and execution of stabilizing-torque occur onboard the spacecraft. The last section provides a brief conclusion, discussion of lessons learned, and suggestions for future developers.

II. Attitude Determination

The determination of satellite's orientation, or attitude, has extensive heritage and represents a well-developed field. There are numerous methods for satellite attitude estimation including star trackers, sun sensors, magnetometers, and various other methods. This section describes a system specifically designed for ION that emphasizes utilization of existing sensors, minimization of hardware-control drivers, and lowering power consumption. Numerous choices, as is often the case in such highly constrained systems, have shaped this subsystem and sometimes forced suboptimal decisions. The authors highly recommend reading the lessons

*Graduate Research Assistant, Department of Aerospace Engineering.

learned section of this chapter for suggestions on how to improve the existing design—many of which will be implemented on ION II.

The attitude of ION is determined using a combination of onboard magnetometer, four solar panels distributed on each of the side faces of the spacecraft, and a solar sensor located on the small face of ION opposite the photometric sensors. Because of previously mentioned limitation on physical space and power capabilities, the majority of attitude determination is accomplished on the ground. The sampling times and durations for all necessary sensors are prescheduled with the necessary commands uploaded from the ground station during a previous communication opportunity. Once the sensors complete their predesignated sampling, the data are downloaded during the next pass and processed.

The attitude determination from the downloaded data occurs in two steps. First, the orbital position of ION must be determined in order to calculate the nominal magnetic field at that location and a vector pointing towards the sun. Then, the measurements taken from the four solar panels, the sun sensor, and the magnetometer are used to determine the attitude at that location.

A. Spacecraft Orbital Position Determination

All data downloaded from the onboard sensors are time-stamped using the onboard clock, which uses the standard UTC time and is synchronized at every pass to correct for clock drift. To estimate the attitude of ION, the orbital position at each of these times must first be known. This is done by first obtaining the orbital position of ION in the two line element (TLE) form data available online at http://celestrak.com at a designated epoch and then propagating this attitude to the desired (data-stamped) times. The specifics of TLE format and explanation of the orbital elements is omitted here for conciseness, but the Celestrak website and [1] provide good explanations of both concepts respectively. Reference [1] also offers convenient conversions between orbital elements and radius/velocity representations of the orbital position.

The TLEs provided by Celestrak are usually updated once a day for recently launched LEO spacecraft, conveniently limiting the maximum propagation times to 24 hours. As a result, the orbital position propagation algorithms can be greatly simplified by assuming only the first-order approximation of a two-body system—an assumption that will be validated later in this chapter. Thus, perturbation to the orbital position as a result of third-body effects such as the moon or the sun, Earth's oblateness, and atmospheric drag are all neglected. For the purposes of validation and testing, the ION attitude determination and control system (ADCS) team has developed detailed satellite tool kit (STK) models that are capable of high-accuracy orbit propagation. These, however, are not designed to be used during regular satellite operation in order to preserve versatility of having control of the satellite from virtually any location—including ground stations without such sophisticated software as STK. This approach is highly recommended to schools with access to software such as STK for validation purposes.

B. Orbit Propagation Algorithm

Orbit propagation is usually accomplished by knowing the position and velocity, or equivalently the orbital elements, of a satellite at a given time and then projecting these values to the desired time. This process is greatly facilitated by relating true anomaly f to an auxiliary angle E, called the eccentric anomaly, through Kepler's Second Law as follows [1]:

$$\tan\left(\frac{E}{2}\right) = \tan\left(\frac{f}{2}\right)\sqrt{\left(\frac{1-e}{1+e}\right)} \tag{9.1}$$

and, further relating the elapsed time to the eccentric anomaly as

$$\Delta t = \sqrt{\frac{a^3}{\mu}}(E - e\sin E) \tag{9.2}$$

Here Δt is the time elapsed since last perigee passage, e is eccentricity, a is the semimajor axis, and μ is the gravitational constant, which, for bodies in Earth's orbit, is equal to 3.986×10^5 km^3/s^2.

Figure 9.1 shows typical orbit propagation logic. In case the orbit information is given as the position and velocity vectors, the first step consists of converting those vectors to orbital elements. This step is shown only for completeness and will usually be unnecessary because the TLE provides a set of independent elements. The next step is to use Eq. (9.1) to convert the true anomaly at the initial time to the eccentric anomaly E_0. Then, using Eq. (9.2), it is possible to calculate Δt_0, the time that has passed since last perigee passage. The epoch time t_0 is then used to calculate t_p, the time of perigee passage, as shown in the diagram. This result, along with the desired time t_d, is then used to calculate the time elapsed from last perigee passage to the desired time Δt_d. It is now possible to reverse the previous process and calculate the eccentric anomaly at the desired time E_d by solving Eq. (9.2). Then, convert E_d to true anomaly at the desired time f_d using Eq. (9.1).

To minimize the computational burden associated with attitude propagation, a simple, two-body gravitational model is assumed. As a result, effects such as Earth oblateness, sun and moon effects, and other third-body perturbations are neglected. Nonetheless, this assumption proves to be sufficiently accurate for the intended propagation durations. Figure 9.2 demonstrates the difference between the Earth-centered inertial (ECI) position vector obtained using the two-body

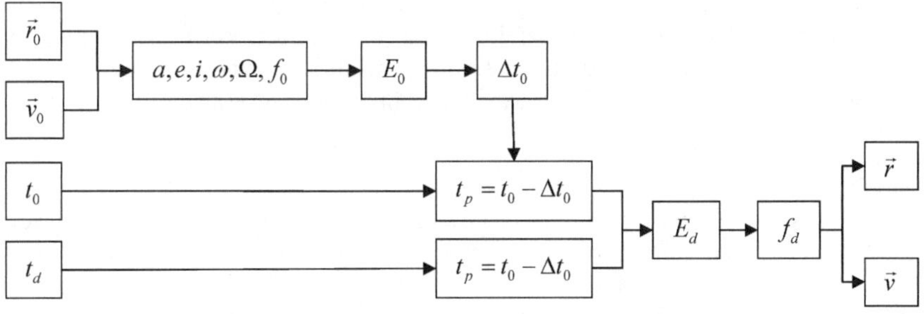

Fig. 9.1 Orbit propagation algorithm.

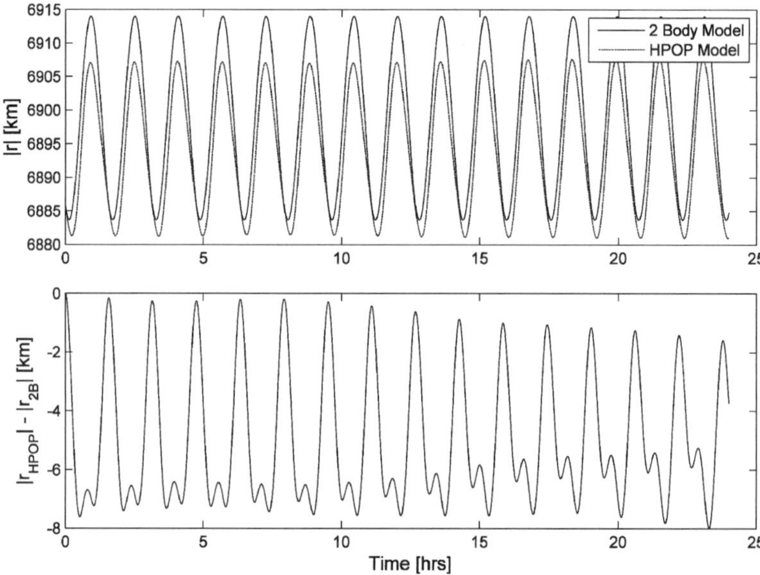

Fig. 9.2 Accuracy of propagated position vector using a two-body model.

model and the same vector calculated using high-precision orbit propagation (HPOP) function in STK. This particular simulation was run for 24 hours with maximum error in magnitude of approximately 7 km. Although this might seem large, it will be shown later that the error in position of this magnitude affects the magnetic field only slightly. Also, the preceding orbit propagator was developed in order to simplify development and testing of the attitude determination and control subsystem. In the event that the two-body assumption is deemed inaccurate, there exist extensive STK simulations capable of propagating ION's position with high precision. Alternatively, perturbation effects can be added to the existing custom-built orbit propagation algorithm.

C. Geomagnetic Field Model

The next crucial step in estimating ION's attitude is the calculation of the geomagnetic field vector at the satellite's position. Figure 9.3 shows total intensity contour lines of Earth's magnetic field projected on a Mercator map. Although it is readily obvious that the field is nonuniform, what is not shown is that the field is also time-varying—a fact that makes its computation difficult.

1. *IGRF Model*

The International Association of Geomagnetism and Aeronomy (IAGA) developed a description of the Earth's main magnetic field, now in its tenth generation, called International Geomagnetic Reference Field (IGRF). The IGRF is a series of mathematical models of the Earth's main field and its annual rate of change (secular variation). The main field coefficients are functions of time, and for the

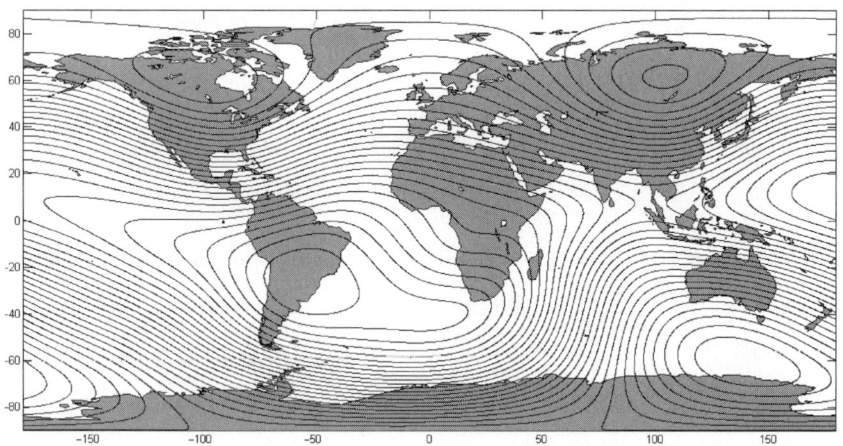

Fig. 9.3 Geomagnetic field lines from IGRF 2005 (surface).

IGRF the change is assumed to be linear over five-year intervals. It is thus possible to accurately calculate the magnetic field up to the IGRF epoch year and, secondly, to extrapolate the coefficients for five years past the IGRF epoch year.

A specific field model is referred to by including either the epoch year or the generation in the name. In addition, there are definitive coefficient sets (DGRF) for which no further revisions are anticipated. The most recently published result is the IGRF2005 (IGRF10) set, which consists of definitive coefficient sets for 1945 through 2000 and preliminary sets for 2005 and for extrapolating from 2005 to 2010. More information on the IGRF models can be found on the IAGA website found in online at http://www.ngdc.noaa.gov/IAGA/vmod/igrf.html.

2. *Magnetic Field Computation*

It is most convenient to compute the geomagnetic field in the Earth-centered fixed (ECF) coordinate system. This can be done by writing the magnetic field as a sum of spherical harmonics, in the following form:

$$\mathbf{b}^E_{\text{calc}} = \sum_{n=1}^{\infty}\sum_{m=0}^{n}\mathbf{b}_{n,m} \tag{9.3}$$

In the preceding equation, the vector $\mathbf{b}^E_{\text{calc}}$ is the calculated magnetic field in the ECF reference frame, and $\mathbf{b}_{n,m}$ is the spherical harmonic of degree n and order m. The components of $\mathbf{b}_{n,m}$ are found by the following equation:

$$\mathbf{b}_{n,m} = \frac{K_{n,m}a^{n+2}}{r^{n+m+1}}\left\{\begin{array}{l}\dfrac{g_{n,m}C_m + h_{n,m}S_m}{r}\left[\left(uA_{n,m+1}(u) + (n+m+1)A_{n,m}(u)\right)\hat{\mathbf{r}} - A_{n,m+1}(u)\mathbf{z_E}\right] \\ -mA_{n,m}(u)\left[\left(g_{n,m}C_{m-1} + h_{n,m}S_{m-1}\right)\mathbf{x_E} + \left(h_{n,m}C_{m-1} - g_{n,m}S_{m-1}\right)\mathbf{y_E}\right]\end{array}\right\} \tag{9.4}$$

The parameters $g_{n,m}$ and $h_{n,m}$ are the Gauss coefficients of degree n and order m, as published by the IGRF. The parameter a is the mean radius of the Earth and is equal to 6371.2 km. The parameter r is simply the magnitude of $\mathbf{r}$, the desired position in the ECF frame. The unit vector $\hat{\mathbf{r}}$ is in the direction of $\mathbf{r}$. The parameter u is the third component of $\hat{\mathbf{r}}$, which can be found as follows:

$$u = \hat{\mathbf{r}} \bullet \mathbf{z}_E \tag{9.5}$$

The remaining terms in Eq. (9.4) can be determined recursively through the following recursions:

$$\begin{aligned}
K_{n,0} &= K_{1,1} = 1, \quad \forall\ n \in \{1, \ldots, \infty\} \\
K_{n,m} &= \left(\frac{n-m}{n+m}\right)^{1/2} K_{n-1,m}, \quad \forall\ m \in \{1, \ldots, \infty\},\ n > m \\
K_{n,m} &= \left[(n+m)(n-m+1)\right]^{-1/2} K_{n,m-1}, \quad \forall\ m \in \{2, \ldots, \infty\}, n \geq m
\end{aligned} \tag{9.6}$$

$$\begin{aligned}
&A_{0,0}(u) = 1 \\
&A_{n,n}(u) = (1)(3)(5)\cdots(2n-1) \quad \forall\ n \in \{1, \ldots, \infty\} \\
&A_{n,m}(u) = \frac{1}{n-m}\left[(2n-1)u\,A_{n-1,m}(u) - (n+m-1)A_{n-2,m}(u)\right], \\
&\qquad \forall\ m \in \{0, \cdots, \infty\}, n > m
\end{aligned} \tag{9.7}$$

$$\begin{array}{lll}
S_0 = 0 & C_0 = 1 & C_m = C_1 C_{m-1} - S_1 S_{m-1} \\
S_1 = \mathbf{r} \bullet \mathbf{y}_E & C_1 = \mathbf{r} \bullet \mathbf{x}_E & S_m = S_1 C_{m-1} + C_1 S_{m-1}
\end{array} \tag{9.8}$$

$K_{n,m}$ are known as the Schmitt coefficients, while $A_{n,m}$ are derived Legendre polynomials. S_m and C_m are related to the Schmitt coefficients, but are separated as a means of simplifying the recursions.

The Schmitt coefficients are independent of the desired position. Thus, if the magnetic field is to be calculated for a number of points, the Schmitt coefficients need only be calculated once, and can be reused for each subsequent calculation. However, the rest of the parameters are location dependent, and must be recalculated for each desired location [2]. A useful reference concerning the derivation of Eqs. (9.4–9.8) can be found in [3] and [4].

3. *Accuracy of Model*

To ensure that the accuracy of the IGRF reflects the high quality of available data, in 2001 the IAGA decided that the main-field coefficients of the IGRF from the year 2000 onwards should extend to degree $\boldsymbol{n}_{\max} = 13$ and be quoted to 0.1 nT precision. Pre-2000 coefficients extend to degree 8 or 10 and are quoted with 1-nT

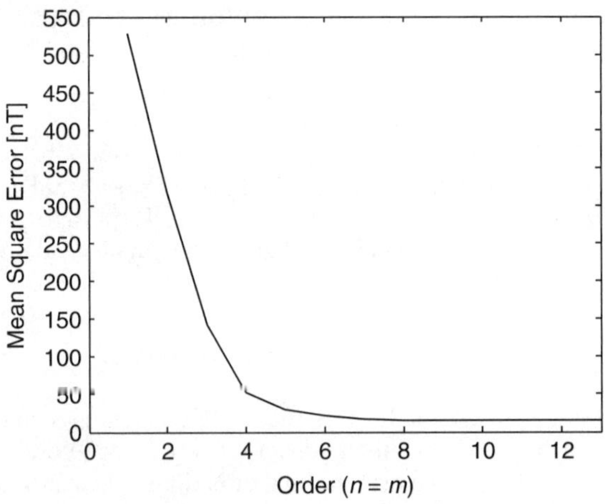

Fig. 9.4 Mean square error vs order ($n = m$).

precision. The new predictive secular variation coefficients for the upcoming five-year epoch are given to degree 8 with a precision of 0.1 nT/yr.

To test the accuracy of the preceding algorithm, the magnetic field is computed for various values of $\boldsymbol{n}_{\max}$ and compared to the magnetic field obtained using STK, which uses the thirteenth-order summation. The results are summarized in Fig. 9.4, which illustrates how the mean square error behaves as a function of order n. Interestingly, the accuracy improvements are minimal once the summation is of the order $n = 10$.

The added degree of accuracy comes at price of increased computational time as can be seen in Fig. 9.5. However, because the majority of the ADCS computation

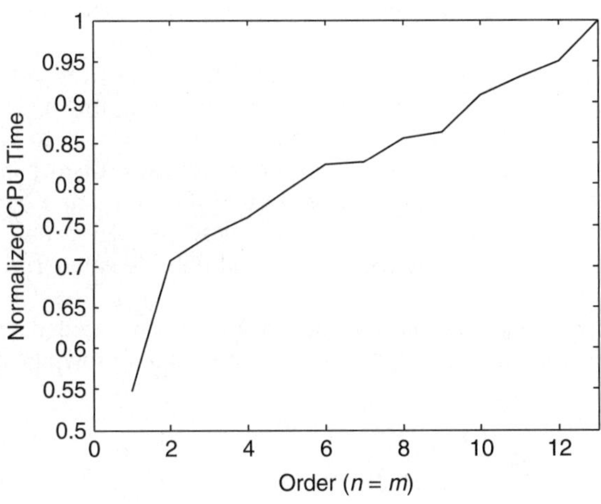

Fig. 9.5 Normalized CPU time vs order.

is performed on the ground, it is possible to utilize the added precision, and hence ION uses the full $n = 13$ degree model of the tenth-generation IGRF.

D. Onboard Sensors

The attitude determination is accomplished using six onboard sensors. A combination of a three-axis magnetometer, four solar panels (primarily used for power generation), and a simple solar cell located on the smaller side of ION, facing away from Earth, is used.

1. *Magnetometer*

The magnetometer used on ION is the HMC2003 model manufactured by Honeywell. It is a three-axis magnetic sensor that uses magnetoresistive transducers to measure the field in three orthonormal directions. This particular sensor can detect magnetic fields between −2 and +2 Gauss with accuracy of 40 μG.

If possible, the magnetometer should be mounted such that its axes are aligned with the satellite's body-fixed frame, providing a magnetic field reading $\mathbf{b}^{B}_{\text{Sensor}}$. Special care should be taken to mount the magnetometer as accurately as possible in order to avoid skewed readings. In case the magnetometer cannot be mounted with its axes coinciding with the satellite body axes, a standard orthogonal rotation can be implemented in software. Also, an obvious precaution that should be implemented in the controller logic (software or hardware) is never to sample the magnetometer while operating the torque coils or other field-generating devices.

Prior to launch the magnetometer must be calibrated. The calibration is a two-step process that ensures interference-free measurements. The first step consists of implementing a set/reset correction that removes any temperature offsets and past magnetic effects. This portion of the calibration can be done without the satellite being finished or assembled. To do this, one reading is taken with the magnetometer sensor set, and then another reading is taken with the sensor reset; the two readings are then subtracted, which removes any temperature and past magnetic effects.

The second part of calibration removes any hard and soft metal effects caused by the satellite's components and must be performed on a finished and assembled spacecraft. The metallic components cause the field readings to be both skewed and offset from their nominal values. To correct for these disturbances, the satellite must be rotated one full revolution around the z axis and once around the x axis. The rotation around the z axis provides the scaling and offset factors for the x and y axes, while the rotation around the x axis determines the z-axis correction factors.

The minimum and maximum values from the collected data are used to compute the scaling and offset factors according to the following equations:

$$X_{\text{SF}} = \max\left\{\frac{Y_{\max} - Y_{\min}}{X_{\max} - Y_{\min}}\right\}, \quad Y_{\text{SF}} = \max\left\{\frac{X_{\max} - X_{\min}}{Y_{\max} - Y_{\min}}\right\},$$
$$Z_{\text{SF}} = \max\left\{\frac{Y_{\max} - Y_{\min}}{Z_{\max} - Z_{\min}}\right\} \tag{9.9}$$

$$X_{\text{OFFSET}} = \left[\frac{X_{\max} - X_{\min}}{2} - X_{\max}\right] X_{\text{SF}} \quad Y_{\text{OFFSET}} = \left[\frac{Y_{\max} - Y_{\min}}{2} - Y_{\max}\right] Y_{\text{SF}}$$

$$Z_{\text{OFFSET}} = \left[\frac{Z_{\max} - Z_{\min}}{2} - Z_{\max}\right] Z_{\text{SF}} \tag{9.10}$$

Figure 9.6 demonstrates an intuitive representation of the calibration results. The rotation of the satellite around a particular axis, say x, causes the readings along that axis to be constant. It is thus possible to plot the remaining two axes in two dimensions as shown in Fig. 9.6. The various components, which are generally nonuniformly distributed around the satellite, distort the magnetic field readings as the satellite is rotated. This effect is clearly shown by the lighter dotted data that are noncircular and offset from the origin. The heavier solid curve represents the same data after the calibration has been applied; the pattern is clearly much more circular and is truly centered at the origin.

2. *Solar Panels*

The solar cells used on ION are the advanced triple-junction (ATJ) high-efficiency solar cells made by EMCORE. Each cell is 4 × 7 cm in size, and five individual cells are placed on each of the four side panels. The cells operate at 27.5% efficiency and contribute up to 5-W peak power generation per panel. In addition, the power point tracking (PPT) circuit ensures that the panels always operate near their peak.

Although the primary function of the solar cells is power generation, the ADCS team utilizes them for attitude determination. The current from a solar cell is dependent on the intensity of sunlight to which the cell is exposed. This intensity, in turn, is dependent on the angle of incidence between a sun ray and a panel. It is thus possible to obtain two angles of incidence from two side panels that are exposed to the sun by simply measuring the current that they are generating. These two angles are instrumental in calculating ION's attitude, as will be shown in

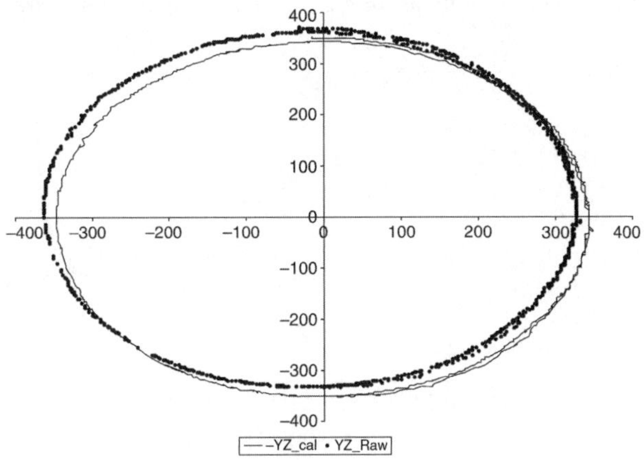

Fig. 9.6 Magnetometer calibration–rotation around *x* axis.

detail in a later section. A simple software check eliminates the panels that are facing the Earth and are receiving albedo effects.

3. *Top Solar Cell*

The top solar cell used on ION is a simple silicone cell capable of producing 0.33 A at 0.55 VDC in full sunlight. The exact cell behavior was characterized by exposing it to sunlight on a clear day and varying the sun incidence angle. To ensure accurate measurement, an online Java applet found at http://www.jgiesen.de/azimuth/index.html was used to compute the elevation of the sun at noon of the specific test day and the specific site location. A simple instrument was constructed in order to vary the cell's position while measuring the incidence angle, voltage, and current through a 1-kΩ resistor. The addition of the resistor makes the measurements less prone to temperature effects, and it distributes the angle effects more evenly over the voltage range giving more resolution at the near normal angles.

The resultant I-V curve has to then be adjusted for the atmospheric effects that reduce the solar intensity on the Earth's surface. Based on the historical data of sun intensity at air mass zero (AM0), these effects are approximately 1.35 times greater in space than on the surface. After making the necessary adjustments, the data were plotted and can be seen in Fig. 9.7. A curve fit was added and adjusted ensuring high accuracy of the fit in the higher voltage range where small voltage variations correspond to large angle changes. The resultant fit is given by Eq. (9.11) and is used to obtain the sun incidence angle from a voltage reading on top of the satellite.

$$\text{Angle} = \frac{32.987 + 4105.715 * V + 8180.705 * V^2 - 231527.176 * V^3}{1 + 121.702 * V - 709.659 * V^2 - 798.071 * V^3} - 32.574 \tag{9.11}$$

E. Attitude Estimation

Attitude determination is a unique problem in which a measurement of one vector is not enough to calculate the attitude, making the system underdetermined. At the same time, two vector measurements provide too much information, forcing

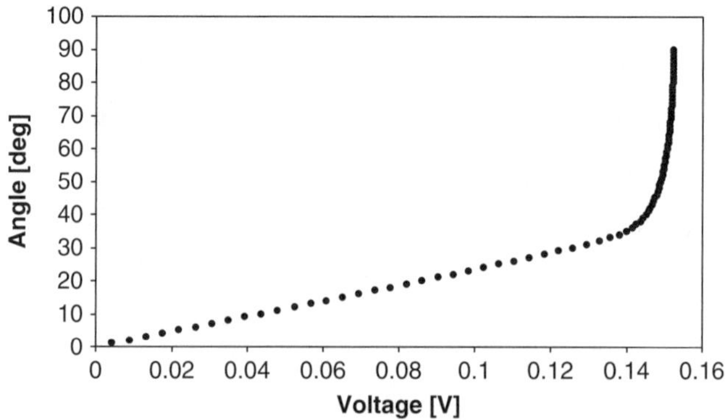

Fig. 9.7 Angle vs voltage for top solar diode.

the problem to be overdetermined. As a result, most attitude determination algorithms are really attitude estimation algorithms, with ION being no exception.

ION's attitude estimation algorithm uses a combination of information from its three sensors and calculation based on its orbital position to find its attitude at discrete points along the orbit. The process implemented in this design is a deterministic attitude determination that uses two sets of vectors. The first set of vectors consists of the magnetic field and sun direction in the satellite's body frame obtained from the magnetometer, sun sensor, and the solar panels. The second set is the magnetic field and sun direction vectors in the ECI frame, calculated based upon orbital position of the satellite. The details of this method are presented in the following section, and more extensive derivation can be found in [5].

1. Sun Direction Vector from Onboard Sensors

Figure 9.8 shows how the solar panels and the top solar sensor are illuminated when ION is in the daylight portion of its orbit. The angle α_3 can be computed immediately from the voltage reading of the top cell, by substituting it into Eq. (9.11).

The dependence of the generated current on the angle of incidence can be written as

$$I = I_{max} \sin \alpha \tag{9.12}$$

where α is defined in Fig. 9.9 and I_{max} is the current induced in the solar panel when the sun shines directly at it.

Applying Eq. (9.12) to both sides of the satellite and creating a ratio of the resulting equations leads to the following result:

$$\alpha_1 = \arctan \frac{I_1}{I_2} \tag{9.13}$$

The remaining angle α_2 can be found from the following relation [6]:

$$\sin \alpha_2 = \cos \alpha_1 \tag{9.14}$$

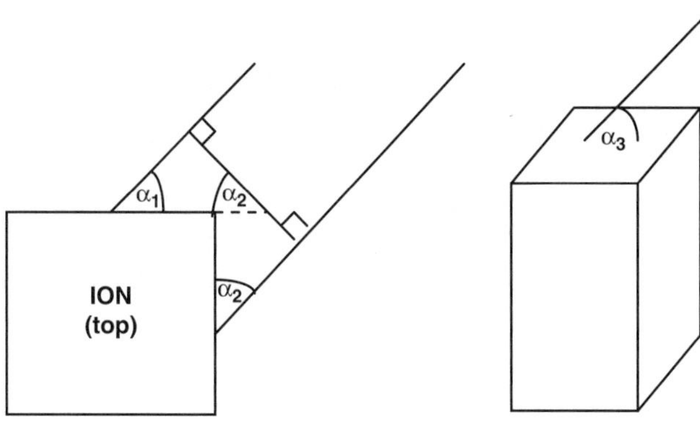

Fig. 9.8 Sunlight incidence angles.

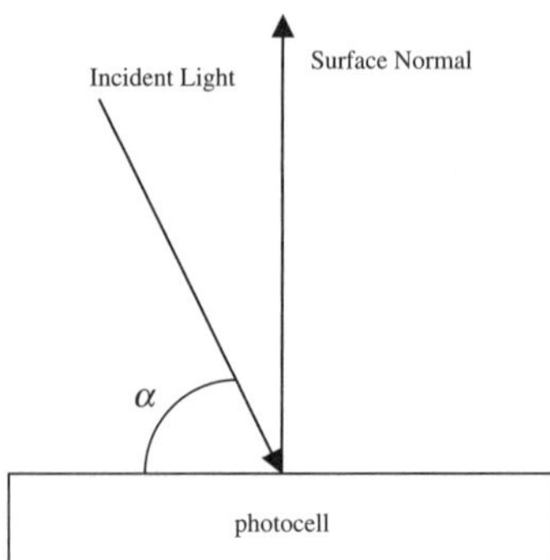

Fig. 9.9 Photocell diagram.

Finally, using the preceding equations, the geometry in Fig. 9.8 can be represented in terms of the three incidence angles as shown in Eq. (9.15). In this form, the equation describes the sun vector in the satellite's body-fixed frame assuming that the incident beam intersects the positive x and y axes, and negative z axis.

$$\mathbf{r}^{\mathrm{B}}_{\mathrm{Sun,Sensor}} = \begin{bmatrix} \sin\alpha_1 \cos\alpha_3 \\ \sin\alpha_2 \cos\alpha_3 \\ -\sin\alpha_3 \end{bmatrix} \tag{9.15}$$

2. Sun Direction Vector from Julian Date

The next step after obtaining the sun position from the onboard sensors is to compute the sun vector based upon the Julian day (JD). Julian day is a universally adopted means of expressing "global" time of any astronomical event. It is defined as number of days since Greenwich noon (12:00 UT) on 1 January 4713 B.C.

However, because JD is not the most intuitive way of scheduling satellite tasks and displaying attitude history, the following format is often used throughout the ADCS software: [year; month; day; hour; minute; second]. Because both time formats are used, Eq. (9.16) gives a simple conversion between the two:

$$\begin{aligned} \mathrm{JD} = 367 * \mathrm{year} - \mathrm{INT}\left\{\frac{7 * \left[\mathrm{year} + \mathrm{INT}\left(\dfrac{\mathrm{month} + 9}{12}\right)\right]}{4}\right\} \\ + \mathrm{INT}\left(\frac{275 * \mathrm{month}}{9}\right) + \mathrm{day} + 1{,}721{,}013.5 + \frac{\mathrm{hr}}{24} + \frac{\mathrm{min}}{1440} + \frac{s}{86{,}400} \end{aligned} \tag{9.16}$$

When formulating the sun vector in the ECI frame, one must remember that the ecliptic plane is inclined to the equatorial plane by an angle ε, known as the

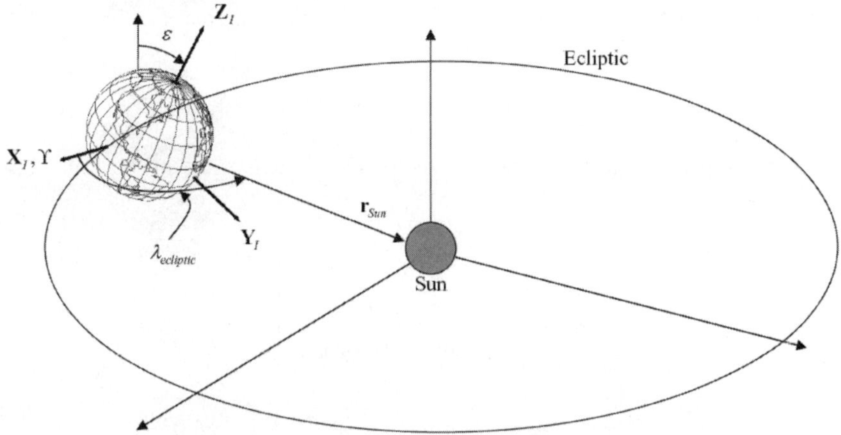

Fig. 9.10 Ecliptic plane with respect to the ECI frame.

obliquity of the ecliptic. This can be seen in Fig. 9.10. Displayed is also the longitude of the sun vector $\lambda_{\text{Ecliptic}}$, which is measured from the direction of vernal equinox along the ecliptic plane.

As an intermediate step, it is useful to define a quantity called mean longitude of the sun as

$$\lambda_{M_{Sun}} = 280.4606184° + 36{,}000.77005361 * T_{\text{UT1}} \tag{9.17}$$

where T_{UT1} is the desired time expressed in Universal Time and which can be computed from the Julian day using the following equation:

$$T_{\text{UT1}} = \frac{\text{JD}_{\text{UT1}} - 2{,}451{,}545.0}{36{,}525} \tag{9.18}$$

The subscript 1 refers to the form of the Universal Time that corrects for polar motion and is independent of station location. For more in-depth explanation on the differences in UT, refer to [7].

Next, it is necessary to compute the mean anomaly of the sun M_{Sun} using Eq. (9.19):

$$M_{\text{Sun}} = 357.5277233° + 35{,}999.050 * T_{\text{TDB}} \tag{9.19}$$

where T_{TDB} is the barycentric dynamical time, which can be acceptably estimated as T_{UT1}. Combining the preceding equations, it is possible to write the ecliptic longitude of the sun $\lambda_{\text{Ecliptic}}$ as

$$\lambda_{\text{Ecliptic}} = \lambda_{M_{\text{Sun}}} + 1.914666471° \sin M_{\text{Sun}} + 0.918994643 \sin 2M_{\text{Sun}} \tag{9.20}$$

Using the same assumption as in Eq. (9.19), namely, that $T_{\text{TDB}} \approx T_{\text{UT1}}$, the obliquity of the equator can be accurately estimated as

$$\varepsilon = 23.43991° - 0.0130042 T_{\text{TDB}} \tag{9.21}$$

It is finally possible to write the sun vector in the ECI frame in the following form:

$$\mathbf{r}^{I}_{\text{Sun,JD}} = \begin{bmatrix} \cos \lambda_{\text{Ecliptic}} \\ \cos \varepsilon \sin \lambda_{\text{Ecliptic}} \\ \sin \varepsilon \sin \lambda_{\text{Ecliptic}} \end{bmatrix} \tag{9.22}$$

3. *Deterministic Attitude Determination*

The attitude determination problem is in essence a problem of finding a rotation matrix between the satellite's body-fixed reference frame and some inertial frame—such as the ECI frame. It can be shown that only three numbers are necessary to fully determine a direction cosine matrix. At the same time, each unit vector provides only two pieces of information because of the unit vector constraint. It is thus necessary to use two vectors, which will provide four known quantities, making the problem overdetermined.

In the ideal case and having the four vectors $\mathbf{b}^{B}_{\text{Sensor}}$, $\mathbf{b}^{I}_{\text{calc}}$, $\mathbf{r}^{B}_{\text{Sun,Sensor}}$, and $\mathbf{r}^{I}_{\text{Sun,JD}}$, it is possible to write Eqs. (9.23) and (9.24), with both $A^{B/I}$ matrices being equal. However, because the system is overdetermined, it is not generally possible to find such $A^{B/I}$.

$$\mathbf{b}^{B}_{\text{Sensor}} = A^{B/I}\mathbf{b}^{I}_{\text{calc}} \tag{9.23}$$

$$\mathbf{r}^{\text{B}}_{\text{Sun,Sensor}} = A^{B/I}\mathbf{r}^{I}_{\text{Sun,JD}} \tag{9.24}$$

To remedy this difficulty, a triad algorithm is used to discard one piece of information. Note, however, that this method does not simply throw away one of the measured components. Also note that $\mathbf{b}^{I}_{\text{calc}}$ is the calculated magnetic field vector in the ECI frame, which is obtained by converting the $\mathbf{b}^{E}_{\text{calc}}$ vector into the inertial frame.

The triad algorithm constructs two triads of orthonormal unit vectors using the vector information gathered. The two triads are the components of the same reference frame t expressed in the body and inertial frames. The reference frame is constructed assuming that one of the body/inertial vector pairs is correct. In the following derivation, it is assumed that the sun vector is the correct one. The process is started by defining the following vector:

$$\mathbf{t}^{B}_{1} = \mathbf{r}^{B}_{\text{Sun,Sensor}} \tag{9.25}$$

$$\mathbf{t}^{I}_{1} = \mathbf{r}^{I}_{\text{Sun,JD}} \tag{9.26}$$

The next step consists of constructing a second base vector as a unit vector in the direction perpendicular to the two observations.

$$\mathbf{t}_2^B = \frac{\mathbf{r}_{\text{Sun,Sensor}}^B \times \mathbf{b}_{\text{Sensor}}^B}{\left|\mathbf{r}_{\text{Sun,Sensor}}^B \times \mathbf{b}_{\text{Sensor}}^B\right|} \tag{9.27}$$

$$\mathbf{t}_2^I = \frac{\mathbf{r}_{\text{Sun,JD}}^I \times \mathbf{b}_{\text{calc}}^I}{\left|\mathbf{r}_{\text{Sun,JD}}^B \times \mathbf{b}_{\text{calc}}^I\right|} \tag{9.28}$$

The third base vector is computed to complete the orthogonal triad as

$$\mathbf{t}_3^B = \mathbf{t}_1^B \times \mathbf{t}_2^B \tag{9.29}$$

$$\mathbf{t}_3^I = \mathbf{t}_1^I \times \mathbf{t}_2^I \tag{9.30}$$

It is now possible to construct two rotational matrices by placing the **t** vector components into the columns of two 3×3 matrices as follows:

$$\begin{bmatrix}\mathbf{t}_1^B & \mathbf{t}_2^B & \mathbf{t}_3^B\end{bmatrix} \quad \begin{bmatrix}\mathbf{t}_1^I & \mathbf{t}_2^I & \mathbf{t}_3^I\end{bmatrix} \tag{9.31}$$

Upon closer inspection, the preceding matrices are simply $A^{B/t}$ and $A^{I/t}$, respectively, and hence the rotational matrix between reference frames I and B can be calculated as follows:

$$A^{B/I} = A^{B/t} A^{t/I} = \begin{bmatrix}\mathbf{t}_1^B & \mathbf{t}_2^B & \mathbf{t}_3^B\end{bmatrix}\begin{bmatrix}\mathbf{t}_1^I & \mathbf{t}_2^I & \mathbf{t}_3^I\end{bmatrix}^T \tag{9.32}$$

Equation (9.32) then defines the sought-after attitude of the satellite with respect to the ECI frame.

4. *Rotational Rate Estimation*

The preceding section describes the methodology of computing the attitude at a single instant in time. In real operation, however, it is necessary to know the satellite's attitude at any time and do so without resampling of the sensors. Fortunately, this can be accomplished by knowing the rotational rate of the satellite, which, in turn, can be easily calculated from a minimum of three separate attitude measurements using the Lagrange interpolation formula. Only the final results are presented here, but a more in-depth derivation and discussion of this method can be found in [6].

Recall that the triad algorithm produces a rotation matrix $A^{B/I}$ according to Eq. (9.32). It is convenient to first transform this attitude matrix into Euler angles ϕ, θ, and ψ using the following equations (for a 3-2-1 Euler rotation):

$$\phi = \arctan\left(\frac{a_{32}}{a_{33}}\right) \quad \theta = -\arcsin(a_{13}) \quad \psi = \arctan\left(\frac{a_{21}}{a_{11}}\right) \tag{9.33}$$

This process is then repeated for all three angles at three discrete sample points, corresponding to measurement times t_1, t_2, and t_3. The angular rates can then be computed using the following equations:

$$\dot{\phi}(t_3) = \frac{t_3 - t_2}{(t_1 - t_2)(t_1 - t_3)}\phi(t_1) + \frac{t_3 - t_1}{(t_2 - t_1)(t_2 - t_3)}\phi(t_2) + \frac{2t_3 - t_1 - t_2}{(t_3 - t_1)(t_3 - t_2)}\phi(t_3) \tag{9.34}$$

$$\theta(t_3) = \frac{t_3 - t_2}{(t_1 - t_2)(t_1 - t_3)}\theta(t_1) + \frac{t_3 - t_1}{(t_2 - t_1)(t_2 - t_3)}\theta(t_2) + \frac{2t_3 - t_1 - t_2}{(t_3 - t_1)(t_3 - t_2)}\theta(t_3) \tag{9.35}$$

$$\dot{\psi}(t_3) = \frac{t_3 - t_2}{(t_1 - t_2)(t_1 - t_3)}\psi(t_1) + \frac{t_3 - t_1}{(t_2 - t_1)(t_2 - t_3)}\psi(t_2) + \frac{2t_3 - t_1 - t_2}{(t_3 - t_1)(t_3 - t_2)}\psi(t_3) \tag{9.36}$$

Note that the preceding equations are tailored to compute the angular velocities at the last time t_3. As it will become clear in the next chapter, the control algorithm reads in an attitude and rotational rate at a discrete point and computes the optimal actuating torque for that time. Although it is possible to obtain the attitude and rotational rates for any one of the three times, it is not desirable to compute optimal torque based on outdated information—as would be the case for t_1 and t_2.

F. Impact of the Two-Body Gravitational Model on Attitude Propagation

A useful result is obtained when investigating the implications of the errors in the two-body gravitational model when applied to attitude propagation during torquing. A two-step simulation was set up by computing two sets of the position vectors, one using a two-body model and the second using STK's high-precision orbit propagator (HPOP) model. In the first step, the two-body position vector was used to find the magnetic field. Then the optimum torquing profile, given initial tip-off attitude and rate, to stabilize the spacecraft was determined. Next, in the second step, the HPOP position vector was used to recompute the magnetic field at the satellite's true location while applying the previously found torque. As a result of the errors in the position found using the two-body model, the applied torque is no longer optimal because of the spatially varying magnetic field. The purpose of this simulation is to investigate the extent that these errors have on the accuracy and stabilization time of the ADCS algorithms.

The errors in the magnitude of the position vector were estimated earlier in this chapter to have a maximum value of approximately 7 km. This result is reproduced in Fig. 9.11 along with the error in the magnetic field obtained using the two position propagation models. The relatively large errors in the position have, by themselves, little effect on the magnetic field. For example, the first maximum error in the position occurs approximately half an hour after the start of the simulation. At the same, the magnetic field computed at both locations is nearly the same as indicated by the top plot of this figure. The converse observation can be made at

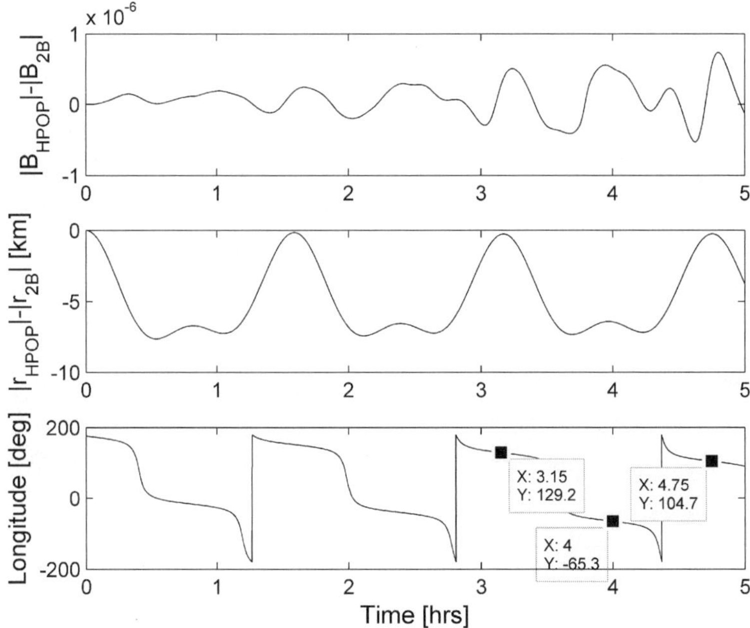

Fig. 9.11 Sources of error in the computed magnetic field.

approximately 1.5 hours into the simulation when the error in the position is nearly zero, but there is a slight increase in the error in the computed magnetic field.

To understand these nonintuitive results, the corresponding longitude of the spacecraft was plotted and can be seen in the same figure. Indicated on the longitude subplot are three points that correspond to three large peaks in the magnetic field error plot. These peaks occur at approximately 3.15, 4.00, and 4.75 hours after the start of the simulation and match up to longitudes of approximately 125, –62, and 103 degrees, respectively. The significance of these longitudes becomes apparent when looking at the contours of the magnetic field of Fig. 9.3. At approximately 125 deg longitude, there exist two large magnetic field anomalies; one resides in the sub-Australian region and another in the North Siberian region. The ground track of ION, whose inclination is 98 deg, passes through both of those anomalies. The first large error in the magnetic field magnitude of approximately 0.5×10^{-7} is directly associated with the steep gradient between the sub-Australian anomaly and the magnetic field at the Equator. Once the satellite enters the equatorial regions where the magnetic field is nearly constant, the error in the magnetic field associated with errors in the position is reduced to nearly zero. However, immediately afterward, the satellite enters the North Siberian anomaly, and even small errors in the position cause large errors in the magnetic field magnitude. Similar results can be seen at 300 and 100 deg longitude where the South Atlantic and North American anomalies are located as seen in Fig. 9.3.

The consequence of the just-described error between computed magnetic field and the actual field is the application of a nonoptimal torque to the spacecraft.

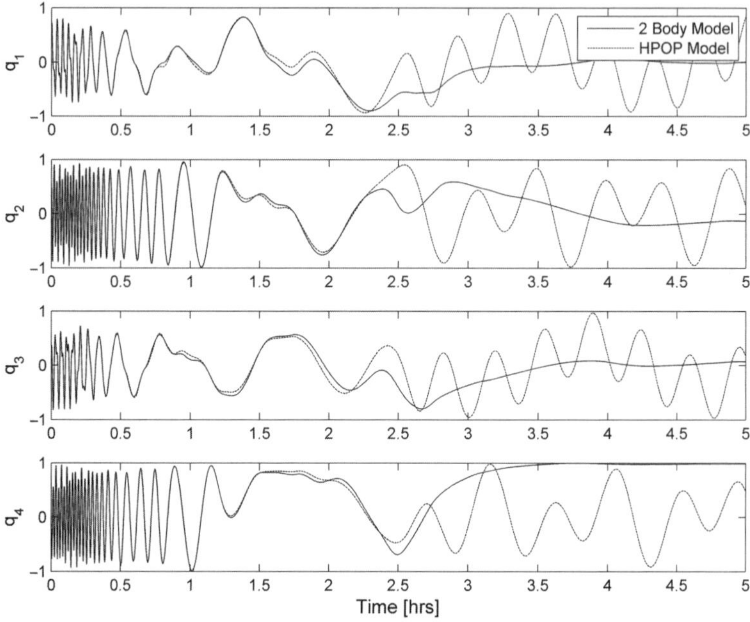

Fig. 9.12 Evolution of the quaternions for two-body and HPOP models.

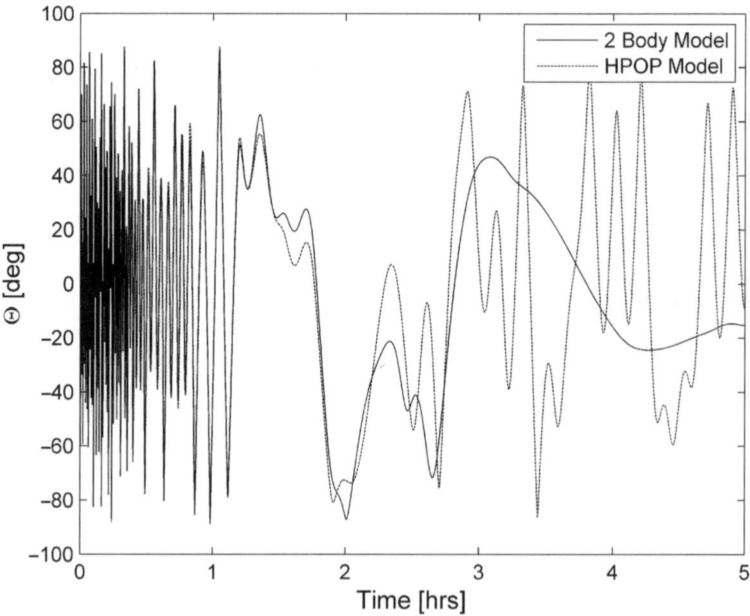

Fig. 9.13 Evolution of nadir offset angle for two-body and HPOP models.

This effect is initially small, but as the spacecraft's attitude deviates from the predicted state, the applied torque becomes less and less optimal. By plotting the predicted and actual response to the applied torque, as is done in Fig. 9.12, it is possible to establish an approximate time for which the torquing is accurate. Although results can vary depending on the orbital location, time of torquing, whether the satellite is within a magnetic field anomaly or strength of the applied torque, the following results are believed to be representative of a typical error encountered by ION during its mission. As seen in Fig. 9.12, the attitude starts deviating approximately 1.5 h after the start of the simulation and reaches unacceptable levels after 2 h. This result instills confidence in the ability of the two-body propagation model to be used successfully because the predicted power budget allows for a maximum of 1 h of continuous torquing. In all of the scenarios presented in this section, the satellite is given the maximum initial tip-off rate of 5 deg/s and is forced to apply the magnetic torque while flying through the magnetic anomalies. Additional results confirming the limit on accurate torque duration can be seen in Fig. 9.13, which compares the predicted and actual offset from the nadir orientation. The results confirm that after 2 h of torquing, the satellite's z axis starts deviating by several degrees from the predicted attitude.

III. Attitude Control

This section describes the methodology used to stabilize ION. The goal of the control algorithms is to drive the rotational rate to near-zero state and to orient the satellite into a nominal, nadir-facing attitude. The linear quadratic theory is used in conjunction with a feedback design system to achieve the preceding goals. Three magnetic torquers, each capable of producing a time-varying magnetic field, are implemented on all three of the satellite's body-fixed axes. Although the initial tip-off rates are unknown prior to launch, the worst-case scenario of 5 deg/s is assumed.

The presented design has the added benefit of reusability on other spacecraft with magnetic torque actuation. The control algorithms are decoupled from the attitude determination software and require only the inputs of initial quaternions and body rotational rates, which can be obtained by any method. The modularity of this design also allows for easy changes to the properties of the satellite (mass, moments of inertia, etc.) and torque coil parameters (size, field strength, magnetic dipole moment, etc.) making adaptation to other spacecraft straightforward. The existence of a complete simulator enables the user to test stability, robustness, and performance of the control algorithms on a particular system.

A. Actuators

ION uses a system of three magnetic torque coils, each placed along the satellite's body axes, to control its attitude and rotational rate. The torque coils produce a magnetic field of variable strength and direction and interact with the Earth's magnetic field to produce a torque on the spacecraft. By knowing the exact magnitude and direction of the Earth's field at every location along the orbit and the satellite's attitude at those positions, it is possible to solve for the field needed to be generated by the torquers in order to correctly orient the satellite. The next two sections describe some of the necessary electromagnetic theory as well as the magnetic torquer design particular to ION.

1. *Torque Coil Theory*

Current flowing through a rectangular loop of wire creates a magnetic field. At the center of the loop, the magnetic field can be characterized by the following equations:

$$\mathbf{b} = \frac{2\mu_0 NI\sqrt{(a^2 + b^2)}}{\Pi A}\mathbf{a}_n \tag{9.37}$$

where μ_0 is known as the permeability of free space and is equal to $4\pi \times 10^{-7}$ H/m, A is the cross-sectional area, I is the current flowing through the wire, N is the number of turns, a and b are the sides of the rectangle, and $\mathbf{a}_n$ is the direction normal to the coil and is found according to the right-hand rule. Additionally, the same loop generates a magnetic dipole moment defined as

$$\mathbf{m} = NIA\mathbf{a}_n \tag{9.38}$$

A key element that enables the attitude control of ION is the fact that when the coil is immersed in a magnetic field, say that of the Earth, it produces a torque. The torque is generated by the interaction of the two magnetic fields and is governed by the following equation:

$$\mathbf{t} = \mathbf{m} \times \mathbf{b} \tag{9.39}$$

Intuitively, the magnetic moment tends to align itself in the direction of the external magnetic field. When the two vectors align, the cross product in the preceding equation is equal to zero, and no torque is exerted on the coil. The coil thus reaches a state of stable equilibrium and rejects any deviations as long as the coil maintains its dipole moment.

2. *ION's Torque Coil Design*

ION's initial torquer design consisted of several loops of copper wire wound in a circular pattern. Although this classical design is inexpensive and easily manufactured, it also proved difficult to mount securely inside the satellite. In addition, the winding process, which was done by hand, produced nonuniform results between the three coils. The coils had the tendency to bend out of shape (become noncircular) and had to be secured in several points along the circumference.

To avoid the preceding difficulties, a new design was proposed and tested. Based on a typical electronics board design, 7-mil traces were etched onto a standard printed circuit board in a wind-down pattern. It is possible to place consecutive traces 7 mils apart on a single layer and to combine multiple layers to form a single torque coil. ION's design, seen in Fig. A.22, uses 30 loops on each of four layers for a combined total of 120 loops. Because of the boards' rigid design, the coils are very easy to mount and handle. Moreover, because of the thin size of the traces, the total resistance per coil is only 99.3 Ω per coil.

The equation for the magnetic field at the center of the coil presented in the preceding section must be corrected for the varying size of the loops. Given the dimensions of the most outer loop as a_{outer} and b_{outer} as well the thickness and separation of the traces, which in this case are equal, as w, it is possible to write the magnetic field at the center of the magnetic torquer as

$$b = 4 \cdot \sum_{N=0}^{29} \frac{2\mu_0 I \sqrt{(a_{\text{outer}} - 4 \cdot N \cdot w)^2 + (b_{\text{outer}} - 4 \cdot N \cdot w)^2}}{\pi (a_{\text{outer}} - 4 \cdot N \cdot w) \cdot (b_{\text{outer}} - 4 \cdot N \cdot w)} \tag{9.40}$$

The magnetic dipole moment is adapted in a similar fashion to obtain the following form:

$$\mathbf{m} = 4 \cdot I \cdot \sum_{N=0}^{29} \left[(a_{\text{outer}} - 4 \cdot N \cdot w)(b_{\text{outer}} - 4 \cdot N \cdot w) \right] \mathbf{a}_n \tag{9.41}$$

The preceding equations are used to calculate the magnetic field and dipole moment of the new coils. Their values, along with other essential parameters, are summarized in Table 9.1.

Although this design offers superior manufacturing quality and desirable mounting characteristics, it suffers from reduced dipole moment as compared to the hand-wound design. This is solely because of the reduced number of turns—a drawback which can be remedied in future designs. In theory, more loops translate to stronger magnetic dipole moment, but at the unfortunate penalty of longer wire with larger total resistance. This increased resistance often becomes the limiting factor in real-life, power-limited systems. ION's torquer design consists of only 120 turns, which is on the lower end of a typical small system such as ION. As a result, the system suffers from a reduced magnetic moment, which correlates directly to reduced torques. Although it is possible to increase the number of board

Table 9.1 Torque Coil Parameters

Coil	X	Y	Z
Equivalent cross-sectional area, m^2	0.0485	0.0485	0.0485
Max/min length of longer side, mm	58.42/38.10	58.42/38.10	58.42/38/10
Max/min length of shorter side, mm	43.18/22.86	43.18/22.86	43.18/22.86
Number of turns (4 layers)	4 × 30	4 × 30	4 × 30
Coil resistance, Ω	99.3	99.3	99.3
Max current, A	0.114	0.114	0.114
Max magnetic dipole moment, Am^2	0.0222	0.0222	0.0222
Max magnetic field, G	4.161	4.161	4.161
Distance between successful traces, mm	0.1778	0.1778	0.1778
Trace thickness, mm	0.1778	0.1778	0.1778

layers to eight or even sixteen, in effect increasing the torque, simulations with the current design have shown very acceptable performance. Particulars of the simulations will be shown in Section IV.D, but even in the worst-case scenario with the satellite tumbling at 5 deg/s on all three axes, the coils were able to despin ION in little over an hour.

With such good performance and added ease of mounting and handling, this design was implemented on ION and is highly recommended for future missions with magnetic torque actuation.

B. Linear Quadratic Regulator

Attitude control via magnetic torque actuation is a challenging problem made difficult by the fact that the control torque, given in Eq. (9.39), is orthogonal to both the Earth's magnetic field as well the magnetic dipole moment of the coil. As a result, it is impossible to generate any control torque in the direction of the geomagnetic field, in effect creating an uncontrollable subspace. Luckily enough, this subspace is time varying, because of the time-varying nature of the geomagnetic field as the satellite changes its orbit. Thus, the satellite is controllable for highly inclined orbits, such as the sun-synchronous orbit of ION.

ION's attitude control system utilizes an asymptotic quasi-periodic linear quadratic regulator (LQR) similar to the control law proposed in [8]. This section first presents the theory associated with the linear quadratic regulator and then proceeds to describe ION's actual design necessary to despin and stabilize the satellite.

1. Linear Quadratic Regulator Theory

The linear quadratic control problem is an optimal control problem. It is covered in most modern control texts, such as [9] or [10]. In this problem, it is assumed that there is a system in which the state dynamics are constant, but the input dynamics vary with time:

$$\dot{\mathbf{x}} = \mathbf{A}\mathbf{x} + \mathbf{B}(t)\mathbf{u}, \quad \text{given } \mathbf{x}(t_0) \tag{9.42}$$

For this problem, the following cost function is assigned:

$$\mathbf{J} = \frac{1}{2}\int_{t_0}^{t_f} \left(\mathbf{x}^T \, \mathrm{Q}\mathbf{x} + \mathbf{u}^T \mathbf{R}\mathbf{u}\right) \mathrm{d}\tau + \frac{1}{2}\mathbf{x}(t_f)^T \, \mathbf{P}_T \, \mathbf{x}(t_f), \quad \text{given } \mathbf{x}(t_0) = \mathbf{x}_0 \tag{9.43}$$

The matrix $\mathbf{Q}$ can be thought of as a penalty on the state vector to discourage the state from deviating too large from the desired state, $\mathbf{R}$ as a matrix to penalize using excessive control effort, and $\mathbf{P}_T$ as a penalty on the final state. All of these are constant matrices. From the point of view of the designer, unless these matrices have some a priori designation, they can be viewed as a means of changing the characteristics of the system dynamics. Thus, these matrices provide a quantitative way of trading off state deviation with control energy. All of $\mathbf{Q}$, $\mathbf{R}$, and $\mathbf{P}_T$ are assumed to be positive definite.

The goal of the linear quadratic regulator is to find the optimal control to minimize the cost function given these various matrices and an initial state condition.

It is a well-known result that the optimal solution to this problem is a full-state feedback controller of the following form:

$$\mathbf{u}^* = \mathbf{Fx} = -\mathbf{R}^{-1}\mathbf{B}(t)^T\,\mathbf{P}(t)\mathbf{x} \tag{9.44}$$

The matrix $\mathbf{P}(t)$ is found by the following differential equation:

$$\dot{\mathbf{P}}(t) = \mathbf{Q} + \mathbf{P}(t)\mathbf{A} + \mathbf{A}^T\mathbf{P}(t) - \mathbf{P}(t)\mathbf{B}(t)\mathbf{R}^{-1}\mathbf{B}(t)^T\,\mathbf{P}(t), \text{ given } \mathbf{P}(t_f) = \mathbf{P}_T \tag{9.45}$$

In principle, this equation can be evaluated in closed form or analytically, to find the time-varying matrix $\mathbf{P}(t)$, which in turn specifies $\mathbf{u}^*$.

A special case of the quadratic linear regulator is where the matrix $\mathbf{B}(t)$ is periodic. In this case, for some value of T and all t,

$$\mathbf{B}(t) = \mathbf{B}(t + T) \tag{9.46}$$

In this case, if $\mathbf{P}_T$ is properly chosen, it can be shown that the feedback gain matrix is also periodic in T. This periodicity in the feedback matrix can be used to show that, in the case of some very general conditions on the penalty matrix $\mathbf{Q}$, $\mathbf{P}(t)$ approaches a steady-state matrix $\mathbf{P}_{ss}$ as the minimum eigenvalue of $\mathbf{R}$ approaches infinity. Thus, for large values of $\mathbf{R}$, it can be expected that this $\mathbf{P}_{ss}$ matrix can be used as a reasonable approximation for $\mathbf{P}(t)$ for all values of t. In this case, the optimal control law becomes

$$\mathbf{u}^* = \mathbf{Fx} = -\mathbf{R}^{-1}\mathbf{B}(t)^T\,\mathbf{P}_{SS}\mathbf{x} \tag{9.47}$$

To find $\mathbf{P}_{ss}$, note that the following long average over one period can be made:

$$\mathbf{B}(t)\mathbf{R}^{-1}\mathbf{B}(t)^T \approx \mathrm{C} = \frac{1}{T}\int_0^T \mathbf{B}(\tau)\mathbf{R}^{-1}\mathbf{B}(\tau)^T\,\mathrm{d}\tau \tag{9.48}$$

These approximations greatly simplify the implementation for the linear quadratic regulator. Inserting this approximation into the differential equation found in Eq. (9.45), it is possible to solve the following algebraic Riccati equation for $\mathbf{P}_{ss}$:

$$\mathbf{0} = \mathbf{Q} + \mathbf{P}_{SS}\mathbf{A} + \mathbf{A}^T\mathbf{P}_{SS} - \mathbf{P}_{SS}\mathbf{C}\mathbf{P}_{SS} \tag{9.49}$$

2. *ION's Asymptotic Periodic Linear Quadratic Regulator Design*

To apply the linear quadratic regulator to ION's attitude control system, the state vector consists of the attitude and angular velocity of the fixed-body coordinate system with respect to the orbital reference frame. Only three elements are necessary to represent the attitude as the fourth element is redundant. Thus, the six-element state is as follows:

$$\mathbf{x}_{\text{lqr}} = \begin{bmatrix} \mathbf{q}^{\mathbf{B/R}} \\ \omega^{\mathbf{B/R}} \end{bmatrix} \tag{9.50}$$

The first three elements of the state are the three-element attitude vector. The final three elements are the angular velocity vectors. The input is the magnetic moment generated by the magnetorquers is as follows:

$$\mathbf{u} = \mathbf{m} \tag{9.51}$$

The system dynamics are clearly nonlinear. Furthermore, the state dynamics are time varying. Thus, to apply the linear quadratic regulator theory, it is first necessary to linearize the system about a nominal trajectory. This derivation is available in [6], and only the results are presented in the following. The system takes the form of Eq. (9.42) with the following definitions:

$$\mathbf{A} = \left\{ \begin{array}{l} \qquad\qquad\qquad 0 \qquad\qquad \frac{1}{2}\mathbf{I} \\ \omega_0 \tilde{\mathbf{I}}^{-1} \begin{bmatrix} 8\omega_0(I_z - I_y) & -6\omega_0 I_{xy} & 2I_{xz} & 0 & -2I_{yz} & I_x - I_y + I_z \\ -8\omega_0 I_{xy} & 6\omega_0(I_z - I_x) & -2\omega_0 I_{yz} & 2I_{yz} & 0 & -2I_{xy} \\ 8\omega_0 I_{xz} & 6\omega_0 I_{yz} & 2\omega_0(I_x - I_y) & -I_x + I_y - I_z & 2I_{xy} & 0 \end{bmatrix} \end{array} \right\} \tag{9.52}$$

The parameter ω_0 is the magnitude of the orbital velocity of the reference coordinate system with respect to the inertial coordinate system. This is assumed to be a constant throughout the derivations. There is some slight deviation of the angular velocity for noncircular orbits. However, the average orbital velocity can be used in the calculations. Also, it is important to distinguish between parameter **I**, which represents the identity matrix, and $\tilde{\mathbf{I}}$, which represents the moment of inertia matrix. Lastly, the preceding system dynamics take into account the gravity-gradient effects when computing the linearized system.

The contribution from the input can be taken into account as

$$\mathbf{B}(t) = \begin{bmatrix} 0 \\ -\tilde{\mathbf{I}}^{-1}\mathbf{X}[\boldsymbol{b}(t)] \end{bmatrix} \tag{9.53}$$

The vector $\boldsymbol{b}$ is the geomagnetic field vector in the fixed-body coordinate system. The matrix **X** is the normal skew-symmetric matrix associated with the cross product. It is defined as

$$\mathbf{X}\left(\begin{bmatrix} b_x \\ b_y \\ b_z \end{bmatrix}\right) = \begin{bmatrix} 0 & -b_z & b_y \\ b_z & 0 & -b_x \\ -b_y & b_x & 0 \end{bmatrix} \tag{9.54}$$

Note that if the spacecraft remained in its nominal attitude, and the changes in the geomagnetic field due to the Earth's rotation are ignored, then the linear matrix $\mathbf{B}(t)$ would be periodic. Thus, it is reasonable to assume that a steady-state matrix solving the algebraic Riccati equation would be a reasonable approximation to the optimal solution.

Integral control. Integral control has many known benefits. Notable are its stabilizing tendencies and noise reduction. To introduce integral control into this design, the preceding system is augmented as follows. Introducing a new state vector consisting of the integral of the three-element quaternion and the previous state,

$$\tilde{x}_{\text{lqr}} = \begin{bmatrix} \int \mathbf{q}^{B/R} \\ \mathbf{q}^{B/R} \\ \boldsymbol{\omega}^{B/R} \end{bmatrix} \tag{9.55}$$

The modified system

$$\dot{\tilde{\mathbf{x}}}_{\text{lqr}} = \tilde{\mathbf{A}}\tilde{\mathbf{x}}_{\text{lqr}} + \tilde{\mathbf{B}}(t)\mathbf{u} \tag{9.56}$$

The modified matrices are

$$\tilde{\mathbf{A}} = \begin{bmatrix} \mathbf{0} & \mathbf{I} \;\; \mathbf{0} \\ \mathbf{0} & \mathbf{A} \end{bmatrix} \tag{9.57}$$

$$\tilde{\mathbf{B}}(t) = \begin{bmatrix} \mathbf{0} \\ \mathbf{B}(t) \end{bmatrix} \tag{9.58}$$

Calculating the C matrix. In the preceding discussion, **C** was calculated by averaging the effect of the matrix $\mathbf{B}\mathbf{R}^{-1}\mathbf{B}^T$ over one orbital period. However, this only works if the **B** matrix is periodic. Because the real matrix will not be exactly periodic, this technique must be modified slightly by performing the average over several orbits to include approximately one day's worth of data or approximately 15 orbits. Performing this longer integration will average out the effect of the rotation of the Earth. Thus, **C** is calculated as

$$\mathbf{C} = \frac{1}{15T}\int_0^{15T} \tilde{\mathbf{B}}(\tau)\mathbf{R}^{-1}\tilde{\mathbf{B}}(\tau)^T \,\mathrm{d}\tau \tag{9.59}$$

In this calculation, the magnetic field vector is calculated in the reference frame because the calculation assumes the satellite is in the nominal orientation. It is only necessary to calculate **C** once. This matrix is then used to solve the algebraic Riccati equation found in Eq. (9.49) for $\mathbf{P}_{ss}$.

Control algorithms. There are two control algorithms. The first is an offline calculation to find the matrix $\mathbf{P}_{ss}$. This algorithm, which is seen in Fig. 9.14, must be run only once, prior to implementing the control algorithm. It begins by calculating the nominal magnetic field for 15 orbits and then uses Eq. (9.53) to find the

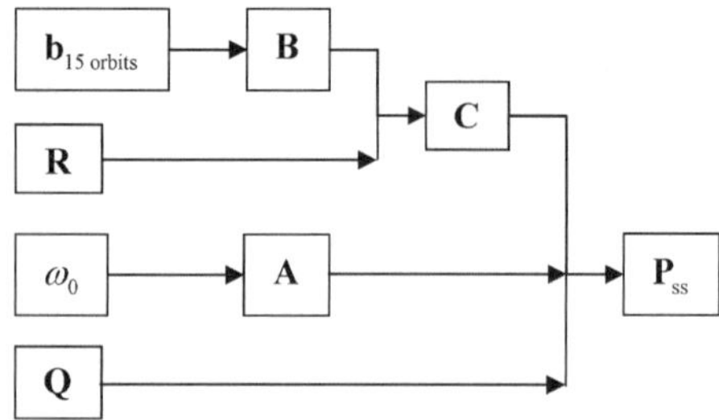

Fig. 9.14 $\mathbf{P}_{ss}$ calculation.

B matrix. Next, using the computed value of **B** and selected value of the penalty matrix **R**, matrix **C** is calculated using Eq. (9.59). Congruently, using the average orbital angular velocity ω_0, matrix **A** is found using Eq. (9.52). These values are then input into Eq. (9.49), which is finally evaluated for P_{ss}.

Before stating the main control algorithm, it is necessary to convert the state vector obtained from the sun sensor and magnetometer measurements to a form used by the LQR. The attitude determination algorithm computes the attitude and rotational velocity of the satellite with respect to the inertial frame, but the linear quadratic regulator requires these state quantities to be with respect to the orbital reference frame [6]. To convert the angular velocity, use the following equation:

$$\boldsymbol{\omega}^{B/R} = \boldsymbol{\omega}^{B/IR} - \boldsymbol{\omega}^{R/I} \tag{9.60}$$

The angular velocity $\boldsymbol{\omega}^{R/I}$ is obtained from the following equation:

$$\omega^{R/I} = -\frac{\|\mathbf{v}\times\mathbf{r}\|}{r^2}\mathbf{y}_R \tag{9.61}$$

To convert between quaternions in different reference frames, the following equation must be used:

$$\boldsymbol{q}^{B/R} = (\boldsymbol{q}^{R/I}*) \otimes \boldsymbol{q}^{B/I} \tag{9.62}$$

It is crucial to note that the symbol $\otimes$ in Eq. (9.62) represents quaternion multiplication. The quaternion $\boldsymbol{q}^{R/I}$ is found by converting the attitude matrix $\mathbf{A}^{R/I}$. After addressing these conversions, it is possible to state the linear quadratic regulator control algorithm as shown in Fig. 9.15.

The magnetic field at the desired location is used to compute the matrix **B** using Eq. (9.53), which in turn is used to find the new matrix $\tilde{\mathbf{B}}$, using Eq. (9.58). At the same time, the estimated state **x** is used to find the augmented state defined in Eq. (9.56) with the aid of Eqs. (9.57) and (9.58). These results, along with the already computed $\mathbf{P}_{ss}$ matrix and the penalty matrix **R**, are then used to solve Eq. (9.47) for **u**.

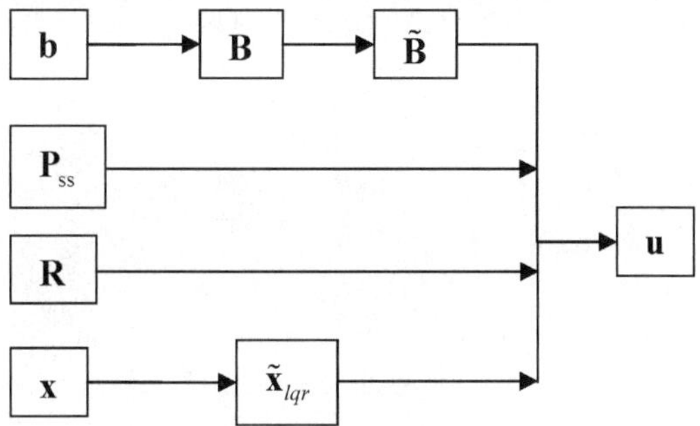

Fig. 9.15 Optimal control calculation diagram for the linear quadratic regulator.

3. Q and R Penalty Matrices from Genetic Algorithm

The selection of values for the **Q** and **R** matrices can be done in several ways. In most cases, the user is searching for a specific range of system responses defined by available power, maximum stabilization time, initial conditions, etc., rather than general stability criteria. As a result, the weighting matrix selection should be performed on a system with specified spacecraft properties, orbit parameters, and user-defined response limits.

The first estimates for ION's LQR parameters were obtained by intuitively adjusting two initial guesses presented by Psiaki in [8] for two different spacecraft configurations. Although this trial-and-error method provided acceptable system response, the team decided to utilize previous experience with genetic algorithms (GA) to improve upon these results.

The following section presents an abbreviated discussion of the GA method used to compute the **Q** and **R** matrices. The full discussion can be found in [6].

GA fitness determination. The GA operates on a finished simulator discussed in detail in the next chapter. The simulator has been designed to operate the satellite in two different modes. The initial mode, called the detumbling mode, emphasizes slowing down the initial high rotational rates of the satellite as it separates from the deployer. The second mode, called the tracking mode, penalizes deviation from the desired state. As a result, separate sets of weight matrices **Q** and **R** must be found for each mode.

The detumbling mode is in effect when the spacecraft's angular velocity along all three axes is greater than 0.1 deg/s. When complete, the tracking mode begins, and new weighting matrices are used by the controller. Although detumbling has no requirement on the actual attitude of the spacecraft, only the rates at which they change, the tracking mode attempts to keep the satellite in a fixed-attitude optimal for instrumentation to work. When all three Euler angles are within set limits for twenty consecutive time steps, it is assumed stability is reached.

Although the weight matrices are different for the two modes, their objectives are the same: stabilize the spacecraft as quickly as possible to the states of minimum rotation for detumbling and nadir-pointing state for tracking. To convert the minimum-time problem to a maximum fitness problem, the time to detumbling

Table 9.2 GA Properties

Parameter	Variable	Value
Crossover probability	$\mathbf{p_c}$	0.9
Mutation probability	$\mathbf{p_m}$	0.01
Selection pressure	$\mathbf{S}$	2
String length	L	32
Population size	$\mathbf{n}$	100
Generations	$\mathbf{G_{max}}$	100

and tracking were subtracted from the maximum time until simulation shutoff, set at fifteen hours. As a result, the **Q** and **R** schemes, which are unable to converge in fifteen hours, have a fitness of zero, and the ones that produce minimum time have the greatest fitness for selection purposes.

GA design. The GA code, building on historical GA LQR designs, uses the parameters found in Table 9.2.

The algorithm also takes advantage of the fact that the **Q** and **R** matrices are diagonal with often equal diagonal entries, in effect reducing the problem size. In particular, the first three diagonal entries of **Q** are weights on the integral of the spacecraft quaternions, the middle three diagonal values of **Q** weights on the actual quaternion vector elements, and the last remaining values are the weights for the time rate of change of the quaternion vector elements. Similarly, the three **R** values are weights on the three angular velocities along the spacecraft axes.

A pair-wise tournament selection with replacement was used, giving a selection pressure of **s**. This operator took two individuals within the population, compared fitness values, and carried the best individual to the child population. This selection type allows the assumption that every member of the population had at least one chance, with an average of two chances, to be represented in the child population. This therefore ensures that the best individual in the population, which cannot be defeated by tournament selection, will be passed to the new population with good odds.

The crossover applied is a single-point crossover that occurs with probability $\mathbf{p_c}$. The operator mates two individuals inserted into the child population by selection to produce switching bits at the point of crossover, chosen randomly. This type of crossover minimizes disruption to the values represented by the individual, causing change in at most one actual weighting matrix value, and instead acting as a shuffler of matrix values between different individuals.

To accomplish mutation, a matrix of identical size to the population was first created. Each value of this matrix is produced as a random number between 0 and 1. If a value were less than $\mathbf{p_m}$, the population value corresponding to that location was mutated. This entails replacing the corresponding value with a random integer not equal to the replaced value. The value of $\mathbf{p_m}$ desired to maintain a minimum diversity at each position by using the smaller value between $\mathbf{p_m} \leq 1/\mathbf{n}$ or $\mathbf{p_m} \leq 1/\boldsymbol{l}$, that is, $\mathbf{p_m} = 0.01$ [11]. After trials with a mutation probability lower than $\mathbf{p_m}$, the convergence rate increased drastically indicating a poor search of the solution space.

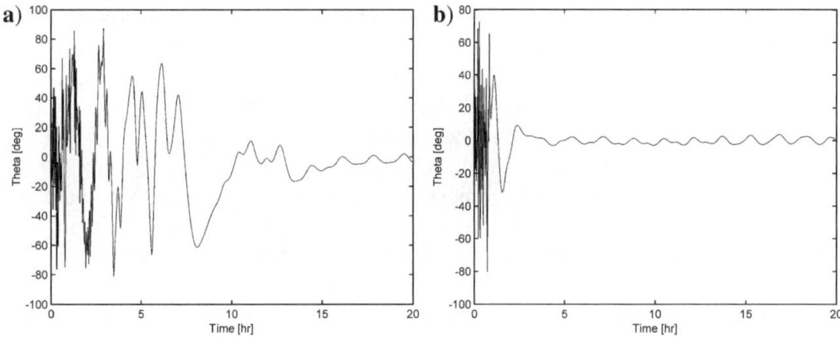

Fig. 9.16 Comparison of offset from nadir pointing direction from a) historical guess and b) GA.

The presented values of $\mathbf{p_m}$, $\mathbf{p_c}$, and $\mathbf{n}$ require large number of generations to reach a convergent solution, which in turn leads to large computation times. At first, the rate of convergence might seem to be a variable desired to be completely maximized, but a closer look shows this is not necessarily true. A faster rate can cause a decrease in the evaluated solution space. With a larger population size and fairly generous $\mathbf{p_m}$ and $\mathbf{p_c}$ values, the GA has a greater solution search space to locate the best possible solution. Although this leads to increased computation time, this was seen as an acceptable drawback because the main objective was to find a better solution to the spacecraft stabilization problem, and not to find the solution faster over some equally successful search algorithm.

GA results and analysis. The fitness found using the GA is significantly higher than historical values. The detumbling algorithm using the weighting matrices determined by historical guessing takes approximately 6 h, whereas the best case that results from the GA could achieve the same end result in as little as 1.16 h. This is more than a four-fold improvement in time. To better understand

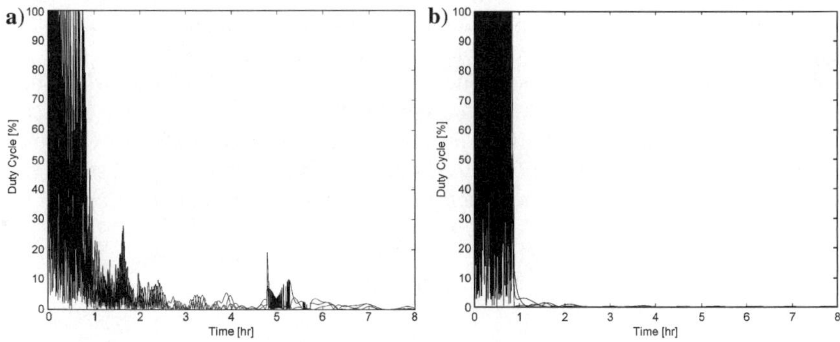

Fig. 9.17 Comparison of duty cycle from a) historical guess and b) GA.

the dynamics of the result, Simulink was used to output statistics on power usage, rotational rates, and orientation angles. Selected outputs compared to historical runs can be seen in Figs. 9.16 and 9.17. A closer inspection of these results shows that the results created from the genetic algorithm might have practical limitations. Although the GA solution does indeed show the satellite reaching the desired state, a best fitness case for the ION LQR controller might require that the torque coils run at full power for a period of time too large for ION to manage. In other words, the GA took a somewhat obvious approach to decreasing the angular rate by applying maximum available torque without considering power limitations.

To solve this problem, the simulator continuously computes how much power it has used while operating the torquers and stops when it reaches a certain threshold. The amount of power available for torquing comes directly from the power team's sampling of the batteries and takes into account operational safety margins. If the safety limit is reached, the torquing stops, the batteries are allowed to recharge, and the sensors are sampled again. Based on the new sensor readings, the attitude is recomputed, and a new duty cycle is uploaded to the satellite for execution. An alternative method consists of modifying the fitness function in the GA, rather than the simulator, to take into account the constraint on the amount of power available. If the constraint is violated, then the fitness would become zero.

IV. Design Implementation on ION

The power and computational limitations imposed on ION by its small size force the majority of attitude determination and control algorithms to be performed on the ground. As a result, a unique combination of open- and closed-loop feedback control algorithms is implemented to achieve all of the ADCS goals. The process starts by sampling the sensors over a limited (order of minutes) time period prior to the next communication opportunity. As soon as contact is established, the power and magnetometer data are downloaded, and the initial attitude is obtained. This constitutes the first step of the open-loop system. The initial attitude is then fed into a closed-loop control algorithm that computes optimal control for stabilization at that particular time step and then uses satellite dynamics to predict the resultant attitude after the control has been applied for a set duration. The controller then recomputes an optimal control torque based on the new attitude. This process is repeated until the satellite is either stabilized or the combined duty cycle exceeds available power onboard the satellite. This closed-loop computation of the torque sequence takes approximately 30 s after which the computed duty cycle is uploaded to the satellite for execution. This entire procedure is largely automated and occurs during a single pass, which, for ION's orbit, varies between 1.5 and 10 min. Included in the uplink is the scheduled sampling of the sensors prior to the next contact time, which completes the next update in the open-loop system.

A. Closed-Loop Simulator

Figure 9.18 illustrates the sequence of events for the attitude determination and control subsystem. The sensor sampling prior to contact is indicated by the

Fig. 9.18 ION operational plan.

dash-dot-dash line, the actual communication time is indicated by the solid line, and the execution of the stabilizing torque is marked by the dashed line.

The sampling of the sensors occurs immediately before the scheduled contact time. This is done in order to minimize the attitude propagation time in the ADCS algorithms, which is the largest source of error. Although it would be most desirable to compute the attitude based on the sampled data at every time step, the just-listed limitations force the software to predict the evolution of the attitude based on an initial set of sensor data. A variety of disturbance torques including gravity gradient, solar pressure, atmospheric drag, third-body perturbations, and residual magnetic effects in the satellite's components contribute to changes in the attitude not accounted for in the torque-free attitude evolution. Although the attitude propagation software tries to account for several of these effects, the predicted attitude will start to deviate from the actual attitude as the propagation time increases.

The extent of this effect will not be known until after the launch, when the attitude determination calibration will be performed. In this phase, all of the attitude determination sensors will be sampled for an extended period of time. The information at a few initial steps will be used to compute the initial attitude, and then the attitude will be propagated for the entire duration of the sampling. The remaining data will be used to compute the "actual" attitude at every time step. The results from both computations will provide a reasonable estimate for how long the propagated attitude is within the user-specified error bound. It is predicted that this "buffer duration" will decrease somewhat during torquing; however, the high-precision magnetic field estimator and precise control on the power levels sent to the torque coils will minimize most of the unwanted effects.

All of the ADCS algorithms are implemented in Matlab® and Simulink in order to utilize the extensive math and control libraries included in these packages. Figure 9.19 represents the closed-loop logic used during the intercept time to compute the optimal torquing profile. The initial values of attitude and rotational rate q_0 and ω_0 are computed from the magnetometer and sun sensor data using techniques presented in Chapter 2.

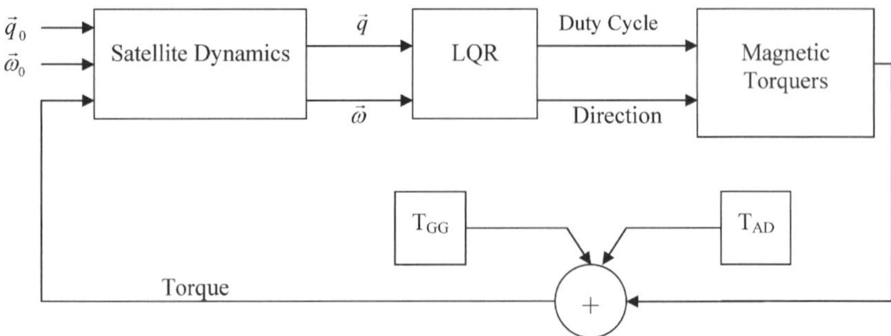

Fig. 9.19 Top-level attitude control simulation diagram.

The satellite dynamics block computes the resultant attitude and rotational rate after an application of external torque. Based on this attitude, the linear quadratic regulator computes an optimum torquing profile to be applied for a user-specified duration, currently set at 15 s. The exact duration that each torque profile will be held was chosen based on simulation results. In theory, the shorter the duration that each duty cycle is held, the better the performance of the system because the LQR is updating the optimal solution using a recomputed attitude. On the other hand, increasing this duration causes the satellite to apply a torque, found for one attitude, to a state that has evolved in time. In other words, as soon as the torque is applied to the system it, in fact, ceases to be an optimum solution—a result that is exacerbated by high rotational rates. Although very short durations would be theoretically desirable, practical limitations such as computational time, system response time, increased power usage for larger number of switching functions, and even the length of the control file that needs to be uploaded, all contribute to the duration being finite and relatively large. Extensive testing performed on a finished simulator showed that the 15-s duration level produced good results.

The optimum duty cycle computed by the LQR is then fed into the magnetic torquers block that combines specific torque coil parameters such as coil resistance, maximum current, and effective surface area with the magnetic field of the Earth at the satellite's location to compute the effective magnetic torque according to Eq. (9.39).

In addition to the applied magnetic torque, the software computes the gravity gradient and aerodynamic drag disturbance torques as outlined in the next section. All three torques are then added together and input into the satellite dynamics block to complete the closed-loop feedback system.

B. Disturbance Torques

The low-Earth-orbit (LEO) environment presents a variety of effects neglected in the assumption that the total torque $\mathbf{t}_T$ is simply the magnetic torque created by the magnetic torquers. In fact, effects such as the atmosphere, gravity gradient, solar pressure, or residual magnetic fields exert finite torques on the spacecraft causing variations in spacecraft's attitude. Although these and other effects contribute to the overall attitude of ION, only gravity-gradient and aerodynamic torques are of significant consequence and will be briefly discussed here. The complete derivation of these results and full explanation of all of the terms can be found in [6].

1. Gravity-Gradient Disturbance Torque

A rigid body in a circular orbit will tend to be aligned by the gravity-gradient effect so that the axis of maximum moment of inertia is normal to the orbit plane and the axis of minimum moment of inertia is aligned along the local vertical. The torque on the spacecraft arises as a result of the inverse relationship between the gravity gradient and the distance to the center of the Earth. As a result, sections of the satellite that are closer to the Earth experience a larger gravitational force than parts farther away. The result is a restoring torque that tends to return the longitudinal axis to alignment with the local vertical. The magnitude and exact direction of the gravity-gradient torque can be computed using the following equation:

$$\mathbf{t}_{gg} = \frac{3\mu}{r_e^3}\hat{\boldsymbol{r}}_e \times (\tilde{\mathbf{I}}\hat{\boldsymbol{r}}_e) \tag{9.63}$$

The parameter μ is the aforementioned gravitational constant, $\boldsymbol{r}_e$ is the vector from the ION's center of mass to the Earth's center of mass in the ECF coordinate system, and the matrix $\tilde{\mathbf{I}}$ is the moment of inertia matrix.

2. Aerodynamic Disturbance Torque

The residual atmosphere present at the LEO altitudes causes the spacecraft to experience aerodynamic drag. This effect decreases with increasing altitude and can be neglected for higher-altitude orbits such as a geostationary orbit. However, in order to increase the accuracy of the attitude propagation algorithms for a satellite in a 500-km orbit, this effect should not be neglected.

The governing equation for the aerodynamic torque can be written as

$$\begin{aligned} \mathbf{t}_a &= \tfrac{1}{2}C_D\rho v_0^2 \int (\boldsymbol{a}_n \bullet \hat{\mathbf{v}}_0)(\hat{\mathbf{v}}_0 \bullet \boldsymbol{r}_s)\, dA \\ &+ \tfrac{1}{2}C_D\rho v_0 \int\Big[\boldsymbol{a}_n \bullet (\omega^{\mathbf{B/I}} \times \boldsymbol{r}_s)(\hat{\mathbf{v}}_0 \times \boldsymbol{r}_s) \\ &+ (\boldsymbol{a}_n \bullet \hat{\mathbf{v}}_0)\big\{(\omega^{\mathbf{B/I}} \times \boldsymbol{r}_s) \times \boldsymbol{r}_s\big\}\Big]\, dA \end{aligned} \tag{9.64}$$

where C_D is the drag coefficient with a typical value between 1 and 2 and ρ is the atmospheric density, which for an altitude of 500 km is approximately 3.614×10^{-14} kg/m^3. The remaining parameters include the unit vector normal to the surface of the satellite $\boldsymbol{a}_n$, the velocity of the satellite in the fixed-body reference frame $\hat{\mathbf{v}}_0$ and the vector $\boldsymbol{r}_s$ directed from the center of mass of the satellite to the infinitesimal area dA. The preceding equation must be evaluated for each surface of the satellite where $\boldsymbol{a}_n \cdot \hat{\mathbf{v}}_0$ is greater than 0. In other words, the atmospheric torque only acts on those sides of the satellite that are in the same general direction as the velocity.

The first term in Eq. (9.64) governs the torque produced by the general translation of the satellite. The second integral is caused by the spacecraft's angular velocity. As long as $\omega \ll v_0$, which is the case for satellites with small angular velocities such as ION, then the second term can be ignored. Equation (9.64) can be greatly simplified as a result of ION's simple, rectangular shape; however, in the interest of space, these results are omitted here. For complete derivation and results of the aerodynamic torque that eliminate the evaluation of the integral, refer to [6].

3. *Disturbance Torque Results*

Using the preceding results, it is possible to write the total disturbance torque $\mathbf{t}_d$ as

$$\mathbf{t}_d = \mathbf{t}_{gg} + \mathbf{t}_a \tag{9.65}$$

It is then possible to rewrite the dynamical equation for the change of angular momentum vector $\boldsymbol{h}$ that includes disturbance torques as

$$\left\{\frac{\mathrm{d}\boldsymbol{h}}{\mathrm{d}t}\right\}_B = \mathbf{t}_m + \mathbf{t}_d - \omega^{B/I} \times \mathbf{h} \tag{9.66}$$

C. Postlaunch Operational Logistics

The three-stage burn of the SS-18 Dnepr launch-vehicle and orbit insertion maneuvers are completed approximately 15 min after the liftoff. The primary payload, Belarusian BelKA satellite, is deployed first followed by the Russian Baumanets satellite developed by the students of Bauman State Technical University. The remaining 16 satellites, including 14 cubesats contained in five PPODs, are launched next with ION's insertion time being predicted at 903.5 s after launch.

Immediately after ION is deployed from its PPOD, the computer initiates the boot-up sequence and performs antenna deployment and battery charging procedures. The omnidirectional antenna sends beacons every five minutes and allows for communications regardless of the initial attitude. The first priority after establishing contact is verification of the overall health of the satellite and testing of the critical systems such as power generation, small intelligent datalogger (SID) operations, integrity of the file system, and uplink/downlink capability. Once completed, all of the remaining onboard sensors will be sampled and the results downloaded for verification. Provided that all systems are performing nominally and the power generation is sufficient, the attitude determination and stabilization will commence.

The first step is verification of the attitude determination algorithms and establishment of the aforementioned "buffer duration" as described earlier in the chapter. The next step is testing of the actual attitude control algorithms and torquer performance. The initial magnetic control is designed to be short duration, one-axis torques to test the attitude propagation algorithm, which includes the magnetic field model, the disturbance torques, and the generated magnetic torque. This phase is not meant to stabilize the spacecraft, but only to establish that the predicted result of an applied torque corresponds to the actual attitude change. In addition, because the applied torques are uniaxial, this phase is used to find appropriate weight factors that can be applied to the attitude simulator to correct for real-life effects that cause the discrepancies between actual and predicted torque response. The duration of these simple torques is increased incrementally until the desired level of propagator accuracy is achieved, which is based on the predicted duration of torquing—usually thirty minutes to one hour depending on available power. This calibration process is then repeated for the remaining two axes. Also, even though these short-duration, single-axis torques are nonoptimal, they are computed in such a way as not to increase the rotational rate of the satellite.

At this point the initial ADCS calibration is completed, and the actual stabilization begins. The just-mentioned combination of open- and closed-loop control systems is used to achieve the final goal of orienting the PMT towards nadir with zero spin. This phase is divided into two stages: the detumbling and tracking modes as described in the section on genetic algorithm. The detumbling mode occurs first and is designed to despin the satellite as quickly as possible with little emphasis on the attitude at the end of this mode. Once the rotational rates on all three axes are less than 0.1 deg/s, the algorithms switch to a different set of **Q** and **R** weight matrices in the LQR to bring the final attitude into a nadir-facing direction.

The attitude determination and control portion of the mission concludes with the stabilization of ION and the demonstration of the three-axis control algorithms using magnetic torque actuation. The remainder of ION's lifetime will be utilized to perform other mission objectives, including sampling of the PMT to study oxygen emission, operating the camera, firing of the electric thrusters, and training future satellite operators.

D. Simulation Results

Because of launch-vehicle failure, only the simulation results are available and will be presented in this section. Limited results were already presented in the section on genetic algorithms, and selected additional cases will be displayed in this chapter. To test the robustness of the control algorithms, large-scale testing was performed by varying the initial attitude conditions. Several thousand cases were run by randomly choosing the initial tip-off quaternion and by selecting a rotational rate between 0 and 5 deg/s. The results from each run, including the despin and stabilization times, evolution of the quaternions and body rotational rates, duty cycle, disturbance torques, and other necessary information were saved for postprocessing. Presented next are two selected cases: one representing a large initial rotational rate with an unfavorable initial attitude, and a second case showing the converse stabilization scenario.

One of the worst-case scenarios encountered throughout the testing process occurred with the initial tip-off quaternion equal to [0.5465, 0.3847, 0.3770, 0.6412] and the maximum rotational rate of 5 deg/s. The attitude represented by this quaternion was the most difficult to stabilize in the specific swath along ION's orbital path and is not, necessarily, the "global" worst case. The best-case scenario

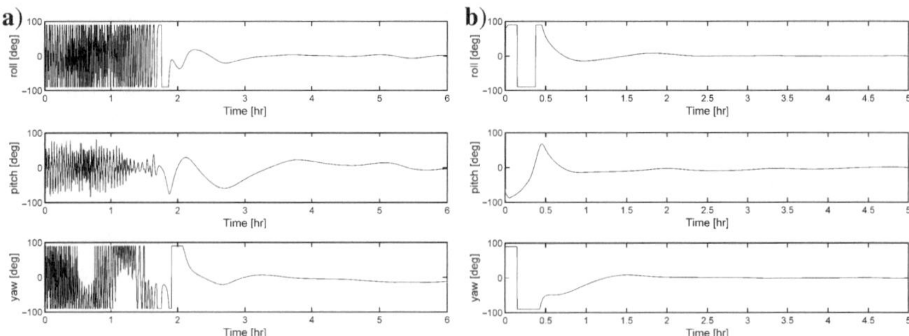

Fig. 9.20 Evolution of Euler angles for a) worst and b) best tip-off scenarios.

occurred with the initial tip-off quaternion equal to [0, 0, 0, 1] and initial rotational rate of 0.1 deg/s. The results for both of these scenarios are presented in Figs. 9.20–9.24. The first two sets of figures display the evolution of the Euler angles and the rotational rates on all three axes of the satellite during the stabilization process. During the detumbling mode, the rotational rates for the worst case (left-side figures) are reduced from their initial rate of 5 deg/s to 0.1 deg/s in approximately 1.25 h. During this time, the control algorithm applies little emphasis on the nadir-facing attitude as seen in the widely varying evolution of the Euler angles. It is only after the detumbling mode is completed that the satellite starts to stabilize and eventually reaches the desired orientation in approximately 3.5 h. In contrast, the best-case scenario (right-side figures) demonstrate that a fully stabilized orientation can be achieved in approximately 1 h. Interestingly, the rotational rate briefly increases from its starting value because of the disturbance torques, as seen in Fig. 9.21. As a result, instead of entering the tracking mode as expected, the system initially engages the detumbling mode until the rates are below the threshold level. During this time, the satellite performs approximately one full rotation on both the roll and yaw axes as seen in Fig. 9.20.

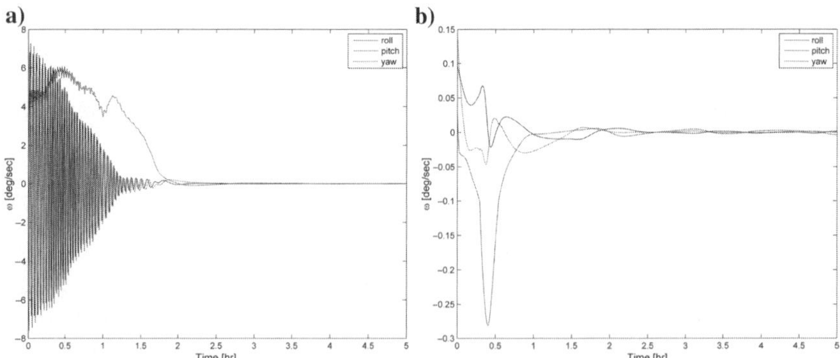

Fig. 9.21 Evolution of body rates for a) worst and b) best tip-off scenarios.

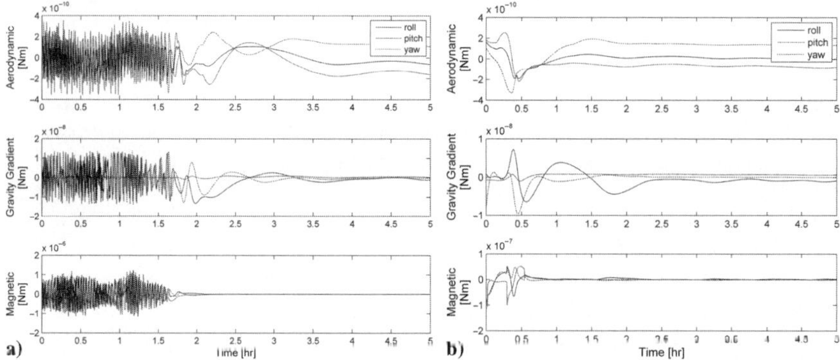

Fig. 9.22 Disturbance and applied torques for a) worst and b) best tip-off scenarios.

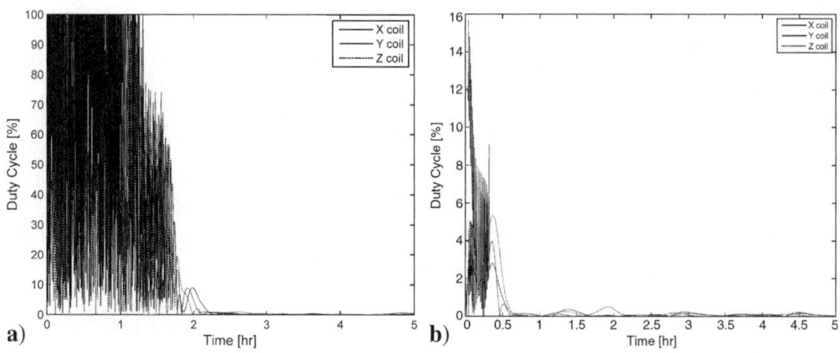

Fig. 9.23 Duty cycle for a) worst and b) best tip-off scenarios.

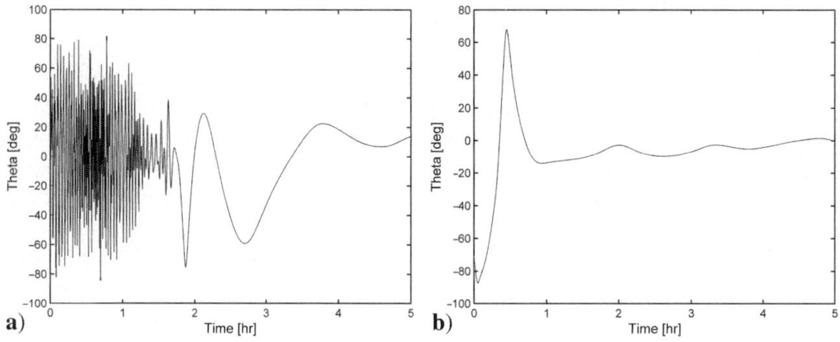

Fig. 9.24 Offset from nadir for a) worst and b) best tip-off scenario.

The resultant gravity-gradient and aerodynamic drag disturbance torques as well as the applied magnetic torque are shown in Fig. 9.22. For both scenarios, the aerodynamic torque is approximately four orders of magnitude less than the applied torque, while the gravity gradient is approximately two orders of magnitude lower. Also, it can be observed that for the best-case scenario, the aerodynamic torque reaches its minimum at the exact time when ION has rotated to present its top face into the direction of motion. Then, once the satellite completes the turn and stabilizes into a nadir-facing attitude, the aerodynamic drag increases and becomes constant.

Because of the initially high rotational rates of the worst-case scenario, the optimum torquing profile found by the LQR, as seen in Fig. 9.23a, uses 100% of available torquing capability for nearly the entire duration of the detumbling mode. It is not until the rotational rates are reduced to below 0.1 deg/s that the applied torque is also reduced until it reaches levels below 10% of its maximum value in approximately 2.0 h after torquing was initiated. The temporary use of the detumbling mode for the best-case scenario can also be seen in Fig. 9.23b, where the controller applies a relatively large magnetic torque on the Y coil in the first few minutes of torquing.

Lastly, Fig. 9.24 demonstrates how the angle between the local vertical and the satellite's pitch angle varies as the magnetic torque is applied. The final attitude reaches the PMT-prescribed bounds in approximately 3.5 and 1 h for the worst- and best-case scenarios, respectively.

V. Conclusion

The top solar cell used on ION does not provide the desired angular resolution for accurate attitude determination. The dual utility of solar panels as power-generating devices and as attitude sensors requires software checks to ensure that batteries are sufficiently charged to avoid inaccurate readings. This, at times, causes loss of valuable data used for attitude determination. To simplify this process, implementation of a Kalman filter that uses exclusively the magnetometer data for attitude estimation is highly recommended. The filter can be implemented in the ground-based software without drastic increase in the computational time. In addition, there exists extensive literature on Kalman-based attitude determination for spacecraft with field-sampling abilities.

The pc-board torque coils greatly simplified overall testing and integration and are highly recommended as actuators. In case larger magnetic moment is required, the torquers can be scaled up at an additional, but still reasonable, cost.

The great majority of the computation time of the presented ADCS is spent on attitude propagation. Simply by removing this necessity and because of high modularity of the software design, the algorithms can be adapted as an onboard system. The real-time attitude data would greatly minimize errors and remove the need for complicated sequencing of sampling events prior to contact, rushed downloading, computation of optimum duty cycles, and upload of the torquing profiles. New developers looking to utilize digital signal controllers as ADCS flight computers should develop their algorithms in hardware-specific language, such as MPLAB C30 and not Matlab®/Simulink.

References

[1] Prussing, J., and Conway, B., *Orbital Mechanics*, Oxford Univ. Press, Oxford, England, U.K., 1993, pp. 46–54.

[2] Gregory, B., "Attitude Control System Design for ION, the Illinois Observing Nanosatellite," M.S. Thesis, Dept. of Electrical and Computer Engineering, Univ. of Illinois, Urbana, May 2004.

[3] Campbell, W. H., *Introduction to Geomagnetic Fields*, Cambridge Univ. Press, New York, 1997.

[4] Roithmayr, C., "Contributions of Spherical Harmonics to Magnetic and Gravitational Fields," NASA Tech. Rept. EG2-96-02, 23 Jan. 1996.

[5] Hall, C., AOE 4140 Lecture Notes, Virginia Polytechnic Inst. and State Univ., Dept. of Aerospace and Ocean Engineering, data available online at http://www.aoe.vt.edu/~cdhall/courses/aoe4140/ [retrieved Oct. 2004].

[6] Pukniel, A., "Attitude Determination and Three-Axis Control System for Nanosatellites with Magnetic Torque Actuation," M.S. Thesis, Aerospace Dept., Univ. of Illinois at Urbana-Champaign, Urbana, Oct. 2006.

[7] Vallado, D., *Fundamentals of Astrodynamics and Application*, Microcosm Press, Hawthorne, CA, 2001.

[8] Psiaki, M., "Magnetic Torquer Attitude Control via Asymptotic Periodic Linear Quadratic Regulation," *Journal of Guidance, Control, and Dynamics*, Vol. 24, No. 2, 2001, pp. 386 – 394.

[9] Bryson, A. E., and Ho, Y., *Applied Optimal Control*, Hemisphere, New York, 1975.

[10] Brogan, W. L., *Modern Control Theory*, Prentice – Hall, Englewood Cliffs, NJ, 1991.

[11] Goldberg, D. E., *Genetic Algorithms in Search, Optimization, and Machine Learning*, Addison Wesley Longman, Boston, 1989.

Chapter 10

Microthruster Propulsion

Lance K. Yoneshige,* Lynnette E. S. Ramirez,† and Carlos F. M. Coimbra‡

Hawaii Space Flight Laboratory, University of Hawaii at Manoa, Honolulu, Hawaii

THE concept of small low-Earth-orbit (LEO) satellites for educational and scientific applications is becoming increasingly popular because of their relatively low cost and short developmental period. Because of the mass and volume restrictions associated with small satellites, this often necessitates the scaling down of existing technologies or the development of new technologies that fulfill the duties of the various subsystems. One such subsystem is propulsion. Although propulsion has not always played an important role in the development of small satellite programs, the reason for this has as much to do with the inherent complexity involved with propulsion subsystems as it does with the selection of first-generation mission objectives of educational programs. As programs evolve and attempt more diverse and complex missions, the need for attitude control and/or orbital maneuvers often emerges as one of the bottleneck technologies. Thus, micropropulsion systems are one of the core enabling technologies for the next generation of small satellites. Current efforts employ a variety of well-tested propulsion subsystems and new concepts that are tailor made for small satellites. In some recent design efforts, such as with the University of Illinois' Illinois Observing NanoSatellite (ION), the primary mission is to demonstrate and qualify a propulsion system [1]. In others, as with the momentum wheels of Hankuk Aviation University's HAUSAT-2 [2], nonpropulsion means are utilized. The focus of this chapter is on propulsion systems that are primarily, but not exclusively, used for attitude control.

---

*Research Assistant; currently Integration and Launch Support Mechanical Engineer, Hawaii Space Flight Laboratory.

†Ph.D. Candidate (awarded Dec. 2009).

‡Associate Professor; currently Associate Professor, Department of Mechanical Engineering and Applied Mechanics, University of California, Merced.

The propulsion systems we are interested in are often referred to as microthrusters. This chapter is not meant to be a comprehensive review of all microthruster technologies available. The following pages are focused on technologies that have demonstrated value or are likely to play a major role in small-satellite applications—in particular, technologies intended for the nanosatellite class (≤10 kg). A special emphasis is placed on those microthrusters that have already been used in small-satellite applications. Also, technologies that are discussed in more detail elsewhere in this book are not included at all (e.g., the microvacuum arc thrusters, etc.).

This chapter begins with a discussion of some of the basic fundamentals of propulsion and the definitions of relevant microthruster characteristics. Following that section is a brief description of different microthruster technologies. We conclude the chapter with an exploratory discussion on a novel microthruster design concept that is under development at the Hawaii Space Flight Lab (HSFL).

I. Definitions and Terminology

We begin by introducing a very short selection of basic principles and the terminology used in propulsion and astronautics.

A. Astronautics

When a thruster is fired, it is often with the aim of altering certain characteristics or behaviors of a spacecraft. These characteristics and behaviors are often the main focus of astronautics. In this discussion, it will be assumed that our orbiting spacecraft or satellite is always primarily under the gravitational influence of the Earth in a simplified two-body system. The satellite's orbit is defined in terms of its altitude relative to the Earth's surface.

A LEO is usually defined to be at an altitude less than 1000 km. However, the majority of satellites in this type of orbit are typically at an altitude in the range of 250–350 km. These orbits are often nearly circular in shape (low eccentricity). Although the atmosphere is much thinner at these altitudes, there is still enough of an atmosphere for a satellite to experience a drag force. If nothing is done to compensate for this drag force, a LEO satellite will eventually *fall* from orbit and disintegrate during reentry. Countermeasures against such an event are often referred to as *stationkeeping*.

Ideally, a satellite is launched directly into the orbit required to accomplish its mission objectives. Direct orbital launch is, however, not always possible, especially for small satellites that are routinely launched as secondary payloads and consequently do not dictate the launch orbit. In such scenarios, a satellite might also need to execute a series of orbital maneuvers to enter a more favorable orbit as defined by the mission objectives.

Although these maneuvers will vary from mission to mission, they all involve a change in velocity Δv. Using Newton's physical interpretation of Kepler's laws, to analyze our idealized two-body system, it can be shown that v for a common elliptical orbit (as in Fig. 10.1) is dependent on the satellite's orbital position with

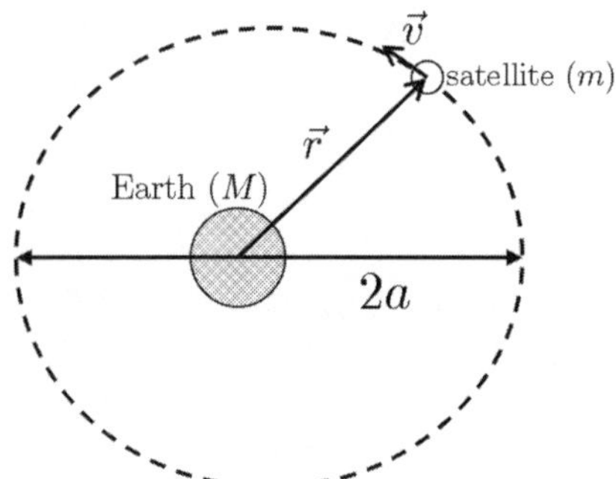

Fig. 10.1 Diagram of a common elliptical geocentric orbit. Note that if the orbit were circular, $r = a$.

respect to the center of the Earth r, the orbit's semimajor axis a, and a combined gravity term $\mu = G(M + m)$ in the following form:

$$v = \sqrt{\frac{2\mu}{r} - \frac{\mu}{a}} \tag{10.1}$$

In the gravity term μ, G is Newton's gravitational constant, M is the mass of the Earth, and μ is the mass of the satellite. The Earth's mass M is understood to be over 20 orders of magnitude greater than m, and so μ can then be simplified to $\mu = GM$. Consequently, v depends only on the orbit and the satellite's position in that orbit. Some maneuvers can also involve the escape velocity v_{esc}

$$v_{\text{esc}} = \sqrt{\frac{2\mu}{r}} \tag{10.2}$$

In the determination of v and v_{esc}, it is clear that Δv is independent of the satellite's physical characteristics. When evaluating the capabilities of different thruster technologies, Δv can be used as a criterion independent from the satellite it is integrated in.

B. Propulsion

The foundation of propulsion is the Momentum Principle. By expelling mass from a body in one direction, that body experiences a net force in the opposite direction. This force is called the thrust force (F) and may vary with time. By definition, impulse is the integrated force acting on a body over time. The total impulse of a thruster (I_t) is then

$$I_t = \int_0^t F \mathrm{d}t \tag{10.3}$$

To measure the efficiency of thrust production, the concept of specific impulse I_{sp} is introduced. The I_{sp} is the total impulse per unit of weight of propellant, where

g_0 is the acceleration of gravity *at sea level* and $\dot{m}$ is the mass flow rate of propellant,

$$I_{sp} = \frac{\int_0^t F\,dt}{g_0 \int \dot{m}\,dt} = \frac{I_t}{g_0 \int \dot{m}\,dt} \tag{10.4}$$

In general, higher values of I_{sp} are desired as this translates into the generation of a greater impulse with less propellant. However, a higher I_{sp} is also generally associated with low levels of thrust. High I_{sp} systems typically require an extended period of thrusting time to perform orbital maneuvers.

The terms discussed thus far (Δv, F, I_t, and I_{sp}) only take into account the direct performance of a thruster. As just noted, the advantage of small satellites is that their small size and mass make them much more affordable to develop and launch. Therefore, another useful criterion for comparing thruster characteristics is the propellant mass fraction (ξ)

$$\xi = \frac{m_p}{m_f + m_p} \tag{10.5}$$

where m_p is the mass of the propellant and m_f is the dry mass of the propulsion system.

The mass fraction ξ measures the amount of "useful" mass m_p in a propulsion system. The closer to unity ξ is, the more effective the propulsion system is in terms of mass utilization. Although a thruster technology might be able to produce a desirable Δv, F, I_t, and I_{sp}, it might not be the best choice if it requires a considerable amount of system overhead. Once in orbit, additional mass generally does not inhibit the ability to perform maneuvers. However, placing that extra mass in orbit has a significant cost.

II. Microthruster Technologies

What follows is an overview of existing microthruster technologies. Each technology is briefly described along with some of its benefits and disadvantages. Concluding the section is a quantitative comparison of the different technologies.

A. Cold Gas

Cold-gas thrusters offer an inexpensive, reliable, low-power, nontoxic, auxiliary propulsion system for small spacecraft. They have been used extensively in various attitude control systems providing multiple low-thrust pulses for actions, such as attitude control, stationkeeping, orbit adjustments, docking maneuvers, and trajectory control. Whereas nitrogen, helium, argon, krypton, and Freon 14 have been used in operational spacecraft, inert nitrogen is the most commonly used cold-gas propellant [3,4]. An example of such technology is found in Surrey Satellite Technology Ltd.'s (SSTL) UoSAT-12 [5]. The basic schematic of a cold-gas thruster system is shown in Fig. 10.2.

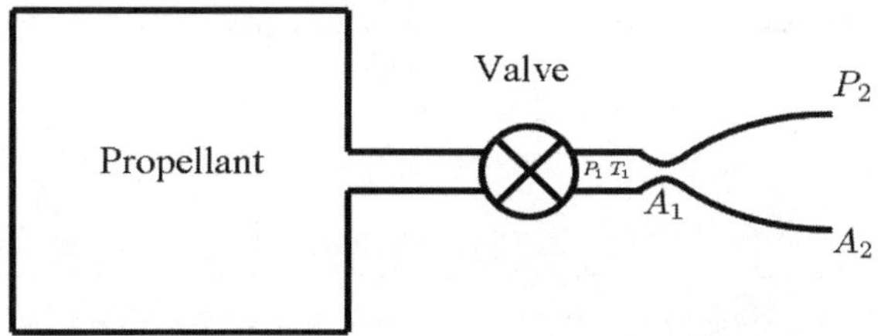

Fig. 10.2 Basic schematic of a cold-gas thruster system.

The pressurized propellant is held in a storage tank and released to the expansion nozzle by a valve. Performance equations for this system are found elsewhere in the literature [6].

Although the simplicity of a cold-gas system seems obvious, actual thruster systems might feature additional components such as pressure regulator assemblies, filters, and relief valves [7]. The need for additional components reduces the theoretical ξ and consumes part of the mass budget that could otherwise be used for other subsystems or additional fuel. Because the propellant is stored at high pressures (~21 MPa), heavier tanks are required, which also negatively affects ξ [4]. The high storage pressure differential with respect to the space "vacuum" also raises concerns over system leakage. Although the high storage pressure does reduce the required volume, the cold-gas system has low storage density (0.28 g/cc for N_2) [8]. Typical performance characteristics can be found in Tables 10.1 and 10.2.

New efforts in cold-gas technology include the usage of microelectromechanical system (MEMS) components for the development of hybrid systems [6]. Hybrid MEMS cold-gas systems, with their reduced volume requirements, offer a viable solution for developing small satellites with full propulsion capabilities, although the cost of development might be prohibitive for many applications. Another extension of the cold-gas concept is an electric propulsion system called a "resistojet," which heats the propellant into a warm or heated gas jet before expulsion through the expansion nozzle [3].

B. Vaporized Liquid

One method of addressing the volumetric inefficiency of cold-gas systems is to use a propellant that can be stored as a liquid (higher density). Compared with cold-gas thrusters, vaporizing liquid thrusters operate similarly but with added complexity because of the liquid storage state of the propellant and the need for additional heating for vaporization. Vaporizing liquid thrusters can also be labeled as resistojets in the literature because of the propellant heating requirement. A concept schematic of a vaporizing liquid thruster system is shown in Fig. 10.3. A vaporizing liquid thruster operates on the same principles of a cold-gas thruster, except the liquid propellant is heated so that it leaves the nozzle in a gaseous state.

The liquids that have been used in the development of this type of thruster include ammonia (NH_3), butane (C_4H_{10}), and water (H_2O) [4,9–11]. Propane (C_3H_8) has also been considered for use as a propellant [12]. In terms of

Table 10.1 Selected Performance Characteristics of Reviewed Thruster Technologies

Technology	Type	Thrust, *mN*	I_{sp}, *s*	Minimum impulse bit, *mN-s*	Required power, *W*
Cold gas	Moog 58 × 125 (N_2)	4.5	65	0.1	2.4
	Marotta (N_2)	50–1000	73	44	0.3
	Hybrid MEMS (N_2)	0.1–10	45	[a]	[a]
Vaporized liquid	Water	50–280	100	[a]	1–1.5
	Butane[b]	40–120	70	1	[a]
Solid chemical propellant	LTCC	446–826	30	0.131–0.279	1–1.5
Pulsed plasma thruster	microPPT	0.002–0.03	[a]	0.002	1–20
	Dawgstar System	0.06–0.264	266	0.06–0.066	15.6–36
Laser plasma thruster	Prototype[c] (NM)	0.25	500	0.001	4
Colloid thruster	Emerald[c] (NaI doped glycerol)	0.4448	500	[a]	4
Field emission electric propulsion	Centrospazio[d] (Cs)	0.04	9000	[a]	2.7
	Centrospazio[e] (Cs)	0.1	9000	[a]	13
	ARC Seibersdorf (In)	0.0001–0.1	8000	0.000005	13
Sublimating solid propellant	SRRS	13.3–73.4	75	[a]	[a]

[a]No value reported.
[b]Based on the SNAP-1 system.
[c]Characteristics are as reported/estimated by the research team.
[d]Qualification model.
[e]Under development.

space-qualified systems, SSTL has successfully flown a butane-based vaporizing liquid thruster on its SNAP-1 nanosatellite [9]. Both ammonia and water have extremely high specific heat capacities, which might offset the density gain by requiring too much energy to induce vaporization.

For most vaporized liquid systems, smaller and lighter propellant tanks can be used for an equivalent mass of inert cold gas as a result of the higher propellant density. Also, because the liquid propellant is stored at a lower pressure, this reduces the likelihood of leakage. With regards to a reduced volume requirement, MEMS technology is being used in some studies to further reduce the system size [10]. As shown in Tables 10.1 and 10.2, the performance of vaporizing liquid thrusters is generally improved when compared with cold-gas thrusters. However, as mentioned before, complexity is introduced into the system as a result of the use of a liquid propellant. There is also the issue of the liquid propellant sloshing

Table 10.2 Selected Mass Characteristics of the Thruster Technologies Covered in Table 10.1[a]

Technology	Type	Required propellant mass, kg	System wet mass, kg	ξ	Ref.
Cold gas	Moog 58 × 125 (N_2)	0.7841	3.3785[b]	0.2321	8
	Marotta (N_2)	0.6982	2.8032[b]	0.2491	8
	Hybrid MEMS (N_2)	1.1326	4.4551[b]	0.2542	6
Vaporized liquid	Water	0.5097	0.5[c]	0.5048	10;24
	Butane[d]	0.7281	3.5005[e]	0.2080	9;12
Solid chemical Propellant	LTCC	1.6989	23.7888[f]	0.0714	13
Pulsed plasma thruster	microPPT	[g]	0.5[h]	[g]	8;25
	Dawgstar System	0.1916	4.2	0.0456	26
Laser plasma thruster	Prototype[i] (NM)	0.1019	0.7219	0.1412	19
Colloid thruster	Emerald[i] (NaI doped glycerol)	0.1019	1	0.1019	8;20
Field emission electric propulsion	Centrospazio[j] (Cs)	0.0057	1.6057	0.0035	8
	Centrospazio[k] (Cs)	0.0057	3.0057	0.0019	8
	ARC Seibersdorf (In)	0.0064	2.5[l]	0.0025	21
Sublimating solid propellant	SRRS	0.6796	1.4371[m]	0.4729	23

[a]The required propellant mass is the amount a microthruster system needs to fulfill an attitude control requirement of $\Delta v = 50$ m/s.
[b]The system mass includes an estimated pressure chamber for the propellant.
[c]The entire system mass is estimated.
[d]Based on the SNAP-1 system.
[e]The system mass includes an extrapolated estimate for propellant storage.
[f]Based on the calculated need of 1888 units.
[g]Not provided due to the lack of an I_{sp} value.
[h]The reported mass does not include the propellant mass.
[i]Characteristics are as reported/estimated by the research team.
[j]Qualification model.
[k]Under development.
[l]The reported system mass is for a 30-g propellant reservoir.
[m]The system mass does not account for additional storage mass.

in its storage tank and disrupting the dynamic behavior of the satellite. In addition, small satellites might not have enough energy available to provide the heating necessary for vaporization.

C. Solid Chemical Propellant

While solid rocket motors have long been in use as a primary propulsion system, solid propellants are only now being used in digital MEMS microthruster arrays

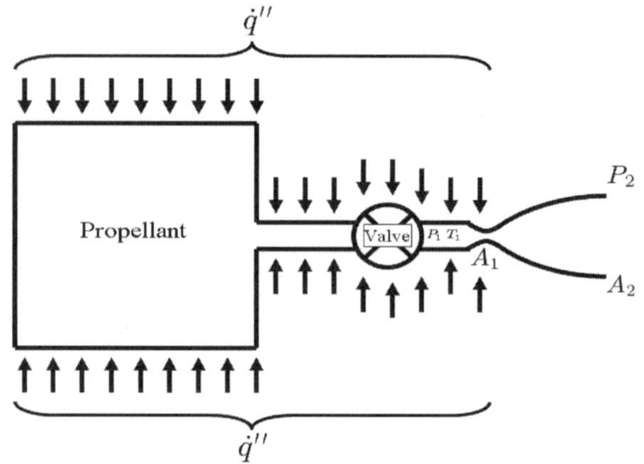

Fig. 10.3 Basic concept of a vaporizing liquid thruster system.

[13,14]. This is illustrated in Fig. 10.4, where the array consists of a number of identical units, each of which is a micromachined combustion chamber filled with solid chemical propellant and an expansion micronozzle.

The MEMS array consists of many single-fire microthrusters. Each microthruster has a combustion chamber loaded with the propellant, which is ignited and allowed to expand through the micronozzle or burst diaphragm. By building an array, solid propellants can produce the necessary small-magnitude bursts required of an attitude control system despite the fact that they cannot be turned off and on like a cold-gas system. Instead of relying on a control system to manipulate the system components to produce known thrusts, this technique depends on the assumed identical performance of each unit in the array. Performance characteristics are listed in Tables 10.1 and 10.2.

Because of the absence of pumps, propellant lines, and valves, the complexity of the system is greatly reduced. The possibility of propellant leakage is small. The main concern with this technology is the reliability of ignition and manufacturing. By nature, solid propellants also are not flexible in terms of on-the-fly

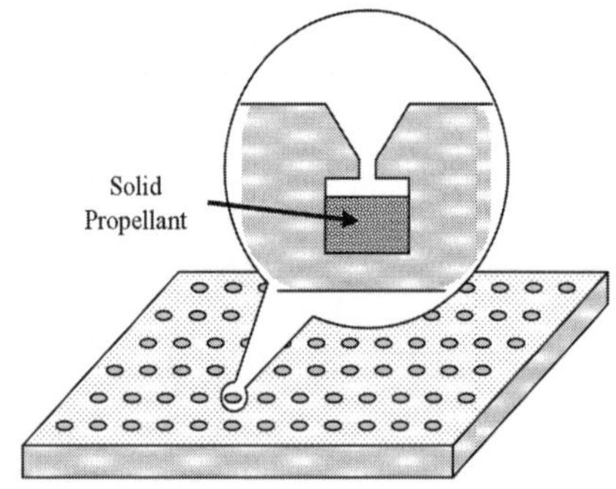

Fig. 10.4 Basic concept of a solid-propellant digital MEMS array.

performance manipulation. The use of arrays is an attempt at circumventing this lack of flexibility.

D. Pulsed Plasma Thruster

The pulsed plasma thruster (PPT) is one of the simplest electric propulsion concepts. PPTs have flown on satellites since the 1960s, but only in recent times have they been scaled down for use as a microthruster on small satellites. Noteworthy is the work done for the University of Washington's "Dawgstar" nanosatellite as part of the ION-F constellation [15]. A conventional PPT generates thrust by ionizing Teflon™ propellant and accelerating the ions electromagnetically. A small contribution also comes from the thermal expansion of nonionized propellant vapor [15].

The propellant bar is supported by the spring so that it is continually replenished between the electrodes as it is consumed. The power processing unit (PPU) charges the discharge capacitor and supplies a high-voltage pulse to the spark plug, which initiates the discharge. This electrical discharge (arc) ablates and ionizes a small amount of the propellant along the surface. As mentioned, the ablated propellant is then accelerated by electromagnetic and gas dynamic forces to provide thrust. Performance characteristics are compared in Tables 10.1 and 10.2.

Overall, PPTs offer an attractive choice for microthrusters because they are self-contained, compact, light-weight, and robust—especially when compared with other forms of electric propulsion. Also favorable is the use of solid Teflon™ as a propellant, which is inert, nontoxic, and requires a reduced feed system. However compared with other forms of electric propulsion, the thrust-power ratio of a PPT is lower. To address this shortcoming, there is research being done on using chemical propellants such as hydroxyl-terminated polybutadiene-ammonium perchlorate in place of Teflon™ [16].

E. Laser Plasma Thruster

A relatively new type of microthruster being touted as an alternative to PPTs is the laser plasma thruster (LPT) [17]. The LPT uses a focused laser to induce ablation and produce plasma jets on a surface. A schematic is shown in Fig. 10.5.

The lens focuses the laser onto a focal point on the target tape where it passes through a transparent substrate layer and heats a propellant layer. This layer ablates and produces a plasma jet perpendicular to the tape surface, which generates the thrust. The LPT can be operated in pulses or in a continuous mode where the tape is fed to keep a steady supply of propellant at the focal point. The transmission mode shown in the figure is favored to protect the optics from contaminants produced by the plasma jet [18]. Thus far polyvinyl chloride and numerous energetic polymer fuels have been tested as propellants, but an ideal material has yet to be found as none have proven durable and/or stable enough. There have also been problems with plume steering where the plasma jet is not perpendicular to the tape surface and deflects in the direction of the tape motion [19]. Performance characteristics for the LPT are listed on Tables 10.1 and 10.2. Despite the favorable performance characteristics and thrust-power ratio produced by the LPT in experiments,

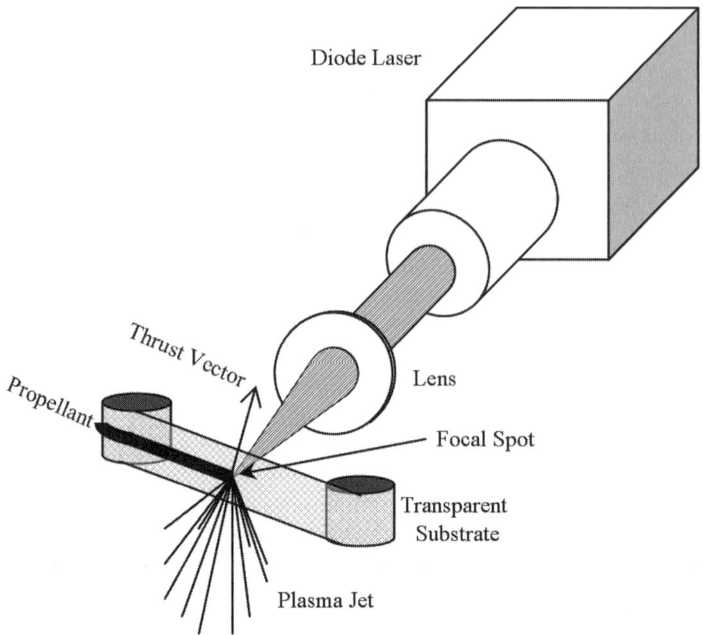

Fig. 10.5 Basic schematic of a laser plasma thruster in transmission mode.

further development is needed to address the issues already mentioned before it can be established as a viable microthruster technology.

F. Colloid Thruster

The colloid thruster is another type of electric propulsion system that uses a charged electrolytic fluid as propellant rather than using the usual method of accelerating ions to produce thrust. Glycerol is commonly used as the fluid propellant and is sometimes doped with salt to increase its conductivity [20]. A simple schematic is shown in Fig. 10.6.

The propellant is channeled out from the reservoir through a positively charged capillary towards the extractor plate. At the orifice a strong electrostatic field is present as a result of the negative charge on the extractor. This causes an imbalance of surface forces as a charge accumulates on the liquid surface. The propellant is pulled into a thin cusp and eventually breaks down into small positively charged droplets, which are accelerated by the same electrostatic field. Thrust is generated by these accelerated droplets.

The colloid thruster generally has lower power requirements for a form of electric propulsion and offers potential for miniaturization through the use of MEMS technologies. A flight demonstration of a colloid microthruster is being developed for flight on the Emerald Nanosatellite [20]. The use of an inert propellant is also advantageous. However, because the accelerated particles are droplets rather than ions, colloid thrusters have lower values of I_{sp}. In addition, the system is somewhat

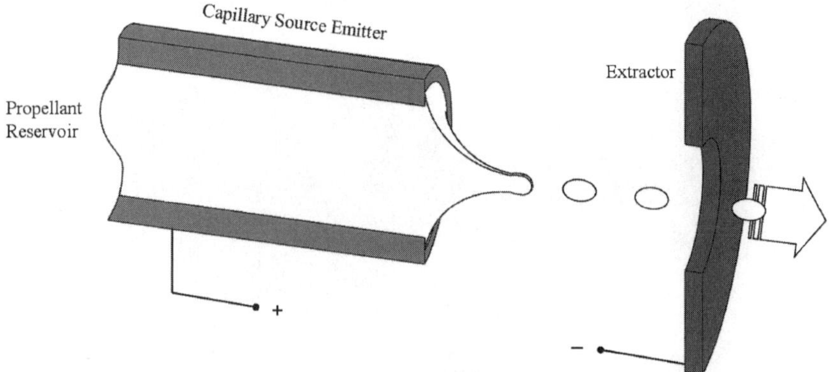

Fig. 10.6 Basic schematic of a colloid microthruster.

complex. As with the other technologies, performance characteristics for a colloid thruster are listed on Tables 10.1 and 10.2.

G. Field Emission Electric Propulsion

Field emission electric propulsion (FEEP) thrusters operate similarly to colloid thrusters but with several key differences. First, ions are generated rather than charged liquid droplets. Also, FEEP thrusters use different types of liquid metals such as indium, cesium, and gallium as propellant [21]. Emitting electrode geometries such as liquid-metal wetted needles or capillaries (tube or long slit) are used to achieve the state where ions can be emitted through surface field ionization or field evaporation. The basic configuration for an indium needle-type FEEP is shown in Fig. 10.7.

The needle-liquid metal pool is positively biased so that the metal wetting the needle is subjected to a high electric field. With a high enough field strength the equilibrium between the surface tension and the electric field at the needle tip forms a Taylor cone with a jet on top. The jet forms as a result of space charge and is shown in the close-up view. At the very tip, atoms are ionized and as with colloid thrusters, accelerated away from the emitter by the same electric field that created them. These expelled ions are then replaced through the hydrodynamic flow of the liquid metal.

Because ionization and acceleration of the propellant take place with the same electric field, FEEP thrusters often have electric efficiencies above 95% [21]. Performance characteristics for FEEP thrusters are listed on Tables 10.1 and 10.2. Also, because FEEP thrusters do not use gaseous discharges to generate ions, they are unencumbered from the associated design restrictions and therefore more readily miniaturized. However, there are concerns over spacecraft contamination and launch safety issues [22].

H. Sublimating Solid-Propellant Thruster

Sublimating solid-propellant thrusters have been used as an auxiliary propulsion system as far back as the 1960s [23]. In general, there are two types of thrusters

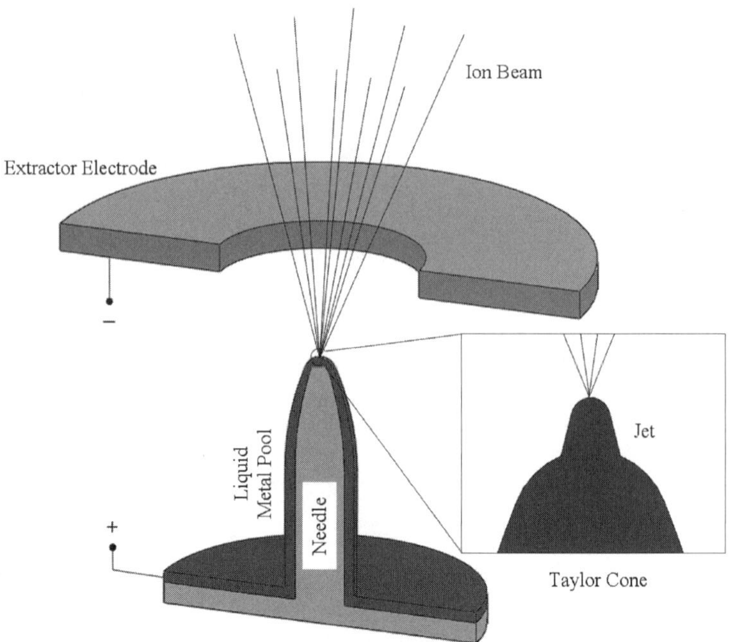

Fig. 10.7 Basic schematic of a field emission electric propulsion microthruster.

for use—valveless and valved. In both types the propellant sublimes into a vapor with a high enough vapor pressure to produce acceptable performance under realistic physical conditions. Common propellants are carbonates $[(X(HCO_3)]$ or carbamates $[X(CO_2NH_2)]$, which sublimate into NH_3, CO_2, and H_2O^4. Specific propellant candidates are ammonium hydrosulfide (NH_4HS) and ammonium carbamate ($NH_4CO_2NH_2$)[8]. The overall simplicity of these thrusters is obvious in Fig. 10.8. Figure 10.8a shows that the propellant is stored in solid-vapor equilibrium and obtains its heat of sublimation from the stored bulk propellant. Figure 10.8b shows that the propellant is self-cooled and therefore needs a heat input to produce a workable vapor pressure.

For the valved thruster, the propellant is stored in solid-vapor equilibrium and obtains its heat of sublimation from the stored bulk propellant. The single valve controls the thrust. For the valveless thruster, the propellant is self-cooled, and thrust is controlled by the heat input needed to produce a workable vapor pressure. In between firings the vapor pressure is considered negligible. Despite the simplicity of these systems, the propellants are often toxic. The two just mentioned can be absorbed by touch through the skin [8]. This leads to higher handling risks and costs.

I. Summary and Comparisons

Thus far, different microthruster technologies have been described along with some of their advantages and disadvantages. In Tables 10.1 and 10.2 we describe

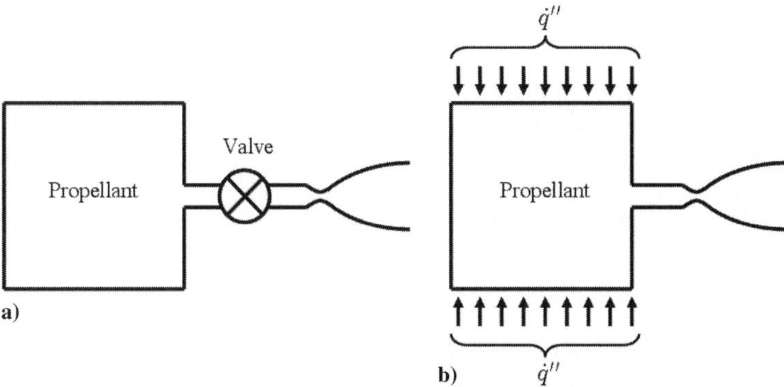

Fig. 10.8 Basic schematics of sublimating solid-propellant thrusters: a) valved and b) valveless.

some performance characteristics for a number of systems that utilize the aforementioned technologies. As much as possible, information was taken as reported in the literature. However, in order to compare the different systems in terms of ξ, certain assumptions were made. First m_p was calculated as

$$m_p = \frac{m_s \Delta v}{g_0 I_{\rm sp}} \tag{10.6}$$

where m_s is the mass of the overall satellite. Because we are interested in nanosatellite-class satellites, this was assumed to be 10 kg. Also, an attitude control requirement of 50 m/s was used for Δv. This is a typical requirement for smaller interplanetary missions at the Jet Propulsion Laboratory (JPL) [8]. Some system masses were also estimated. Those masses are labeled.

Nanosatellite-class satellites have a typical dimension of 0.3 m and a moment of inertia of 0.15 kg-m^2 assuming that the satellite is cubical in shape [8]. These physical properties correspond to required impulse bits that range from 0.0001 mN-s for a 0.02-mrad dead band with 100 s between pulses to 0.43 mN-s for a 17-mrad dead band with 20 s between pulses. Nanosatellite-class satellites are also estimated to be capable of providing a maximum power on the order of 20 W.

As shown in Tables 10.1 and 10.2, the listed microthruster technologies are, for the most part, viable candidates for nanosatellite-class missions. Where a technology does not appear to be qualified, as with the LTCC solid propellant because of its substantial mass, it is important to note that many of these technologies are still under development, and the listed systems do not represent the technology as a whole. Overall, we see that chemical/gaseous microthruster technologies are often one to two orders of magnitude less efficient than electric microthruster technologies in terms of $I_{\rm sp}$. An exception to this is the pair of vaporized liquid microthrusters. However this advantage in $I_{\rm sp}$ comes at the cost of power requirements and ξ. Heavy, power-hungry support systems are generally needed to obtain these higher levels of $I_{\rm sp}$. Finally, here is a note of caution: reported mass figures were often estimated and/or assumed to include all necessary support hardware if a number was obtained from the literature.

III. Sublimating Solid Microthruster Development at the HSFL

As noted in Section II.H, sublimating solid microthrusters are simple in design while also possessing the inherent advantages of storing the propellant as a solid. At the HSFL we are working on the development of a microthruster that retains these qualities while addressing the traditional system's shortcomings. We also aim to keep the system affordable as an economical alternative to other technologies.

A. Design

Our microthruster design consists of a prenozzle pressure chamber that holds the sublimating gas at the equilibrium vapor pressure of the high-density, low-toxicity, nonvolatile naphthalene ($C_{10}H_8$) propellant. Naphthalene only becomes a safety hazard at excessively high quantities and extensive exposure. A basic schematic of the thruster is shown in Fig. 10.9.

The thruster system consists of two chambers, the storage chamber and the charging chamber, and two independent valves. Solid naphthalene is held in the storage chamber, which is kept at the naphthalene equilibrium vapor pressure. In preparation for firing, the sublimated naphthalene gas is fed into the charging chamber. The first valve then isolates the storage chamber from the charging chamber before and during a standard pulse firing. In this charging chamber setup, it is believed that the performance will be more consistent in pulse to pulse firings. During continuous firing, the first valve is left open. The second valve with a submillisecond response time controls the time-dependent discharge of the gas to a low-pressure (or vacuum) environment on demand.

The equilibrium pressure in the prenozzle charging chamber is determined to millibar accuracy through a short-response time temperature control system. Because of the high temperature sensitivity of the vapor, this system may or may not be heated by a low-current electrical resistor. The magnitude of the naphthalene gas thrusting jet is therefore purely controlled by the equilibrium vapor pressure and electrical current (temperature) input. The expected maximum pressure for the system is 1 atm because the maximum vapor pressure due to the sublimation of naphthalene is less than 1 kPa in the operational temperature range. Therefore, thin aluminum or plastic shells can be used.

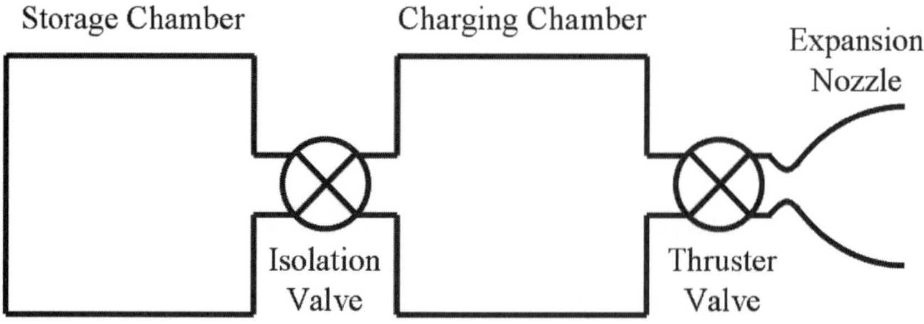

Fig. 10.9 Basic schematic of the microthruster system.

B. Preliminary Analysis

A preliminary analysis of the thruster's expected performance is performed using energy and momentum balances and assuming quasi-steady, isentropic process from the charging chamber and through the exit nozzle. A schematic is shown in Fig. 10.10.

From material property data found in the literature [27], it is assumed that the ratio of specific heats (γ) for naphthalene is a constant 1.05 for the temperature interval of interest (0–50°C). The following relation is then used to determine the critical upstream pressure for choked flow (flow at Mach 1) at the throat:

$$\frac{P_{0,\text{critical}}}{P_{\text{external}}} = \left(\frac{\gamma+1}{2}\right)^{\frac{\gamma}{\gamma-1}} \tag{10.7}$$

With an assumed external pressure of 10^{-7} torr, it was determined that the flow through the throat will remain choked until the upstream pressure is so small that it will produce an insignificant level of thrust. With such a small subsonic contribution, the assumption that the flow at the throat is always choked holds. Choked conditions are labeled with a * in Fig. 10.10. The time-dependent relation for thrust is then determined using the following equation:

$$F_{Th} = |\dot{m}v_2| + A_2(P_2 - P_{\text{ext}}) \tag{10.8}$$

Using the preceding assumptions, the following time-dependent relation for thrust is found along with the time function for the chamber pressure P_0:

$$F_{\text{Th}} = P_0(t)\left[\frac{\gamma A_1 M_2}{\sqrt{1+\frac{\gamma-1}{2}M_2^2}}\left(\frac{2}{\gamma+1}\right)^{\frac{\gamma+1}{2(\gamma-1)}} + A_2\left(1+\frac{\gamma-1}{2}M_2^2\right)^{\frac{\gamma}{1-\gamma}}\right] - A_2P_{\text{ext}} \tag{10.9}$$

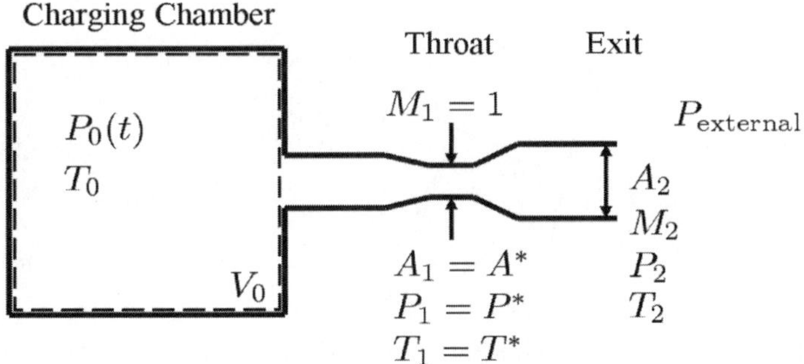

Fig. 10.10 Schematic used in the theoretical analysis.

$$P_0(t) = \left[P_{0,i}^{\frac{1-\gamma}{2\gamma}} - \frac{A_1(1-\gamma)}{2V_0} \sqrt{\gamma RC} \left(\frac{2}{\gamma+1} \right)^{\frac{\gamma+1}{2(\gamma-1)}} t \right]^{\frac{2\gamma}{1-\gamma}} \tag{10.10}$$

The initial chamber pressure ($P_{0,i}$) in Eq. (10.10) is calculated from a vapor-pressure curve and property data [27]. As seen in Eqs. (10.9) and (10.10), the generated thrust is dependent upon a number of parameters. The effect of altering these parameters on the yield thrust is shown in Fig. 10.11 (time-thrust curves). The parameters for the "Original" line in Fig. 10.11 are as follows: initial temperature of 298 K, chamber volume of 1 cc, valve throat diameter of 0.005 mm, and area ratio of 30. A number of key performance characteristics for each set of design parameters is compiled in Table 10.3. Of note is the large influence the valve and nozzle dimensions have over the performance characteristics while consuming the same propellant mass.

As shown in Table 10.3, altering the valve and nozzle dimensions appears to have the most significant impact on performance in terms of F_0, I_t, and I_{sp}. Although a change in the initial temperature influences the initial thrust and the total impulse

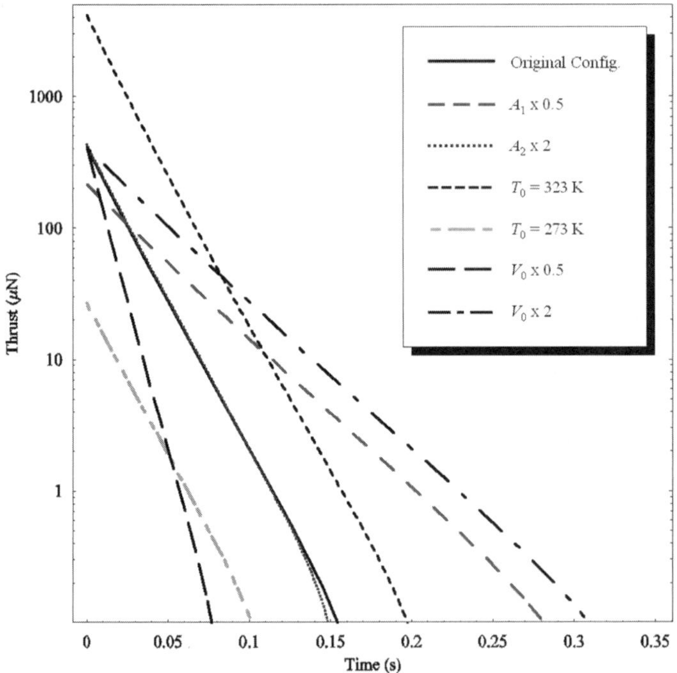

Fig. 10.11 Thrust plotted as a function of time while varying different parameters as noted in the figure's legend. The nearly logarithmic decay of thrust over time is also evident in the plot.

Table 10.3 Initial Thrust (F_0), I_t, I_{sp}, and Consumed Mass (m_c) Calculated for Each Set of Design Parameters

Design parameters	F_0 (μN)[a]	$I_t(\mu N - s)$[a]	I_{sp} (s)	m_c, mg
Original	410.3	7.418	1.427	0.5301
$A_1 \times 0.5$	215.6	7.791	1.498	0.5301
$A_2 \times 2$	431.2	7.791	1.498	0.5301
$T_0 = 323$ K	4151	72.15	1.486	4.948
$T_0 = 273$ K	26.49	0.4955	1.349	0.03743
$V_0 \times 0.5$	410.3	3.709	1.427	0.2650
$V_0 \times 2$	410.3	14.84	1.427	1.060

[a]F_0 and I_t are shown graphically in Fig. 10.11.

to some extent, the initial temperature does not have much of an effect on the I_{sp}. The effect on I_{sp} is limited despite the tangible change in F_0 and I_t because the temperature change also affects the consumed propellant mass. In comparison, changing the valve and nozzle dimensions induces significant changes in F_0 and I_t with the same amount of mass. Also shown at the bottom of Table 10.3 is that changes solely to the amount of propellant used through different chamber sizes affects only the total impulse.

IV. Conclusion

In this chapter we discussed a variety of microthruster technologies that are applicable to attitude control of satellites. Although many of these technologies require further development before being qualified for flight, they all hold great potential in bringing full propulsion capabilities to nanosatellite-type missions. We concluded the chapter with a discussion of a novel microthruster design concept under development at the Hawaii Space Flight Laboratory of the University of Hawaii at Manoa.

References

[1] Rysanek, F., Hartmann, J. W., Schein, J., and Binder, R., "MicroVacuum Arc Thruster Design for a CubeSat Class Satellite," *16th Annual AIAA/USU Conference on Small Satellites*, Paper SSC02-I-2, Aug. 2002.

[2] Chang, Y.-K., Lee, B.-H., and Kim, S.-J., "Analysis of the HAUSAT-2 Attitude Control with a Pitch Bias Momentum System," *5th International Symposium of the IAA*, International Academy of Astronautics, April 2005.

[3] Sutton, G. P., and Biblarz, O., *Rocket Propulsion Elements*, 7th ed., Wiley-Interscience, New York, 2000, Chap. 4.6, 6.8, 7.6, 8.2, pp. 136–139, 228–232, 263–264, 300–304.

[4] de Groot, W., and Oleson, S., "Chemical Microthruster Options," NASA Contractor Report 198531, Oct. 1996.

[5] Wu, S.-F., Steyn, W. H., and Bordany, R. E.,. "In-Orbit Thruster Calibration Techniques and Experiment Results with UoSAT-12," *Control Engineering Practice*, Vol. 12, No. 1, 2004, pp. 87–98.

[6] Köhler, J., Bejhed, J., Kratz, H., Bruhn, F., Lindberg, U., Hjort, K., and Stenmark, L., "A Hybrid Cold Gas Microthruster System for Spacecraft," *Sensors and Actuators A*, Vol. 97–98, No. 1, 2002, pp. 587–598.

[7] Adler, S., Warshavsky, A., and Peretz, A., "Low-Cost Cold-Gas Reaction Control System for Sloshsat FLEVO Small Satellite," *Journal of Spacecraft and Rockets*, Vol. 42, No. 2, 2005, pp. 345–351.

[8] Mueller, J., "Thruster Options for Microspacecraft: A Review and Evaluation of State-of-the-Art and Emerging Technologies," *Micropropulsion for Small Spacecraft*, edited by Michael M. Micci and Andrew D. Ketsdever, Progress in Astronautics and Aeronautics, Vol. 187, AIAA, Reston, VA, 2000, Chap. 3, pp. 51–56, 68–71, 84–123.

[9] Gibbon, D., Underwood, C., and Sweeting, M., "Cost Effective Propulsion Systems for Small Satellites Using Butane Propellant, *Acta Astronautica*, Vol. 51, No. 1–9, 2002, pp. 145–152.

[10] Mueller, J., Ziemer, J., Green, A., and Bame, D., "Performance Characterization of the Vaporizing Liquid Micro-Thruster (VLM)," *International Electric Propulsion Conference*, IEPC 03-237, 2003.

[11] Maurya, D. K., Das, S., and Lahiri, S. K., "Silicon MEMS Vaporizing Liquid Microthruster with Internal Microheater," *Journal of Micromechanics and Microengineering*, Vol. 15, March 2005, pp. 966–970.

[12] Gibbon, D. M., Ward, J., and Kay, N., "The Design, Development and Testing of a Propulsion System for the SNAP-1 Nanosatellite," *14th Annual AIAA/USU Conference on Small Satellites*, Paper SSC00-1-3, Aug. 2000.

[13] Zhang, K. L., Chou, S. K., and Ang, S. S., "Development of a Low-Temperature Co-Fired Ceramic Solid Propellant Microthruster," *Journal of Micromechanics and Microengineering*, Vol. 15, March 2005, pp. 944–952.

[14] Zhang, K. L., and Chou, S. K., "Performance Prediction of a Novel Solid-Propellant Microthruster," *Journal of Propulsion and Power*, Vol. 22, No. 1, 2006, pp. 56–63.

[15] Cassady, R. J., Hoskins, W. A., Campbell, M., and Rayburn, C., "A Micro Pulsed Plasma Thruster (PPT) for the 'Dawgstar' Spacecraft," *Aerospace Conference Proceedings, 2000 IEEE*, Vol. 4, IEEE, 2000, pp. 7–14.

[16] Mashidori, H., Kakami, A., Muranaka, T., and Tachibana, T., "A Coaxial Pulsed Plasma Thruster Using Chemical Propellants," *Vacuum*, Vol. 80, No. 11–12, 2006, pp. 1229–1233.

[17] Phipps, C., and Luke, J., "Diode Laser-Driven Microthrusters: A New Departure for Micropropulsion," *AIAA Journal*, Vol. 40, No. 2, 2002, pp. 310–318.

[18] Phipps, C. R., Luke, J. R., McDuff, G. G., and Lippert, T., "Laser-Driven Micro-Rocket," *Applied Physics A*, Vol. 77, May 2003, pp. 193–201.

[19] Luke, J. R., Phipps, C. R., and McDuff, G. G., "Laser Plasma Thruster," *Applied Physics A*, Vol. 77, May 2003, pp. 343–348.

[20] Pranajaya, F. M., and Cappelli, M., "Progress on Colloid Micro-Thruster Research And Flight Testing," *13th Annual AIAA/USU Conference on Small Satellites*, Paper SSC99-VIII-6, Aug. 1999.

[21] Tajmar, M., Genovese, A., and Steiger, W., "Indium Field Emission Electric Propulsion Microthruster Experimental Characterization," *Journal of Propulsion and Power*, Vol. 20, No. 2, 2004, pp. 211–218.

[22] Mitterauer, J., "Micropropulsion for Small Spacecraft: A New Challenge for Field Effect Electric Propulsion And Microstructured Liquid Metal Ion Sources," *Surface and Interface Analysis*, Vol. 36, Nos. 5–6, pp. 380–386.

[23] Rocket Research Corp., "Development of the Subliming Solid Control Rocket," NASA Contractor Report CR-712, March 1967.

[24] Mueller, J., Chakraborty, I., Bame, D., and Tang, W., "Vaporizing Liquid Microthruster Concept: Preliminary Results of Initial Feasibility Studies," *Micropropulsion for Small Spacecraft*, edited by Michael M. Micci and Andrew D. Ketsdever, Progress in Astronautics and Aeronautics, Vol. 187, AIAA, Reston, VA, 2000, Chap. 8, p. 228.

[25] Burton, R. L., and Turchi, P. J., "Pulsed Plasma Thruster," *Journal of Propulsion and Power*, Vol. 14, No. 5, 1998, pp. 716–735.

[26] Rayburn, C. D., Campbell, M. E., and Mattick, A. T., "Pulsed Plasma Thruster System for Microsatellites," *Journal of Spacecraft and Rockets*, Vol. 42, No. 1, 2005, pp. 161–170.

[27] Lide, D.R. (ed.), *CRC Handbook of Chemistry and Physics*, CRC Press, Boca Raton, FL, 2005, pp. 5–48, 12–195.

Chapter 11

Novel Propulsion System for Nanosatellites

Filip Rysanek*

University of Illinois at Urbana-Champaign, Urbana, Illinois

I. Vacuum Arc Thrusters

THERE is a wide range of available propulsion systems for satellites larger than 1000 kg. Many of these scale down for use in smaller 100 kg, or even 10-kg satellites. However, spacecraft weighing on the order of 1 kg have few propulsion options capable of providing orbit transfer, active attitude control, or even desaturation of momentum wheels.

Among the few options that do exist for satellites smaller than 1 kg are vacuum arc thrusters (VAT) [1–3]. These thrusters provide thrust by utilizing the high-velocity ions ejected from the cathode of a vacuum arc. The reduced mass of these thrusters is achieved by using an inductive energy storage (IES) power processing unit. Figure 11.1 shows a simplified schematic of the thruster circuit.

A semiconductor switch is closed to draw current from the dc, low-voltage (5–35 V) power supply through an inductor. Once the switch is opened, a voltage peak of $L\,dI/dt$ is produced, igniting a plasma by running current through a thin metal film coating between the two electrodes [4,5].

After the initial vacuum arc formation, a fully ionized metal plasma is produced from a macroscopically cold cathode. The metal plasma plume, produced at cathode spots in a way reminiscent of laser plasma production, streams outward to achieve velocities of $5–23 \times 10^3$ m/s over a wide range of elements from Carbon to Tungsten [6]. Time-of-flight measurements of the charge state distributions have shown that the ions are predominantly of charge states 1^+ to 3^+ depending on the metal species used and the arc current density [6]. Arc discharges are produced with arc currents from tens of amperes to many kilo-amperes. Pulse lengths can be from a few microseconds upwards, and the pulse repetition rate can readily be up to a few hundred pulses per second. This implies a very wide dynamic thrust range. The μVAT can be operated in a pulsed mode with an average power of 1–100 W and masses of about 0.1 kg. For orbit transfer and orbit maneuvering of

---

*Ph.D. Candidate (awarded 2007).

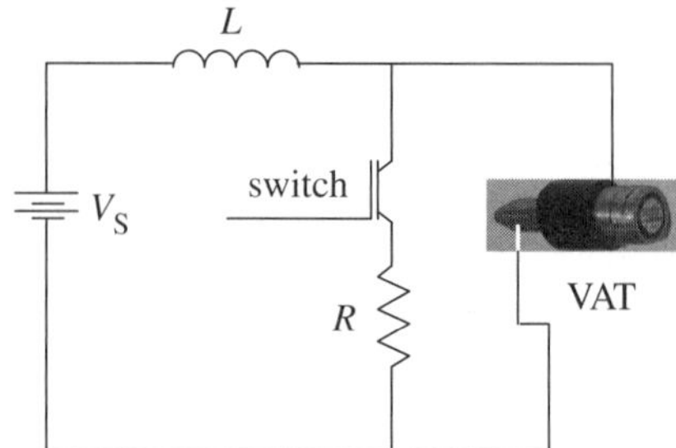

Fig. 11.1 Schematic of the circuit used on a typical vacuum arc thruster.

a satellite, high thrusts are desired to save time, whereas for stationkeeping the required thrusts are relatively low. The μVAT has the potential of providing a wide dynamic range of thrusts without efficiency loss for these missions.

Another factor that makes the VAT a useful and versatile thruster is the range of geometries available. Three main geometries are commonly used in the thruster design. The coaxial geometry shown in Fig. 11.2 has a central cathode and an outer anode, separated by an insulator. In the sandwich geometry shown in Fig. 11.3, the thruster electrodes and insulator material are layered sheets. Finally the ring geometry (Fig. 11.4) has a cylindrical cathode with a ring-shaped anode separated by a ring-shaped insulator.

In each of the geometries, the key to igniting the discharge is a thin metallic film on the insulator between the two electrodes. During operation, the arc

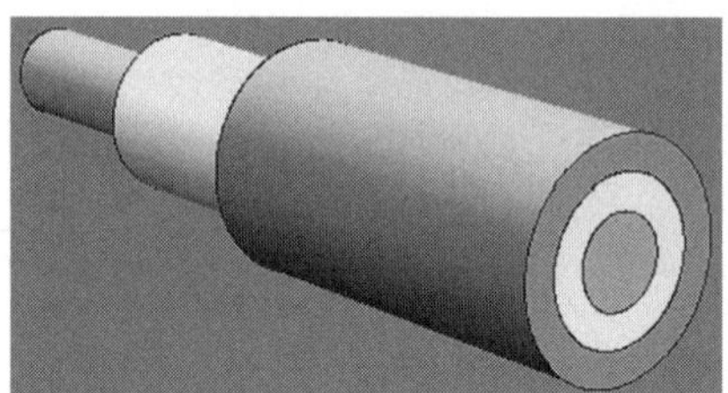

Fig. 11.2 Schematic of a coaxial geometry vacuum arc thruster.

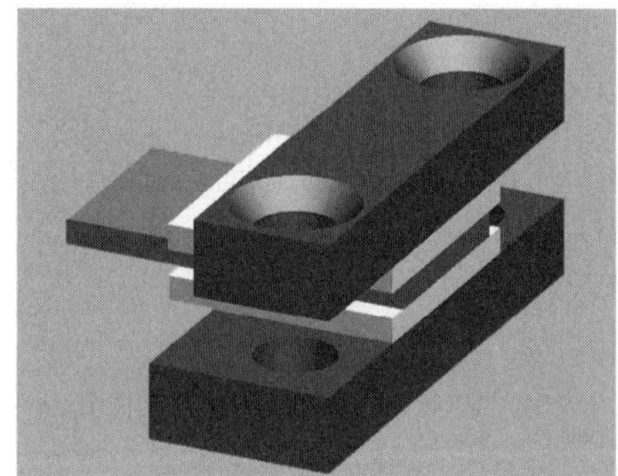

Fig. 11.3 Schematic of a sandwich geometry vacuum arc thruster.

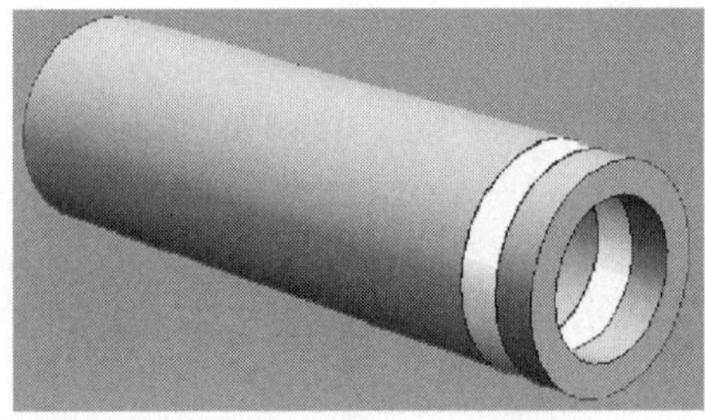

Fig. 11.4 Schematic of a ring geometry vacuum arc thruster.

discharge redeposits a layer of cathode material on the insulator between the two electrodes, maintaining the thin metal film necessary for operation. In the ring geometry, because the insulator is opposite the location of the arc discharge, this redeposition of the metal film on the insulator is enhanced. This makes the ring geometry more robust than the other geometries.

Along with the versatility of the geometry, the VAT has the advantage of using a high-density solid (namely, the cathode metal) as the fuel. This allows the VAT to be more compact and also allows the option of incorporating the thruster into the structure of the spacecraft, possibly even using the spacecraft structure as the fuel itself.

II. VAT System on ION

A. PPU

The University of Illinois power processing unit, designed and built by Alameda Applied Sciences Corporation and shown in Fig. 11.5, incorporates the same inductive energy storage design. The ION PPU was designed to control four thrusters individually. The PPU has a modular design (Fig. 11.6) with one main board housing the inductor and timing circuit. Each thruster is connected to a separate control board that houses the semiconductor switch used to control the current through the inductor. PPU size and mass have been driven by the cubesat requirements and amount to $4 \times 4 \times 4$ cm and 150 g, respectively. The PPU is powered from a 12–24 V power bus.

Four TTL level control signals determine which thrusters are operational. Multiple control signals can be sent in order to fire multiple thrusters simultaneously.

Fig. 11.5 Vacuum arc thruster PPU designed for use onboard ION.

Fig. 11.6 Vacuum arc thruster PPU designed for use onboard ION, open view.

In this case, the arc will randomly initiate on one of the two thruster heads. Another way to fire two thrusters simultaneously is to alternate the control signal. In this fashion, the relative power ratio going to each thruster can be controlled.

The ION PPU can be fired in two modes of operation. First, an onboard timing circuit can be used to fire the thrusters at a preset pulse frequency and power. This mode of operation requires only a single activation control signal. The second mode of operation for the PPU is one where the ION onboard computer sends the PPU a square wave signal generated by a pulse-width modulator circuit (PWM) to directly control the semiconductor switch for each thruster. In this manner, the computer can control the pulse frequency as well as the energy per pulse. A typical pulse duration is approximately 4 μs, with a peak current under 10 A.

B. Thrusters

The original μVAT design incorporated a cylindrical thruster. With this design, the thruster was to be placed in the feet of the satellite. It was determined that the cylindrical geometry would require significant insulation, as well as tight tolerance machining to operate reliably. It was also unclear if a cylindrical design would withstand launch vibration. A sandwich geometry was adopted instead. In testing, the sandwich geometry proved to be more reliable, as well as easier to manufacture. The first design included a sandwich of copper, ceramic, and titanium. The arc forms between the center (titanium) and outer electrodes. This design was then scaled to approximately a 1-cm-wide thruster.

During the initial design, the high density and high I_{sp} of the tungsten electrode were favored. The tungsten could provide the highest ΔV given a limited mass and volume; however, in order to demonstrate a thruster system with a large volume of fuel for possible future missions, the μBLT thrusters were incorporated into the satellite structure in such a way that the aluminum structure became the fuel. The anode is separated from the structure (cathode) with two high-alumina ceramic plates. The arc will either attach to the satellite structure or to the aluminum bar used to clamp the thruster together. When the thrusters were assembled and prepared for operation, the ceramic closer to the aluminum bar was not plated with the conductive layer, increasing the chance that the arc will form between the anode and satellite structure.

The four thrusters are located on the satellite in such a way as to allow both translation and two-axis rotation. Two thrusters are placed on each of the 10×10 cm faces of the satellite. Each thruster is in the opposite corner of that face as shown in

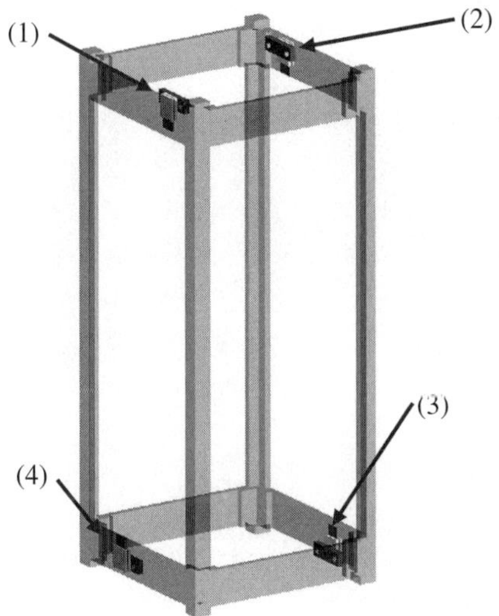

Fig. 11.7 μBLT thruster locations on satellite.

Fig. 11.7. With this layout, when thrusters 1 and 2 or thruster 3 and 4 are fired simultaneously, the satellite experiences translation. When thrusters 1 and 3 or thruster 2 and 4 are fired, the satellite experiences rotation about one axis. Firing thrusters 1 and 4 or thrusters 2 and 3 rotates the satellite about another axis.

To estimate the effect that the thrusters have on the satellite, a theoretical analysis, which provided an estimate of thrust with an aluminum cathode to be 13.5 μN-s/W, was conducted [7]. With this thruster layout, each thruster will have approximately 5 cm of lever arm about the axis of rotation. It is assumed the satellite has the moment of inertia of a uniform density block (8.3×10^{-3} kg-m^2), and the thruster fires at a nominal 4 W, producing 54 μN of thrust. The pair of thrusters produces 5.4×10^{-6} N-m of torque. This results in an angular acceleration of 6.56×10^{-4} rad/s^2. This means that by firing the thrusters continuously for approximately 4 s, the satellite will turn 90 deg in 10 min.

This estimate is only an approximation to the actual motion of the satellite. Because of the location of the thrusters, torque will not be applied through the center of mass, thus likely resulting in something other than pure rotation. The dynamics of this motion are yet to be simulated; however, the simplified model just described is sufficient to estimate an approximate effect that the thrusters will have.

C. Diagnostics

One issue that has often confronted mission designers is whether the exhaust from thrusters will redeposit on solar panels or optical experiments. Such deposition can reduce solar-panel lifetime and efficiency. To estimate this effect, a conductive deposition monitor (CDM) experiment was designed to help determine the amount of deposition accumulating on the satellite due to the thruster.

The CDM is a 1 cm × 2 cm circuit board with two leads. The layout of the circuit board is an intertwined comb shape to maximize the effect of deposited material (Fig. 11.8). As the fuel from the thrusters deposits on the ceramic plate, the

Fig. 11.8 (CDM) collector.

resistance between the two wires will fall from infinity. Because the amount of deposition is expected to be small, a voltage divider circuit is used to measure the resistance between the two leads on the CDM. This circuit is designed to be very sensitive between infinite resistance and 10 MΩ.

A number of methods have been devised to help verify the operation of the thrusters. At the very basic level, the attitude control system is supposed to be able to detect any change in attitude using the onboard magnetometer. Although this is not a direct determination that the thrusters are functioning properly, it is enough to verify that an end has been achieved.

The attitude control system is also capable of compensating for the thrust with the onboard torque coils. The amount of current through the torque coils necessary to compensate for the thrusters can then be converted into a force or torque produced by the thrusters. This method can be used as an in-flight thrust-stand to measure the exact thrust produced by the thrusters.

The ION computer system is capable of measuring the temperature of the satellite in up to 63 locations. The temperature of the inductor as well as one of the IGBT switches and its corresponding thruster are monitored. This will provide a means to protect the PPU and thrusters from overheating as a result of normal operation. In case of a failure resulting in a short circuit, the temperature measurement will help to protect and diagnose the system for future operation.

One final diagnostic tool is a Channel Island Circuits model 711B current probe, used to measure the current through one of the thrusters. This current probe is a torroidal inductive probe that measures the derivative of the current between the PPU and the thruster head. A circuit was designed to integrate the derivative signal and return the actual current to the analog input of the computer. This data will be used as a diagnostic tool and possibly in conjunction with a pulse counter.

III. VAT Issues

The vacuum arc thruster technology was developed by Alameda Applied Sciences Corporation. During the development of the ION vacuum arc propulsion system, a number of design issues were confronted. Thruster startup, lifetime, and circuit protection are among the most important design issues faced.

As just mentioned, the thrusters rely on a thin metallic film between the two electrodes to initiate the plasma breakdown. When the thruster is first built, a high-resistance film (~1 kΩ) must be applied to the insulator between the anode and cathode. To apply this film, a second thruster was pointed at the flight thruster and fired repeatedly. To maximize the amount of material ejected from the second thruster onto the flight hardware, the second thruster is powered using a pulse forming network [8]. This increases the energy per pulse of the plating thruster, thus increasing the amount of material transferred to the flight hardware. This method has the disadvantage of requiring the use of a vacuum chamber as well as a significant amount of time to complete.

To reduce the build time of each thruster, colloidal graphite can be applied instead of a metallic film. If graphite is used, the resistance between the two electrodes will undergo a break-in period. The length of this break-in period is approximately 10,000 pulses, depending on the pulse energy. During this time, the resistance between the electrodes will fall to a minimum as the metallic film is slowly forming. As the thruster continues to fire, the resistance will increase to approximately 1 kΩ. This resistance will remain approximately constant until approximately 100,000 shot. At this point, the resistance will slowly increase to a point where the thruster no longer fire properly.

The design of the vacuum arc thruster circuit shown in Fig. 11.1 requires that a thruster be present, and the voltage required to initiate a plasma on that thruster is less than the voltage rating of the IGBT switch. If the plasma does not initiate properly, or the PPU is fired without a thruster attached, the IGBT switch will be exposed to the entire $L\, dI/dt$ voltage generated upon opening. If that voltage is above the rated voltage of the IGBT, it will cause the IGBT to fail. The IGBT typically fails closed, which results in a short circuit of the power supply.

To reduce the risk of this occurring, a MOV (metal-oxide varistor) is placed in parallel with the thruster. When the voltage across a MOV is below a threshold value, the resistance remains infinite. When the voltage increases above the threshold, the resistance of the MOV decreases quickly to a small value. Metal-oxide varistors are typically found in surge protectors. This protects the IGBT by sending the energy through the MOV instead of the IGBT if the thruster does not fire. The rating on the MOV must be chosen below breakdown voltage of the IGBT, to take into account the time it takes the MOV resistance to fall. It must also be selected as high as possible to reduce the chance that the MOV will divert energy from the thruster and keep the thruster from initiating a plasma breakdown. This MOV selection is dependent on the thruster geometry and pulse energy.

Another issue taken into consideration is the magnetic field produced by the PPU during operation. This magnetic field interacts with the Earth's magnetic field, producing a torque on the satellite. The magnitude of this torque can be on the order of the torque produced by the thrusters. For this reason, the PPU was encased in a mu-metal to reduce this effect. A trade study was done to determine the thickness of the mu-metal necessary.

IV. Conclusion

Vacuum arc thrusters provide a unique opportunity for microsatellite designers. The inductive energy storage PPU eliminates the need for bulky high-voltage power supplies, along with the additional insulation that accompanies them. The solid fuel provides a compact, high-density propellant, requiring no valves or pressure vessels. The ability to use the spacecraft structure as fuel can further reduce system mass. The modular PPU design allows for expansion to any number of thrusters. All of this results in the low system mass necessary for use on 1–5-kg satellites. Few other propulsion systems at this level of development can provide attitude control for a microsatellite.

References

[1] Qi, N., Schein, J., Binder, R., and Krishnan, M., "Compact Vacuum Arc Micro-Thruster for Small Satellite Systems," AIAA Paper 2001-3793, July 2001.

[2] Schein, J., Qi, N., Binder, R., Krishnan, M., Ziemer, J. K., Polk, J. E., and Anders, A., "Inductive Energy Storage Driven Vacuum Arc Thruster," *Review of Scientific Instruments*, Vol. 73, No. 2, 2002, p. 925.

[3] Keidar, M., Schein, J., Wilson, K., Gerhan, A., Au, M., Tang, B., Idzkowski, L., Krishnan, M., and Beilis, I.I., "Magnetically Enhanced Vacuum Arc Thruster," *Plasma Source Science and Technology*, Vol. 14, No. 4, 2005, pp. 661–669.

[4] Anders, A., Brown, I. G., MacGill, R. A., and Dickinson, M. R., "'Triggerless' Triggering of Vacuum Arcs," *Journal of Physics D: Applied Physics*, Vol. 31, No. 7, 1989, p. 584.

[5] Anders, A., Schein, J., and Qi, N., "Pulsed Vacuum-Arc Ion Source Operated with a 'Triggerless' Arc Initiation Method," *Review of Scientific Instruments*, Vol. 71, No. 2, 2000, p. 827.

[6] Anders, A., and Yushkov, G. Y., "Ion Flux from Vacuum Arc Cathode Spots in the Absence and Presence of a Magnetic Field," *Journal of Applied Physics*, Vol. 91, No. 8, 2002, p. 4824.

[7] Polk, J. E., Sekerak, M., Ziemer, J. K., Schein, J., Qi, N., Binder, R., and Anders, A., "A Theoretical Analysis of Vacuum Arc Thruster Performance," 27th International Electric Propulsion Conference, IEPC Paper 01-211, Oct. 2001.

[8] Anders, A., MacGill, R. A., and McVeigh, T. A., "Efficient, Compact Power Supply for Repetitive Pulsed, 'Triggerless' Cathodic Arcs," *Review of Scientific Instruments*, Vol. 70, No. 12, 1999, p. 432.

IV. Electrical Power Systems

Chapter 12

Power Systems for Cubesat-Class Satellites

Purvesh Thakker* and Jonathan W. Kimball†
University of Illinois at Urbana-Champaign, Urbana, Illinois

I. Introduction

CUBESAT deployers [1] typically hold three 10 × 10 × 10-cm cubesats that conform to a simple interface defined in the cubesat specification (data available online at http://cubesat.atl.calpoly.edu/pages/documents/developers.php). These deployers are then carried to orbit using leftover space on existing launches. Thanks to this growing secondary payload launch infrastructure, universities and other organizations have begun utilizing these spacecraft for education [2], scientific measurement [3–7], and space technology testing missions [8,9].

This work provides a complete investigation of power systems for these spacecraft covering all system elements. The work provides a collection of techniques and recommendations that address fundamental power system design considerations with detailed lessons and data from the Illinois Observing Nanosatellite (ION) [10]. The paper walks through the elements of a typical cubesat power system illustrating ION's approach as either the recommended approach or as a counter example depending on the development experience.

Section II outlines requirements for the system. Section III provides an overview of typical cubesat power system architecture. The subsequent sections delve into generation, storage, regulation, and distribution details. The later sections return to the overall system discussing fault-tolerance, power budget analysis, fabrication, and integrated system testing.

II. Requirements

In the Cubesat Spec, some requirements that affect the power system are defined (Table 12.1), whereas other power system requirements are defined internally

*Program Manager, ION Cubesat, Department of Electrical and Computer Engineering.

†Senior Research Engineer, Department of Electrical and Computer Engineering.

Table 12.1 Cubesat Spec Requirements

Mass	1.0-kg per cube (up to 3 cubes)
Size	10 × 10 × 10-cm per cube
Environmental testing	Vibration and bakeout
Wiring	Kill switches, pull pin
Dataport	Battery recharging, diagnostics, external power

(Table 12.2). In addition to providing a simple interface between cubesats and their deployers, the Cubesat Spec's requirements minimize risk to the rest of the launch vehicle and payloads. For example, power to the satellite must be cut while the satellite awaits deployment. Also, the satellite must be completely enclosed by the deployer, and vibration and bakeout procedures must be performed prior to launch.

Figure 12.1 shows ION's hardware layout with power items circled. Kill switches in the satellite's feet cut power to the satellite until deployment. The satellite has "feet" on both the top and the bottom of its frame. The switches are normally implemented on the top feet, which do not contact the table during ground development. In addition, a pull pin cuts power as the satellite is being inserted into the deployer. Satellite owners can remove this pull pin after integration with the deployer. Alternately, the pull pin can be removed just before launch and returned to the satellite owner as proof that the procedure was completed. The pull pin is often the primary method for switching the satellite on and off during development, so a robust mechanical design is important. A single pull pin can activate multiple switches.

In the Spec, the deployer is required to provide small windows for accessing the satellite's data ports. These data ports can be used for recharging batteries and performing diagnostics while the satellite waits in the deployer. Because the deployer cannot be opened after integrated vibration testing, teams can use the data ports to subsequently test their satellite. The kill switches (and possibly

Table 12.2 Internal System Requirements

System	Description
Orbit	Low Earth orbit (000 to 000 km)
Lifetime	Six months
Operating temperature	–20° to +85°
Regulation	for example, 5 V (1-A peak), 9.6 V (3-A peak), 20 V (100-mA peak)
Load	Support all components simultaneously
Generation	Solar cells on most satellite surfaces; some deployable panels possible
Storage	24-h operation without recharging; recover from dead battery
Budget	50% normal/90% sleep mode margins; sustain system without attitude stabilization; automatic sleep mode
Fault-tolerance	Power cycle system when frozen; radiation shielding for Total Ionizing Dose <5 krads

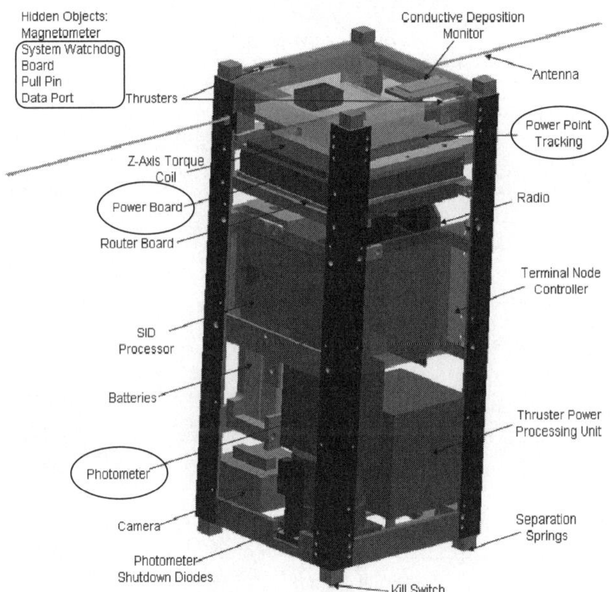

Fig. 12.1 ION hardware.

pull-pin switches) are depressed inside the deployer, so that the flight batteries are disconnected. An external power source can be connected through the data port for diagnostic purposes. Similarly, some data-port lines should feed directly to the batteries to allow an external battery charger to recharge batteries in the deployer. The data port's diagnostic, external power, and external charger lines are also useful during development before the satellite is placed in the deployer. Using an external power source during development prevents wear on the flight batteries and pull pin.

Figure 12.2 shows a possible wiring scheme for the data port, kill switches, and pull-pin switches. The data port can include connections for programming the processor and monitoring certain lines, such as the +5-V regulated line. When all switches are in their normal closed configuration, the batteries and solar cells pass through to the system. When the kill switches are open (depressed in the deployer), the internal batteries and solar cells are disconnected from the system. When the pull-pin switches are open (pull pin inserted), the internal batteries and solar cells similarly cannot reach the system. However, with the pull-pin switches in their open position (pull pin inserted), the data port can access the batteries directly for recharging. Similarly, the data port can power the system directly as if it were a solar panel. The diodes shown in the diagram can be implemented either with Schottky devices (~0.3-V drop) or with current sense resistors (~10 mΩ), MOSFETs, and control circuitry.

For internal requirements, missions are often specifically selected such that they can be completed in a relatively short time frame (e.g., six-month lifetime for ION). This minimizes lifetime concerns and keeps the development time short. Most importantly, reducing lifetime engineering effort keeps the focus on first-day

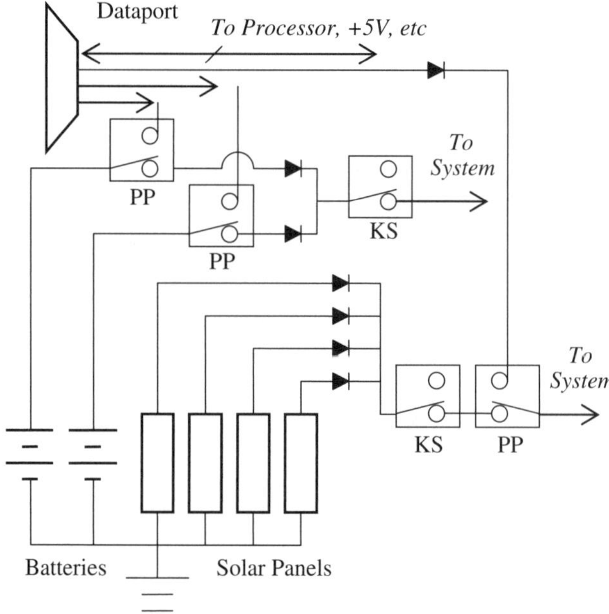

Fig. 12.2 Possible wiring scheme for dataport, kill switches (KS), and pull-pin switches (PP).

functionality as appropriate for inexperienced university satellite teams. Performing lifetime analysis and testing, as well as adding unnecessary fault tolerance, can add tremendously to the complexity of a project that already suffers from limited space, mass, power, time, and expertise. The six-month lifetime requirement limits these lifetime efforts to the simplest, most effective techniques. Battery wear and total ionizing dose (TID) radiation effects are the primary lifetime factors considered in ION's development.

The required operating temperature of the satellite is typically set to a range of about –20°C to +85°C, roughly corresponding to industrial grade components. Military and space grade components offer a wider temperature range, but availability and cost often prohibit their use. Industrial-grade components offer wide availability, and so there is no need to step down to the narrower commercial temperature range. The satellite's thermal design must keep it within this operating range. Satellite thermal analysis is beyond the scope of this work.

The required voltages and corresponding peak loads are determined from the needs of all of the system's elements. For cubesats, the power system can often be designed to support all system loads simultaneously with little extra effort. This approach eliminates the need to understand the time-varying load on the regulators and battery and allows more duty cycle flexibility during testing and flight operations.

Power generation on cubesats is limited by their small surface area. Normally, they have solar cells on all outer surfaces not already occupied by optics or other hardware. Designs can also deploy panels to fixed positions. Cubesats do not normally have solar panels that actively track the sun because of the extra complexity required for hinges, motors, sensors, and attitude stabilization.

Energy storage is required to sustain the satellite through eclipse and through high power consumption bursts from communication transmissions or payloads. The satellite should be designed for shallow discharge cycles because of uncertainty in orbit, generation, and consumption patterns, particularly at the beginning of the project when the battery design is defined. A 15% maximum discharge depth goal is recommended for normal operation. In addition, the satellite should be designed to recover from a dead set of batteries in the event of an operational error or discharge while waiting for launch. For example, ION did not launch for more than a year after being delivered. ION's battery did not discharge, but other battery technologies or a power leak could create problems. In addition to the main battery, there might also be other backup batteries for clocks on the processor or communication system. The system should be able to operate without these backup batteries, which would ideally be removed entirely during development. These batteries are sometimes difficult to replace because of unusual form factors and their location in the integrated satellite. Complex emergency surgery had to be performed on ION just before launch because of a dead clock battery that was required to start up the system.

For the power budget, the satellite should be able to sustain the system without attitude stabilization even if an attitude control system is anticipated because of the risks associated with this system. In addition, the satellite should have a normal mode power margin goal of 50% because of attitude stabilization and power budget analysis uncertainty. Power consumption of components also increases with radiation exposure while power generation from solar cells decreases. In the event that the batteries are drained, the satellite should have a sleep mode with a 90% power margin goal. The power system should automatically trigger this sleep mode and recharge the batteries to a near full state when they fall below a critical charge level of about 35%.

For fault tolerance, the power system should power cycle the system if it freezes up. In addition, radiation protection is required to prevent component failure from cumulative TID radiation effects. Complementary metal-oxide-semiconductor (CMOS) components are typically the first to fail after receiving 100 krad to 10 Mrads of accumulated radiation exposure [11]. A shielding level that allows a maximum of 5 krad of exposure during the satellite's design lifetime is recommended because there are many CMOS components whose failure can bring down the system. The 5-krad recommendation is less than the 10 krad that commercial components can tolerate [8]. The sleep mode provides another fault-tolerance mechanism that allows recovery from drained batteries. The power cycle and sleep mode functions should ideally be implemented entirely in hardware because software can be disrupted by radiation single-event effects (SEEs). Just before exiting sleep mode, the system should be power cycled in case the failure was caused by a SEE or some other anomalous condition that might be repeated.

III. Architecture

Satellite power systems utilize two basic architectures, direct energy transfer (DET) and power point tracking (PPT) [12]. For power sources, spacecraft have four options: static, dynamic, fuel cell, and photovoltaic [13]. See Part I, Section II.B of Chapter 1 for more details on these design alternatives. ION uses a PPT with photovoltaic cells.

IV. Overview

Figure 12.3 provides an overview of a typical cubesat power system. The data port allows an external power source to operate the system by feeding lines directly into the PPT like a solar panel (see Sec. II). All solar panels and batteries feed through the kill switches (see Sec. II), pull pin (see Sec. II), and watchdog timer switches so that power can be cut. Solar panels with PPT circuitry generate power, which recharges the batteries or powers the system directly. From there, unregulated and regulated power is distributed to the system. In the case of ION, the watchdog timer uses relays to cut power to the entire system. The watchdog timer and other fault-tolerance topics are covered in the fault-tolerance section. Power-path control circuitry switches paths on and off to control power consumption and direct power into or out of each battery. Voltage and current sensors throughout the power system provide information for automated software control and manual ground monitoring. The power budget must be carefully characterized through analysis and testing of generation and consumption to ensure that the system does not run short of power. Power budgets, fabrication, and integrated system testing are covered in later sections.

V. Generation (Solar Panels and PPT)

Each of ION's four side panels has of a string of five solar cells in series, and these four panels are tied together using blocking diodes to prevent reverse currents, as shown in Fig. 12.4. In addition, bypass diodes allow current to flow around each cell that is not receiving light to protect the panels.

ION uses a simple hardware-only fixed-voltage PPT circuit. The design is independent of ION's main centralized processor. Cubesats typically have space for only one PPT circuit, and so strings of solar cells from multiple panels are tied together using blocking diodes (Fig. 12.4). Figure 12.5 shows the characteristic I-V curve test data for ION's panels. Lower curves represent less light intensity or shallower sun angles. The current drawn from the panels determines

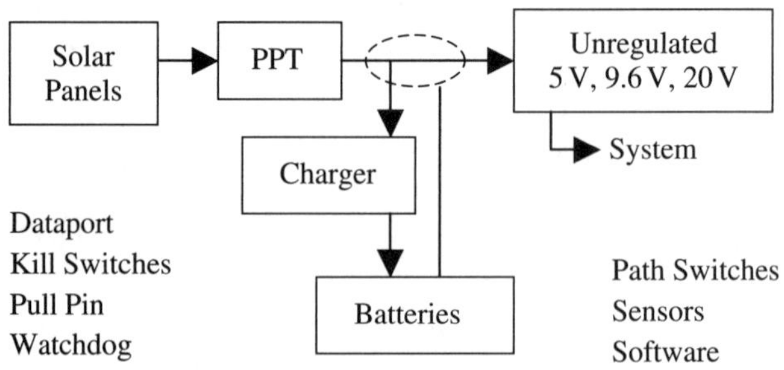

Fig. 12.3 Power system overview.

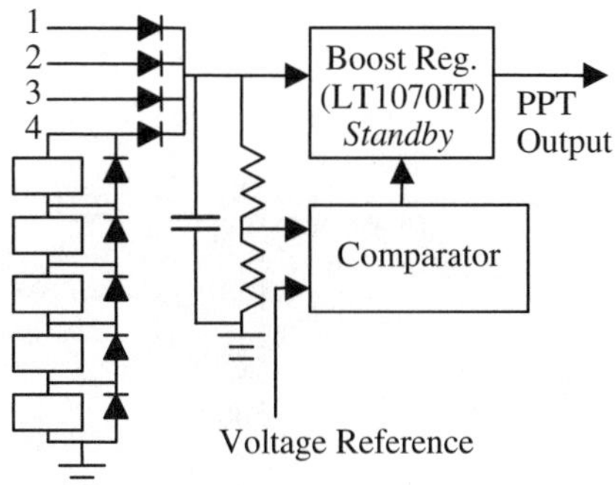

Fig. 12.4 Solar panels and power point tracking.

the operating point on a particular curve. Maximum power is generated near the knee of the I-V curve, and PPT circuitry ensures that the panels operate near this peak power point.

As shown in Fig. 12.4, ION utilizes a simple circuit with a comparator and boost regulator to keep the panels at a fixed operating voltage. Hysteresis on the solar-panel voltage puts the boost converter into and out of standby mode. When the boost converter is operating, the panel voltage drops, whereas in standby mode, the panel voltage rises. The combination of hysteresis and input capacitance allows low-frequency mode switching to maintain a nearly fixed panel voltage. Power generation for these cells follows a sine rule down to shallow angles of about 20 deg, below which reflections can further reduce efficiency. As shown in Fig. 12.6, this simple fixed-voltage PPT design keeps the panels operating at an average of about 95% of peak power down to 20 deg (1700 mW) at room temperature. Solar cells with antireflective coatings would improve performance at shallow angles.

The I-V curves shift left with higher temperature and with radiation exposure, and so the voltage threshold must be set below the first-day room temperature maximum power point. The ideal power point is 11.2 V [(2.3 V/cell × 5 cells) – 0.3 V diode loss]. The solar-cell data sheet provides information about sensitivity to temperature, and experimental data provide information about sensitivity to light intensity. To prevent a rapid dropoff in tracking efficiency at the end of the curve, the voltage is set to 10 V ± 0.25 V. This lower-voltage target allows near-maximum power generation over a wide temperature range. Radiation effects

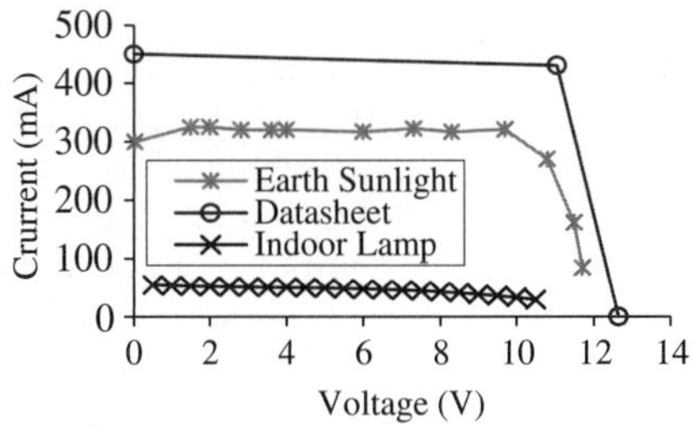

Fig. 12.5 Solar-panel I-V curves with fixed-voltage PPT.

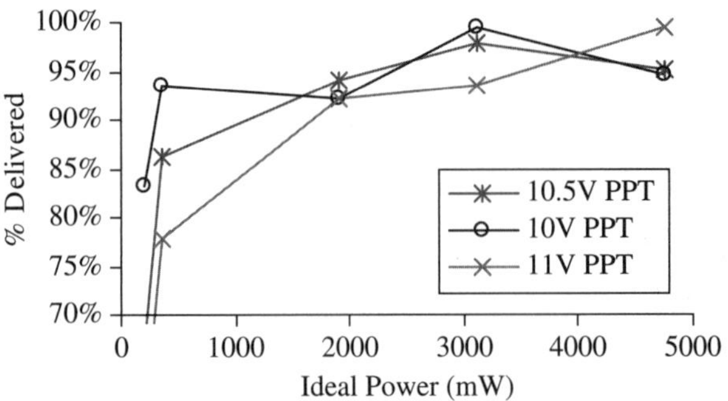

Fig. 12.6 Percent ideal power delivered by fixed-voltage PPT (23°C).

are neglected because ION's lifetime requirement is short compared to the spacecraft for which these cells are designed. For a more sophisticated approach to setting the target voltage, the cells can be characterized using an equivalent circuit model. This level of sophistication was not deemed necessary for ION thanks to the large power margin goal.

ION's hardware-only fixed-voltage PPT design does not require assistance from the processor as with other more complex PPT designs that actively track the peak power point. This simplicity decouples the PPT design from other system elements for faster development and improved robustness. The only drawback is that if one cell is lost in a panel, the remaining cells will be forced to operate at too high a voltage, where little power is generated.

Several other PPT techniques have been proposed over the past few decades [14], most of which require a microcontroller. A fixed-voltage PPT, as with ION, has decreased performance as temperature varies. A slightly better method is fractional open-circuit voltage (V_{oc}), which allows for temperature compensation. However, the fractional-V_{oc} method requires a sample-and-hold circuit. A widely used method is perturb-and-observe (P&O), where the operating point is periodically changed and power generation is monitored. The P&O method requires both a sample-and-hold and a current measurement. Ripple correlation control (RCC) uses a relatively complex analog circuit to continuously adjust the operating point based on power ripple. Again, current measurement is needed. Other methods are possible, all of which require current measurement, complex circuitry, or both. The fixed-voltage method was chosen for its simplicity.

Solar simulators, with the correct intensity and spectrum, can be used for solar-panel testing but are expensive. For ION, a bright car headlamp was used indoors for consistent low-intensity functionality tests. The panels were also tested outdoors in bright sunlight for high-intensity performance. Table 12.3 shows test results for the four completed solar panels with PPT circuit. Each panel is tested individually, oriented normal to the sun, and pairs of panels around the satellite are tested at 45-deg angles each. Testing with real sunlight requires good weather and suffers from uncertainty about solar conditions. Because of atmospheric losses, sunlight intensity is approximately 20% brighter in space for single junction cells [15] and 41% brighter for the advanced triple junction cells [16] used on ION. Because power tracking is approximately 95% at room

Table 12.3 ION Power Generation Tests

Test Setup	Description Bright sunlight, onboard sensors, PPT
One normal panel	2.8 W to 3.4 W (tested) 4.96 W ∗ (0.95/1.41) = ~3.3 W (expected)
Two panels at 45 deg	4.0 W to 4.6 W (tested) 4.96 W ∗ 2 ∗ sin 45 deg ∗ (0.95/1.41) = ~4.7 W (expected)

temperature, expected values from the datasheet are scaled by 0.95/1.41. Results show that the panels are generating approximately the correct amount of power. A large power margin in the power budget (see Sec. IX) accommodates any error in panel assembly and testing.

VI. Storage (Batteries and Charger)

Data on battery options were provided by Jonthan Kimball. Many secondary battery technologies can be used for satellite applications. Two metrics are critical: volumetric energy density and gravimetric energy density. Larger spacecraft generally use NiCd or possibly NiH2 batteries. Cubesats, including ION, often utilize Li-ion batteries [4,6,17,18], which are found in portable devices such as cell phones. These batteries have high gravimetric energy density and high columbic efficiency. They return more than 90% of the milliamp-hours that are put in. Li-ion batteries require a special two-stage charging profile, as shown in the test data in Fig. 12.7. In the first stage, a constant current (CC) is fed into the battery until its voltage reaches a particular value. Then, the battery is kept at a constant voltage (CV) while the current drops until the battery is topped off. Each battery can hold 1950 mAh of charge for a total of 3900 mAh. The satellite consumes about 95 mA in eclipse, 675 mA during communication transmissions, and 132 mA during communications receptions (see Sec. IX for power values and

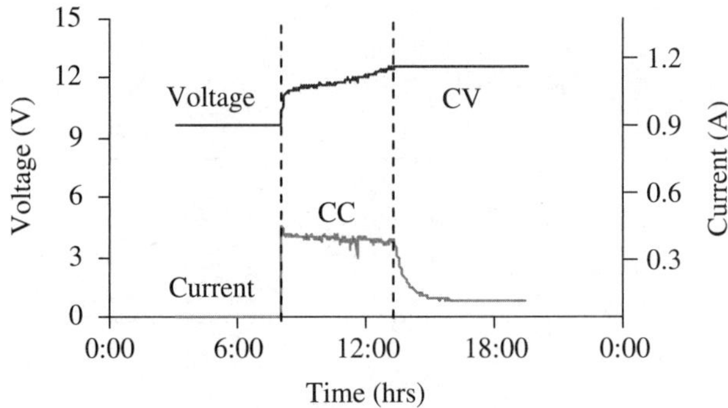

Fig. 12.7 Li-ion battery charging profile.

divide by 12.6 V). A full set of three consecutive communications passes might consist of a maximum of three 35-min eclipses, 20 min of transmission, and 10 min of reception. This results in $(3 \times 35 \times 95 + 20 \times 675 + 10 \times 132)/60 =$ 413 mAh, which is 413/3900 = 11% maximum discharge depth without including any battery recharging. This meets the 15% requirement in Section II. As a whole, the batteries store 3900/95 = 41 h of charge, or nearly two full days of normal operation without any recharging. Li-ion batteries provide 300 to 500 charge cycles with no memory effect and prefer shallow discharges. As a result, the batteries should last (300 to 500) × 41 hrs = 12,300 to 20,500 hrs, or 1.4 to 2.3 yrs. ION's Li-ion batteries also have low leakage, with approximately 90% of charge maintained after sitting on the shelf for one year while waiting for launch. Finally, the batteries were tested in a vacuum to ensure that they do not expand or leak. Li-polymer batteries, such as used on TU-Sat, might also make a good choice for these tiny satellites [4] when adequately tested in a vacuum.

Many software and ground operating decisions require knowledge of the battery charge level. For example, the software needs to know when to switch from one battery to the other, or a ground operator might need a full battery to safely operate certain high-power experiments. However, it can be challenging to obtain a good instantaneous measure of battery level. Coulomb counting circuits can be complex and difficult to properly initialize. A simple battery voltage measurement can provide incorrect information because the voltage drops slowly until the battery is nearly empty (0.75 V change for 12.6-V battery across most of charge range). High charge and discharge currents can perturb the voltage by as much as 1 and 0.3 V, respectively, as a result of resistive effects. Temperature can affect voltage by as much as 0.75 V across a 30°C range. To use the battery voltage as an instantaneous measure of battery charge level, the measured voltage must be adjusted for current load (positive discharge or negative charge) and temperature. A linear model of the resistance, determined from the battery datasheets, eliminates the voltage jumps caused by high currents, as shown in Fig. 12.8. In the first half of the graph, the system is switching between batteries, while in the second half the system is using only the displayed battery. The battery is full at 12.6 V and largely empty at 10 V. Similarly, the voltage is adjusted for temperature using a piecewise linear model, and then it is converted directly to a

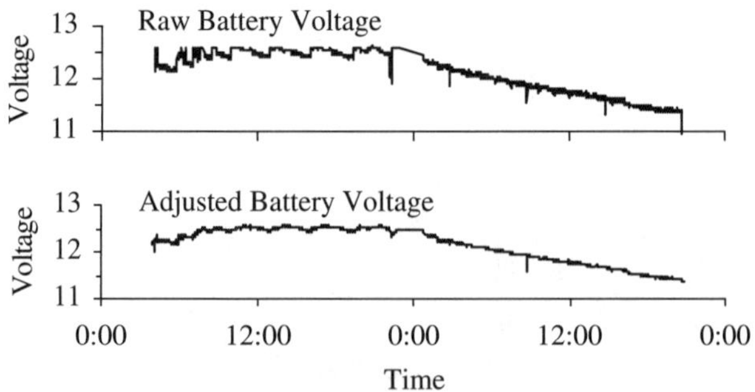

Fig. 12.8 Battery voltage adjustments.

battery charge level using another piecewise linear model. The remaining spikes in the measurement are caused by slight differences in the sampling times and filters for currents and voltages. There are some residual dynamic voltage effects that also affect the results. The described approach provides a good simple instantaneous measure of battery charge level.

VII. Regulation

Battery voltage changes significantly with state-of-charge and current (charging or discharging). Some loads can tolerate significant voltage variation and can run at the 12-V nominal battery voltage, such as ION's torque coils. Other systems, though, need a clean, regulated supply at a specific voltage. Generally, two types of regulators can be used. Linear regulators provide the cleanest output, but efficiency is always the ratio between input and output voltage. For example, a linear regulator with a 12-V input and 5-V output would be 41.7% efficient. This directly impacts the power budget. Switching regulators can be designed with low-ripple output and high efficiency, but are often larger than linear regulators. Switching regulators must be used when the output is greater than the input.

The ION system uses 5-, 9.6-, and 20-V regulators for its loads. The 5-V regulator is a high-efficiency (90%) buck converter that powers the processor and most other electronics. The 9.6-V regulator is a simple linear regulator with a typical efficiency of 85% for the typical 11 to 12.6-V input range offered by the batteries. This output primarily powers the communications system. The linear regulator has a 1-V dropout voltage, so that the output voltage droops whenever the battery voltage is less than 10.6 V. Voltage droop should not be problematic as the batteries are designed to stay full, and the communication system can still operate at lower voltages. The 20-V regulator is a boost converter that is only used by the magnetometer set-reset circuit and draws minimal current.

In addition to basic functionality testing, all regulators were tested for input voltage range support, load-handling capability, and efficiency across loads. Also, all voltage and current sensors were carefully characterized and adjusted for accuracy in software using a linear model.

VIII. Distribution

Once the required regulators and other components have been defined, a scheme must be defined for the distribution and control of system power. For startup issues, the watchdog timer requires its own regulator directly off of the battery. Without its own regulator, the watchdog timer would cut power to itself. The watchdog timer requires a minimal amount of current, and so a simple low-efficiency linear regulator can be used. In addition, the system should be designed to start up using any battery or the solar panels. This way, if one or all batteries are drained, the system can still start up.

Relays or MOSFETs can be used to switch power throughout the system, and each has its own unique properties. Relays must tolerate launch vibrations without inadvertently closing and require higher power than MOSFETs to switch. N-channel MOSFETs require a higher voltage on the gate than on the switched line in order to obtain a low on-resistance. However, processor output control voltages are typically 5 V or less. As a result, these processor outputs must be amplified or level-shifted above the switching line voltage to switch the MOSFET properly.

The distribution system requires sensors to acquire the necessary information. Sensors often require filters to eliminate transients and noise, and a simple RC filter is often sufficient. The sensors also often pass through multiplexers (muxes) so that multiple sensors can be fed into a single analog-to-digital converter port. When multiple sensor lines pass through a single mux and then a single filter, the sampling rate is limited because the filter must settle every time the line is switched. A rule of thumb for the settling time is 10 times the RC filter time constant. Once implemented, the sensors can be characterized, and their accuracy can be adjusted in software.

On ION, a power path controller chip gives the processor a simple interface with sequenced switching to control the flow of power between the batteries, PPT, system, and charger. The chip can also place the power system into a "three-diode mode" that allows any of the three power sources (two batteries, PPT) to start up the system. In the three-diode mode, the batteries do not charge. The software system must load up (a time-consuming process on ION) and select a battery to charge, and so adequate power must be available to run the processor.

Once operating, the software uses one battery while charging the other and swaps the two functions when the battery in use falls below a certain voltage. Logic also exists for dealing with special cases, such as when both batteries are drained. As discussed in Section II, a sleep mode with minimal power consumption is important to allow the batteries to recharge from a low-capacity situation. ION's sleep mode consists of turning off optional systems, such as the payload items and torque coils, while minimizing communication requests. Further power savings would require turning off the processor, which could be incorporated into future designs if further power reduction is needed to reach the 90% sleep mode margin goal.

The design and testing would have been greatly simplified if the two batteries and solar panels were connected through power diodes instead of being dynamically switched (Fig. 12.9). The switching requires extra hardware, extra software, careful timing, and a great deal of integrated system testing. Battery charge level, line voltage, and line currents can be challenging to use as decision inputs because of transient effects and noise. Accounting and testing for all scenarios in the

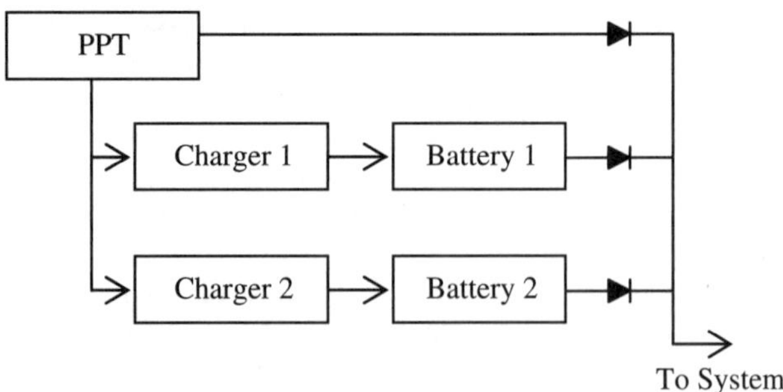

Fig. 12.9 Diode-connected sources.

integrated system can also be challenging, especially as software is developed on a central processor by a separate team. With the diode design, all three sources are backing each other up all of the time, and the one with the highest voltage provides current first. The PPT voltage can be set slightly higher than the batteries, so that solar cells are the primary power source and run the system directly without charger losses. Excess power goes to charging the batteries. Using solar panels directly without battery backup in a switching system is problematic because the solar panels cannot provide adequate current for high-power items. The software must switch over to batteries for high-power items and must continually test for eclipse conditions. ION's design functions in this manner. A diode loss of 0.3 V out of 12.6 V (2.4%) is well worth the simplicity that the diode design offers. As discussed in Section II, techniques exist for eliminating even this small diode loss.

IX. Fault Tolerance

The satellite's design should assume that failure points exist. This section discusses simple fault-tolerance techniques appropriate for university cubesat teams. In general, lifetime-related fault-tolerance distracts the team from first-day functionality and uses valuable space, mass, time, and expertise. Instead, the emphasis is on tolerance against unexpected operating modes, inadequate power generation, and SEEs that disrupt the system. Techniques to consider include watchdog timers, ground-controlled reset switches, resettable fuses, an automatic sleep mode, radiation shielding, tantalum tape, and latch-up protection.

Many failures can result in system freeze-up. The watchdog timer is a simple and effective technique for detecting these conditions and resetting parts or all of the system. ION's watchdog timer was frequently used as a backup plan to deal with anomalous scenarios such as hardware, software, and radiation errors. ION has multiple watchdog timers. The system watchdog timer cuts power to the entire system for two minutes if the processor does not reset it within eleven minutes. Watchdog timers with adjustable timeout and reset times are commercially available. The two-minute off-time allows the large power capacitors in the system to fully drain before restarting the system. In addition, the processor has its own watchdog timer with a shorter timeout period and a pulse reset. The processor can get caught in a loop that resets the processor watchdog, but not the system watchdog. Also, on ION the 5-V regulator automatically shuts down on certain fault conditions, and power must be cycled to reset it. The system watchdog performs this function and generally allows the entire system to be restarted from scratch. In addition, the processor can trigger the watchdog timer directly to reset the system at regular intervals. This prevents less significant failures from continuing to interfere with the system. For additional fault tolerance, resettable fuses can be used to protect individual systems from excessive current draw that could permanently damage them [19]. Resettable fuses automatically resume normal operation when the fault is removed.

As with any complex electronic system, sometimes the only solution is to cut power and reset the whole system with a reset button. To incorporate a reset on the satellite, a direct hardware line is fed from the communication system to the watchdog timer. This line allows a reset switch on the ground to directly reset the watchdog timer independent of TNC (modem) or processor failures. Cubesat programs typically use standard amateur radio equipment that includes audio inputs and outputs. Dual-tone multiple-frequency (DTMF) components used in phone systems can be used to encode and decode the wireless reset signal from the ground directly into the watchdog timer.

As discussed in Section II, the automatic sleep mode allows the satellite to recharge its batteries when they are all nearly drained. This can happen while the satellite is on the ground waiting for launch or while the satellite is in flight because of design errors or operational errors. The sleep-mode functionality should ideally be implemented in hardware as part of the power system. This makes the function independent of the main processor reducing the risk that software complexity or single event effects will interfere with its functionality. The sleep mode can be implemented entirely in hardware with the help of comparators that check the battery voltage level and a timer that establishes a charging time.

Unfortunately, space-grade components resistant to radiation cost too much for the ION project and similar university projects. They also tend be older and have limited availability. For these reasons, ION primarily uses commercial hardware not specifically designed to withstand radiation. Although these components present some extra risk, the benefits outweigh the risks for small satellites in low Earth orbits (LEO). On ION, the SID processor board from Tether Applications is the only part specifically designed for operation in a radiation environment. The SID also consists of commercial parts, but it includes latch-up protection circuitry throughout the board as well as heat dissipation and electromagnetic interference (EMI) control techniques. The SID has been tested for radiation exposure in LEO.

Radiation exposure with shielding is calculated using the online tool, SPENVIS (data available online at http://www.spenvis.oma.be/spenvis/intro.html) with ION's orbit as an example [20]. A minimum amount of radiation shielding is required for the system to operate for its six-month design life. As discussed in Section II, a maximum of 5 krad of cumulative exposure is recommended over the satellite lifetime. Because of strict mass restrictions, ION could only support approximately 1 mm of shielding in the form of one 0.2-mm aluminum plate and four 0.2-mm polyethylene sheets. The polyethylene sheets are expected to provide protection similar to the aluminum at a lower mass. This level of shielding should result in 2.7 krad of radiation exposure for the six-month mission (Fig. 12.10). ION is not expected to have problems reaching the two-week minimum lifetime requirement and should satisfy the longer six-month operational goal. This can be particularly for beneficial for critical software components, like processors, which are also subject to SEEs. Latch-up protection can further extend the life of the components. Latch-up conditions are detected as excess current consumption in a component. If power is cut quickly enough, the component's lifetime can be extended. Additional techniques for addressing SEEs are discussed in [21], which covers software techniques for improving radiation hardness of commercial components.

X. Power Budget

This section provides a power budget calculation template using ION as an example. As discussed in Section II, the satellite must generate more power than it consumes regardless of attitude. Power consumption of each subsystem is carefully characterized and adjusted for supporting regulator efficiency losses and

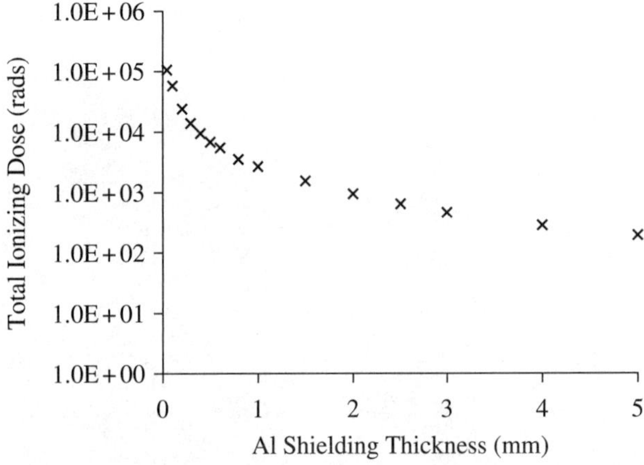

Fig. 12.10 ION's six-month radiation dosage vs shielding level from SPENVIS.

operating duty cycle. As shown in Table 12.4, ION consumes 1120 mW in normal mode. A sleep mode was not incorporated into ION's early design, but a simple ad hoc sleep mode was added later in a way that would not require a hardware design change. In this ad hoc sleep mode, optional systems and ground communication are turned off in software bringing consumption down to 883 mW.

Three power generation values correspond to low, uniform, and high levels of solar incidence (Table 12.5). Uniform power generation is calculated with the assumption that the sun direction is uniformly distributed about the satellite, and

Table 12.4 Power Budget (Consumption)

	Raw power, mW	Source	Efficiency, %	Duty cycle, %	Normal power, mW	Duty cycle, %	Sleep power, mW
Power and router boards	250	Unreg.	100	100	250	100	250
Processor (SID)	383	5 V	90	100	426	100	426
Photometer shutdown circuit	40	5 V	90	100	44	100	44
Magnetometer	180	9.6 V	85	5	11	0	0
Photometer	85	5 V	90	30	28	0	0
TNC and radio receiver	467	9.6 V	85	12.0	66	11.7	64
TNC and radio transmitter	8455	9.6 V	85	2.40	239	1.00	99
Torque coils	1143	Unreg.	100	5	57	0	0
Average consumption					1120		883

Table 12.5 Power Budget (Generation and Margin)

	Generated power (incl. cell efficiency, eclipse, and angle)	PPT board power consumption	PPT tracking effectiveness, 95%	PPT regulator efficiency, 85%	Charger efficiency, 88%	Normal power margin	Sleep power margin
	Calculated/ simulated, mW	minus 150 mW $*$ 66% sunlight	mW	mW	mW	1120 mW cons.	883 mW cons.
Low (w/albedo)	2200 (2860)	2100 (2760)	1995 (2622)	1696 (2229)	1492 (1961)	25% (43%)	41% (55%)
Uniform (w/albedo)	2850 (3705)	2750 (3605)	2612 (3424)	2220 (2910)	1954 (2561)	**42% (56%)**	55% (66%)
High (w/albedo)	3000 (3900)	2900 (3800)	2755 (3610)	2342 (3069)	2061 (2700)	46% (59%)	57% (67%)

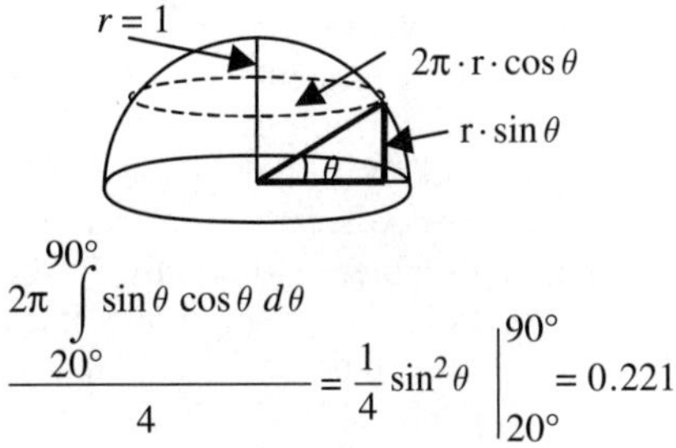

Fig. 12.11 Power generation with uniform angle distribution.

the panels generate power down to 20 deg. In Fig. 12.11, the sin θ term finds the amount of sunlight normal to the panel, whereas the $2\pi \cdot \cos\theta$ term finds the circumference over which this power calculation should be weighted. The calculation is normalized to 1 W of possible generated power ($r = 1$) for simplicity. The normal power is multiplied with its circumference weight, and the angle is integrated from 20 to 90 deg. The result is divided by the surface area of the whole sphere (4π) to find the average power generation after angle losses around the entire sphere (0.221 W). Because ION's four panels each generate 4.96 W when normal to the sun, and the satellite is in sunlight for 65% of its orbit, ION generates an average of 2850 mW (4 panels $\times$ 4.96 W $\times$ 0.221 $\times$ 0.65). In addition to uniform power generation, low (2200 mW) and high (3000 mW) values were found from simulations of some random attitudes in Satellite Tool Kit (STK), orbital simulation software from Analytical Graphics, Inc. (AGI).

In addition to direct sunlight, the Earth's 30% average albedo (sun reflection) can add to the power generated. Whereas the sun provides a point source, the Earth albedo spans a 134-deg field of view that reaches most panels (Fig. 12.12). Because of the difficulty of characterizing Earth albedo [13], separate margins are calculated with and without a simple 30% Earth albedo bonus. ION's normal mode margin is 40% (54% with albedo), which is near the 50% goal recommended in Section II. ION does not have a sleep mode, but its ad hoc sleep mode margin is 51% (63% with albedo). ION should be able to sustain itself with any attitude thanks to a 42% (55% with albedo) sleep mode margin in the worst simulated attitude case. Additional power could be generated by deploying panels out to fixed positions such as in [7]. Panels that actively track the sun are possible, but likely add too much complexity for these simple satellites.

XI. Fabrication

Practical fabrication issues, such as system wiring, staking epoxy, and conformal coating, significantly impact system operation. Wiring can be incredibly

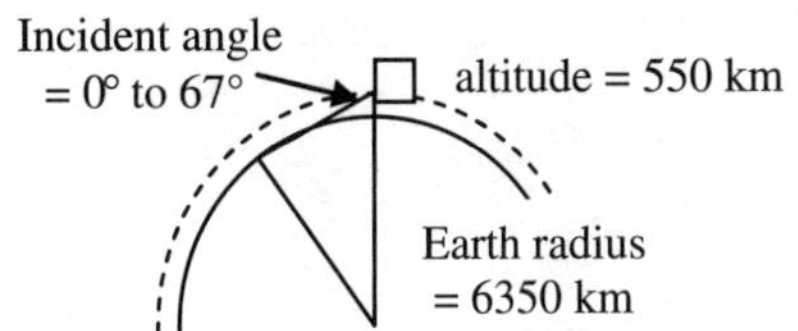

Fig. 12.12 Earth albedo reflection angle.

tedious and error prone despite its conceptual simplicity. On ION, most wiring was routed individually. This allowed the system to be integrated earlier because fitted boards did not need to be created. However, the wiring occupied a lot of space in the middle of the satellite and wiring is a likely failure point during vibration testing and launch. A better design would have involved creating a single board backplane with the other boards fitted into the backplane. Alternatively, the boards could be stacked, but this design is not as flexible because pinouts must change on multiple boards simultaneously. With a backplane design, the backplane can be revised periodically to accommodate all pinout changes. Some wiring is still required to accommodate the switches, payloads, and other hardware that do not plug into the backplane.

Once the satellite has been wired, the large components on the boards must be staked down with epoxy to secure them prior to vibration testing. In addition, the boards can be covered with conformal coating to secure smaller components. Wires should also be staked down before launch. With ION, wires were not staked down until after some vibration testing was completed because the ability to debug the system hardware is lost after this step. In general, staking and conformal coating can peel off for minor fixes, but removing the staking from ION's wiring would have required irreversibly breaking connectors soldered into boards.

XII. Integrated System Testing

Integrated system testing techniques and analysis of onboard sensor data are crucial to cubesat program success. Data analysis can be challenging partly because discrete voltage and current measurements must be used to indirectly measure continuous power flow. One test involves simulating 24 hours of operation. During these tests, the satellite went through some early manual testing, and then it was left to operate on its own normally for many hours. An external power supply simulated the solar panel inputs by turning on and off with approximately a 66% duty cycle and a 1.5-h period. At the end of the test, some additional manual testing was done.

Figure 12.13 shows measured power for various signals during this test. The power values are the product of separate voltage and current measurements. The shark fin pattern indicates that the two batteries alternately accept charge from

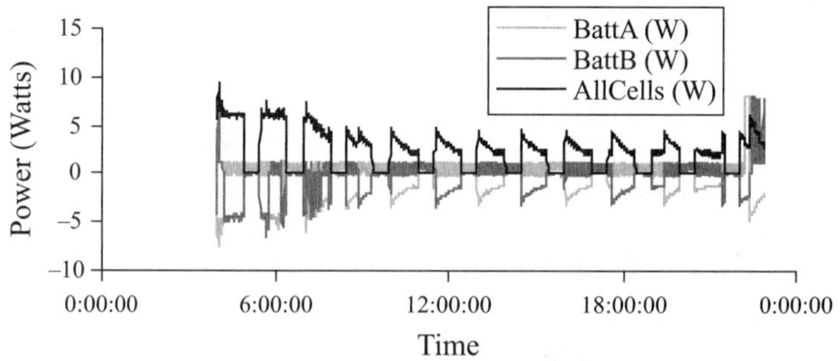

Fig. 12.13 Power over 24 hours.

the solar cells as recharge power. The system is correctly charging the batteries with decreasing power as they fill up with charge. The system also correctly switches between the two batteries. The large spikes at the end are as a result of communications activities.

Table 12.6 calculates average power for a number of items during the time period from 0930 to 1700, when the satellite operated normally. These values are not all explained in detail here because of space restrictions. These measurements can be cross-checked in a variety of ways, such as checking known efficiencies, summing power at a node, or checking power generation against power consumption over long time horizons. The last item in the table (pLost) is a sum of powers into a single node. The sum should be zero, and so the value of pLost provides insight into the accuracy of the other measurements. The lost power should be compared against pSolB, the largest value entering the node. The accuracy of the measurements can be improved by averaging high-frequency samples onboard or filtering outpower spikes.

Another way to check the accuracy of the measurements is to divide power into a regulator by power out of a regulator. This value can be compared against the known regulator efficiency (e.g., Solar Boost Efficiency = pSolB/pAllCells = 90%). The unregulated power consumption provides a direct measure of system power consumption excluding battery charging and power tracking losses (Ave Power Consumption 1 = pUnreg = 0.74 W). The average power distribution of each battery can be found from the average power out of each battery (not shown in the table, pBattAout = 0.186 W, pBattBout = 0.185 W, Sum = 0.371 W). Because the system operates from batteries 33% of the time, the system consumes 1.12 W (Ave Power Consumption 2 = 0.371/33 = 1.12 W). This value is slightly higher than the first power consumption measure because the high-power items operate only on batteries. The system switches over to batteries from solar cells when a high-power item is required. In the long run, the total charging power should also equal the power consumption (pBattAin = 0.719, pBattBin = 0.505, Sum = Ave Charge Power = Ave Consumption 3 = 1.22 W). With this third measure of power consumption, the value is slightly high because there is a net charge entering the batteries during the test period. The total power generated including power

Table 12.6 Average Power over 24 Hours

Item	Power (mW)
pUnreg	741
pBattA	–533
pBattB	–320
pCellA	1830
pCellB	14
pCellC	26
pCellD	26
pAllCells	1896
pSolB	1697
pLost	103

tracking losses and battery charging is equal to (Ave Power Generation Required = pAllCells = 1.90 W). This is the average power generation target for the solar cells. The power margins can be calculated as shown here and are roughly consistent with the predicted values:

$$\text{Ave Power Cons 1/Ave Power Gen} = 0.74/1.90 = 39\%$$

$$\text{Ave Power Cons 2/Ave Power Gen} = 1.12/1.90 = 59\%$$

$$\text{Ave Power Cons 3/Ave Power Gen} = 1.22/1.90 = 64\%$$

XIII. Conclusion

This work presented useful strategies, design options, techniques, and recommendations for cubesat power systems based on development experience from the Illinois Observing Nanosatellite. The PPT technique provides a simple, effective method of operating solar panels. Similarly, the power budget calculation procedure provides an effective template that other teams can replicate. With a complete collection of such techniques spanning all power system elements, this work should help to contribute to the success of these spacecraft and grow the opportunities that they offer.

References

[1] Nason, I., Puig-Suari, J., and Twiggs, R., "Development of a Family of Picosatellite Deployers Based on the Cubesat Standard," *IEEE Aerospace Conference Proceedings*, Vol. 1, IEEE, Piscataway, NJ, 2002, pp. 1-457–1-464.

[2] Lan, W., Brown, J., Toorian, A., Coelho, R., Brooks, L., Suari, J. P., and Twiggs, R., "CubeSat Development in Education and into Industry," *AIAA Space*, AIAA Paper 2006-7296, Sept. 2006.

[3] Thakker, P., Swenson, G., and Waldrop, L., "ION-1 and -2 Cubesat Optical Remote Sensing Instruments," *Emergence of Pico- and Nanosatellites for Atmospheric Research and Technology Testing*, edited by P. Thakker and W. A. Shiroma, Progress in Astronautics and Aeronautics, AIAA, Reston, VA, 2010.

[4] Holmes, W. C., Bryson, J., Gerig, B., Oehrig, J., Rodriguez, J., Schea, J., Schutt, N., Voss, D., Voss, J., Whittington, D., Bennett, A., Fennig, C., Brandle, S., Dailey, J., and Voss, H. D., "TU Sat 1—A Novel Communications and Scientific Satellite," *16th Annual AIAA/USU Conference on Small Satellites*, Paper SSC02-I-1, Aug. 2002.

[5] Gregorio, A., Bernardi, T., Carrato, S., Kostadinov, I., Messerotti, M., and Stalio, R., "AtmoCube: Observation of the Earth Atmosphere from the Space to Study 'Space Weather' Effects," *Proceedings of Recent Advances in Space Technologies*, Aeronautics and Space Technologies Inst., Istanbul, 2003, pp. 188–193.

[6] Obland, M., Hunyadi, G., Jepsen, S., Larsen, B., Klumpar, D. M., Kankelborg, C., and Hiscock, W. A., "The Montana State University NASA Space Grant Explorer-1 Science Reflight Commemorative Mission," *15th Annual AIAA/USU Conference on Small Satellites*, Paper SSC01-III-2, Aug. 2001.

[7] Flagg, S., Bleier, T., Dunson, C., Doering, J., DeMartini, L., Clarke, P., Franklin, L., Seelback, J., Flagg, J., Klenk, M., Safradin, V., Cutler, J., Lorenz, A., and Tapio, E., "Using Nanosats as a Proof of Concept for Space Science Missions: QuakeSat as an

Operational Example," *18th Annual AIAA/USU Conference on Small Satellites*, Paper SSC04-IX-4, 2004.

[8] Simburger, E. J., Liu, S., Halpine, J., Hinkley, D., Srour, J. R., Rumsey, D., and Yoo, H., "Pico Satellite Solar Cell Testbed (PSSC Testbed)," *Conference Record of the 2006 IEEE 4th World Conference on Photovoltaic Energy Conversion*, IEEE, Piscataway, NJ, 2006, pp. 1961–1963.

[9] Rysanek, F., and Hartmann, J. W., "MicroVacuum Arc Thruster Design for a CubeSat Class Satellite," *16th Annual AIAA/USU Conference on Small Satellites*, Paper SSC02-I-2, Aug. 2002.

[10] Thakker, P., Ames, D., Arber, L., Dabrowski, M., Dufrene, A., Pukniel, A., Rein, A., Coverstone, V., and Swenson, G., "Case Study: Overview of ION as Applied to Atmospheric Research and Technology Testing Problems," *Emergence of Pico- and Nanosatellites for Atmospheric Research and Technology Testing*, edited by P. Thakker and W. A. Shiroma, Progress in Astronautics and Aeronautics, AIAA, Reston, VA, 2010.

[11] Barnes, C., and Selva, L., "Radiation Effects in MMIC Devices," *JPL Publication*, 96-25, Jet Propulsion Lab., Pasadena, CA, 1996, Chap. 10, pp. 203–243.

[12] Patel, M. R., *Spacecraft Power Systems*, CRC Press, Boca Raton, FL, 2005.

[13] Larson, W. J., and Wertz, J. R., *Space Mission Analysis and Design*, 3rd ed., Microcosm Press, El Segundo, CA, 1999.

[14] Esram, T., and Chapman, P. L., "Comparison of Photovoltaic Array Maximum Power Point Tracking Techniques," *IEEE Transactions on Energy Conversion* (to be published).

[15] Flood, D. J., "Advanced Space Photovoltaic Technology: Applications To Telecommunication Systems," *19th International Telecommunications Energy Conference*, IEEE, Piscataway, NJ, 1997, pp. 647–652.

[16] Morales, R. A. L., "Performance of Advanced Triple Junction Solar Cells," *Escuela de Ingenieria, Ingeniero Civil Electricista*, Santiago de Chile: Pontificia Univ. Catolica de Chile, 2005.

[17] Obland, M., Klumpar, D. M., Kirn, S., Hunyadi, G., Jepsen, S., and Larsen, B., "Power Subsystem Design for the Montana EaRth Orbiting Pico-Explorer (MEROPE) CubeSat-Class Satellite," *IEEE Aerospace Conference Proceedings*, Vol. 1, IEEE, Piscataway, NJ, 2002, pp. 1-465–1-472.

[18] Young-Keun, C., Byoung-Young, M., Ki-Lyong, H., Soo-Jung, K., and Suk-Jin, K., "Development of the HAUSAT-2 Nanosatellite For Low-Cost Technology Demonstration," *Recent Advances in Space Technologies, 2005. RAST 2005*, Aeronautics and Space Technologies Inst., Istanbul, 2005, pp. 173–179.

[19] Voss, D. L., Kirchoff, A., Hagerman, D. P., Zapf, J. J., Hibbs, J., Dailey, J., White, A., Voss, H. D., Maple, M., and Kamalabadi, F., "TEST—A Modular Scientific Nanosatellite," *Space 2004 Conference and Exhibit*, AIAA Paper 2004-6121, Sept. 2004.

[20] Athanasopoulou, E., Thakker, P., and Sanders, W. H., "Evaluating the Dependability of a LEO Satellite Network for Scientific Applications," *Second International Conference on the Quantitative Evaluation of Systems (QEST'05)*, 2005, pp. 95–104.

[21] Mehlitz P. C., and Penix, J., "Expecting the Unexpected—Radiation Hardened Software," *Infotech@Aerospace*, AIAA Paper 2005-7088, Sept. 2005.

Chapter 13

Picosatellite Power System Design

E. J. Simburger*
The Aerospace Corporation, El Segundo, California

I. Introduction

THIS report covers the various steps in the design process for selecting the specific power system architecture for a picosatellite mission. These steps include 1) the determination of the optimal solution for a specific mission, 2) determination of optimal bus voltage for the main power system bus, 3) selection of circuit configurations and control schemes, 4) determination of the amount of energy storage required, and 5) determination of energy balance or consumption for the mission.

II. Architecture Trade Study

There are two possible power system architectures that can be implemented in the design of a picosatellite. They are 1) energy storage (ES) where a primary battery provided all of the power required for the picosatellite mission and 2) solar generation with energy storage (SGES). The specific architecture that the designer selects will be determined by performing a trade study. This trade study should consider mission duration, cost of implementation, complexity, weight, volume, schedule, reliability, spacecraft real estate, space qualification of components, and power consumption requirements for the mission.

The first picosatellite missions used the ES solution. However, the short duration of the mission possible with an ES solution prompted the development of SGES solutions. This solution, however, increases complexity, cost, schedule, weight, volume, spacecraft real estate, and reliability.

---

*Senior Engineer.

To illustrate the trade study process, we shall define a picosatellite mission and perform a simple trade study to determine which of the two primary power system architectures would be optimal. The requirements are as follows:

1. Mission duration: two weeks minimum with a goal of one month
2. Cost budget: $2000.00
3. Schedule: one year from concept to delivery
4. Volume of picosatellite: 4 in. × 4 in. × 5 in.
5. Mass: less than 1 kg

There are several methods of performing a trade study to determine the optimal solution for the power system for a particular mission. The method that will be used for this example is the 1, 3, 9 weighting factor method. Weighting factors are assigned as follows: a weighting factor of 1 is assigned to a trade category that has little impact on the mission, a weighting factor of 3 is assigned to a trade category that has some impact on the mission, and a weighting factor of 9 is assigned to a trade category that addresses one or more of the mission requirements. Tables 13.1–13.3 comprise a completed trade study analysis that results in the best power system architecture being the ES solution.

III. Energy Storage Solution

The energy storage solution for providing electric power for a picosatellite is the simplest and has a relatively low cost for implementation. The downside to this solution is a relatively short mission duration. That said, let us look at the design and performance of a power system that uses only energy storage for powering a picosatellite.

Table 13.1 Weight Factor Assignment

	Weighting factor	Comments
Mission duration	9	Minimum mission duration is a top-level requirement.
Complexity	9	One year for design & construction will not allow for complex system design.
Cost	9	Cost of $2000.00 is a top-level requirement.
Weight	9	Mass is a top-level requirement.
Volume	9	Volume is a top-level requirement.
Space qualification of components	3	Primary lithium thionyl chloride batteries have flown on picosatellites ejected from shuttle.
Spacecraft real estate	3	External real estate used for power system not available for other mission critical components.
Reliability	3	Short mission duration.
Power consumption requirements for mission	9	Power system must meet minimum mission power requirements.

Table 13.2 Assessment Factor Scoring

Assessment factor	ES assessment	Comments	SGES assessment	Comments
Mission duration	7	The ES solution will meet the minimum mission duration requirement but will not meet the goal.	10	The GES solution will meet both the minimum duration requirement and the goal.
Complexity	9	The ES solution requires a minimum of power processing electronics.	3	The SGES will require more complex power processing electronics when compared to the ES solution.
Cost	10	Cost of primary batteries and associated power processing electronics estimated to be less than target of $2000.00.	3	The SGES solution cannot be implemented within the specified budget of $2000.00.
Weight	10	Power system mass for ES solution estimated to be less than 150 grams.	10	Power system mass for SGES solution estimated to be less than 200 grams.
Volume	10	Power system volume estimated to be less than 3.5 cubic inches out of 80 cubic inches available.	10	Power system volume estimated to be less than 5 cubic inches out of 80 cubic inches available.
Space qualification of components	10	Primary lithium thionyl chloride batteries have flown on pico satellites ejected from shuttle.	10	Li-ion batteries multijunction solar cells have flown on the space shuttle.
Spacecraft real estate	10	No external real estate required.	5	External spacecraft real estate required for mounting solar cells.
Reliability	10	Primary batteries are highly reliable.	10	The combination of solar cells and rechargable batteries provides two separate sources of power.
Power consumption requirements for mission	5	Meets minimum requirement with little margin.	10	Meets all requirements with margin.

Table 13.3 Picosatellite Power System Architecture Trade Study Results

Power architecture	Mission duration	Complexity	Cost	Weight	Volume	Space qualification components	Space-craft real estate	Reliability	Power consumption requirements for mission	Total
Weight factor	9	9	9	9	9	3	3	3	9	–
ES assessment	7	9	10	10	10	10	10	10	5	–
ES score	63	81	90	90	90	30	30	30	45	549
SGES assessment	10	3	3	10	10	10	5	10	10	–
SGES score	90	27	27	90	90	30	15	30	90	489

The first step in designing an energy storage power system is the selection of the battery. The specifics of the remainder of the power system are the selection of the electronics that will act as the interface between the battery and the loads.

The following items are to be taken into consideration in selecting a specific battery for powering a picosatellite:

- Discharge voltage characteristic
- Rated capacity at specified discharge rate
- Maximum rated output current
- Short-circuit current
- Short-circuit protection
- Storage loss (nominal % per year)
- Volumetric energy density
- Spaceflight history
- Operational temperature range
- Cell operating pressure
- Cell pressure vessel burst pressure (test)
- Cell pressure vessel material
- Cell vent

Once the specific battery is selected, it must be qualified for spaceflight by performing a qualification test. The qualification test must verify that the batteries can function correctly and withstand the stresses that it might encounter during its life cycle, including end-of-life performance. Qualification testing should be performed in accordance with TR-2004(8583)-1 [1], which replaces Mil-Std-1540.

In addition to the thermal and mechanical environment testing just specified, the battery qualification testing should include the following tests as a minimum:

- Cell capacity test
- Discharge voltage characteristic test
- Maximum discharge current test
- Short circuit test
- Seal leakage test
- Survival temperature test

The balance of the power system for the energy storage solution would consist of a fuse on the battery's negative leg to protect the battery and spacecraft wiring from short circuits and a blocking diode to ensure that current flow is from the battery to the loads. The individual loads would be connected to a distribution bus as shown in Fig. 13.1.

The bus voltage for the energy storage solution is the battery voltage. The battery voltage should be selected to be within the input voltage operating range of the single largest load on the spacecraft bus. For most picosatellites, this load is the radio used for communications to the ground.

The remaining individual loads would be required to operate on the bus voltage supplied by the battery. To accomplish this, the individual spacecraft loads would include power processing electronics that would convert battery voltage to that

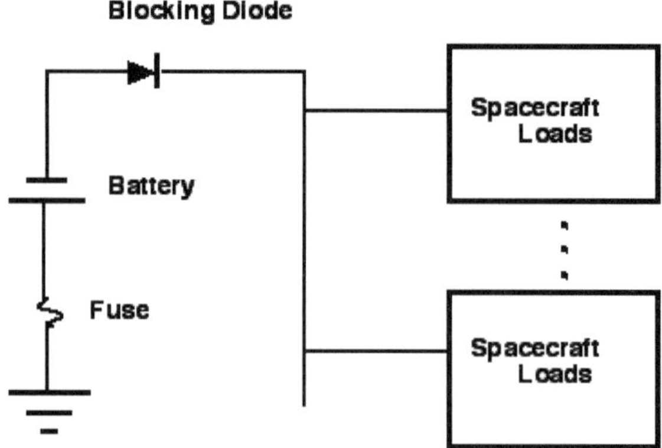

Fig. 13.1 Schematic diagram of picosatellite power system using energy storage solution.

required by the specific load. For example, if the battery voltage were selected to be 30 VDC, and a particular load required a tightly regulated 5 VDC for operation of its electronics, then the load element would be required to provide a 30-V/5V dc-dc converter.

IV. Solar Generation with Energy Storage

If the mission requirements for a picosatellite require a mission lifetime more than a week or two, then the solar generation with energy storage (SGES) solution would be required. There are three flavors to the SGES solution, namely, the unregulated bus, regulated bus, and ring bus.

A. Unregulated Bus

The unregulated bus solution is the simplest to implement from a circuit complexity point of view. The unregulated bus scheme simply adds a string of solar cells to the circuit shown in Fig. 13.1. The resultant power system circuit is shown in Fig. 13.2.

The solar-cell string open-circuit voltage is matched to the battery open-circuit voltage as close as possible. Thus, the bus voltage is regulated by the battery's state of charge. As the battery discharges, the battery voltage drops. This moves the voltage on the solar cells from the open-circuit point toward the maximum power point. Ideally, the battery minimum voltage during discharge would match the peak power operating point for the solar cells. However, this happy set of circumstances is extremely hard to obtain with real-world batteries and solar cells.

The bus voltage selection and power processing requirements for the various loads would be same as the ES solution.

Given the limitations of the unregulated bus solution, if all mission requirements can be met with this type of power system, then it would be optimal for long-duration missions because of its relatively low cost and simplicity.

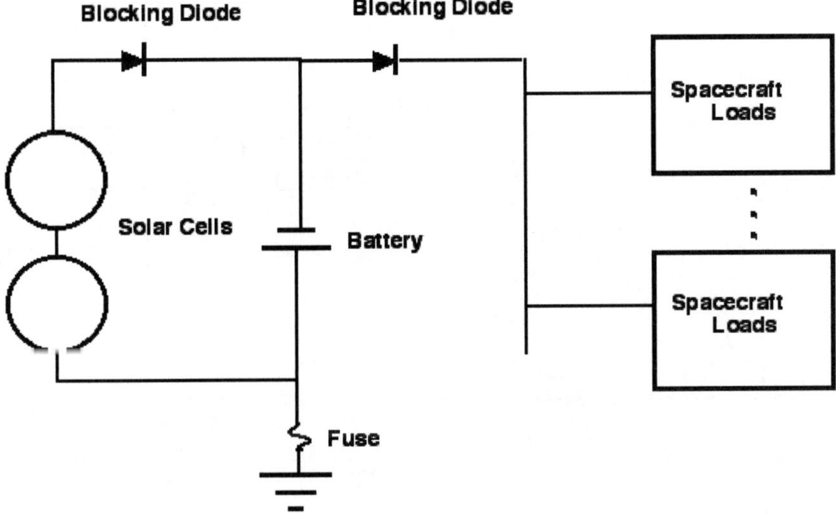

Fig. 13.2 Schematic diagram for the SGES with an unregulated bus.

B. Regulated Bus

The regulated bus configuration will improve overall power system performance with an increase in system complexity. The unregulated bus configuration does not allow for maximizing the output of the solar cells because the battery state of charge defines the operating point for the solar cells. In addition, the mismatch between the battery voltage when fully charged and the open-circuit voltage of the solar cells will either result in overcharging or undercharging the battery. This condition will, over time, result in a degradation of the battery, thus reducing mission life.

The regulated bus configuration addresses these shortcomings in the unregulated bus at the expense of additional power processing electronics. Figure13.3 provides one of many configurations for implementing a regulated bus.

The regulated bus shown in Fig. 13.3 has three components added to the power processing electronics: the solar-array partial shunt electronics, the battery charger and discharge boost regulator, and the power system controller.

The power system controller maintains bus voltage by providing control signals to the solar-array shunt electronics and the battery charger and discharge boost regulator.

The solar-array string is split into two parts. The lower part of the string is designed to provide an open-circuit voltage that is slightly less than the desired bus voltage. The upper part of the string provides enough cells connected in series to deliver the desired bus voltage at end of life at the maximum power point for the solar cells. Because the maximum power point voltage will decrease with time on orbit as a result of radiation damage, the solar array at beginning of life will not be operating at its peak power point for 99% of the mission.

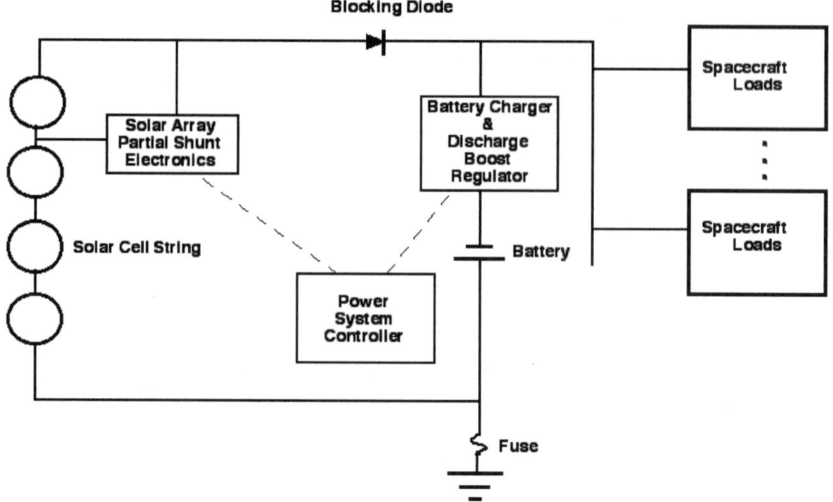

Fig. 13.3 Schematic of a picosatellite power system using a regulated bus configuration.

The solar-array partial shunt electronics control the output current of the solar array by shunting some of the current from the upper section of the solar array such that the current produced by the lower section of the solar array matches the current demand for charging the battery and supplying the loads.

The battery voltage is generally set to be slightly less than the desired bus voltage in a regulated bus configuration. This allows the battery to be isolated from the main power bus, which allows for developing battery charge regimes, which can be optimized for the specific battery chemistry. This will maximize the battery lifetime on orbit. This isolation also requires that a battery discharge regulator be inserted between the battery and the power system bus to increase the battery voltage to the desired bus voltage.

C. Ring Bus

The ring bus was developed to resolve the problem of having various segments of a solar array mounted on different faces of a picosatellite. The problem is that one cannot connect *in series* solar cells that are mounted on different faces of the picosatellite. One can connect a string on different faces *in parallel*, however, providing that the individual strings have blocking diodes to prevent reverse biasing the solar cells when they are in shadow. However, the operating point for the individual string will vary with the different amount of solar energy input as a result of the different angle that they have relative to the sun. The ring bus was invented to resolve this issue.

This architecture was originally developed for the PowerSphere [1–3]. The configuration of the major power system elements for the ring bus is shown in Fig. 13.4. In this configuration, each of the individual power modules senses the voltage on the “ring bus.” For this example, the bus voltage selected is 10 VDC.

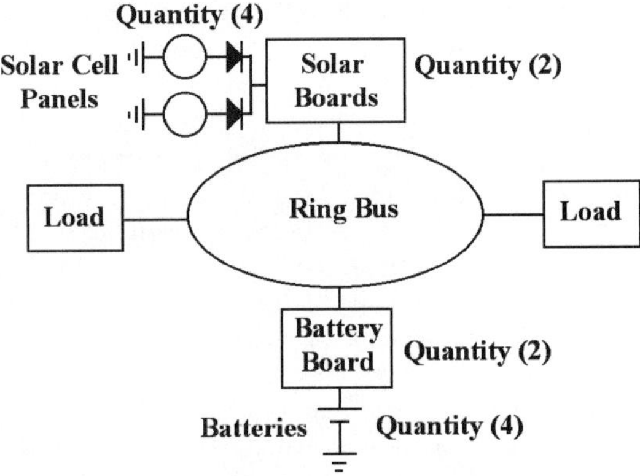

Fig. 13.4 Schematic diagram of ring bus architecture.

Each of the solar-array regulators shown in Fig. 13.5 has a microprocessor and a dc-dc converter. The solar array in this example has solar cells on four of the six faces of a picosatellite. Two of these arrays on opposing faces of the picosatellite are connected to a single solar-array regulation unit through blocking diodes. For operation in the sunlight portion of the orbit, the first level of the control loop is the pulse-width-modulated (PWM) dc-dc boost converter, which provides a regulated bus with voltages between 9.5 and 10.5 V. If left alone, this PWM dc-dc converter would increase the current demand on the controlled solar array beyond the peak power point. If this happens, the power output of the converter would collapse to zero. To prevent this from happening, a microprocessor monitors the bus voltage and output current and implements a peak-power-tracking algorithm. Thus, if an increase in current demand from the PWM dc-dc converter results in a decrease in power output, then the microprocessor commands a lower current demand by the PWM dc-dc converter. The microprocessor also monitors the solar-array voltage and turns the PWM dc-dc converter off if the solar-array voltage drops below 3.0 V and turns it back on when the voltage exceeds 3.2 V. The battery subsystems provide power to the "ring bus" when the bus voltage

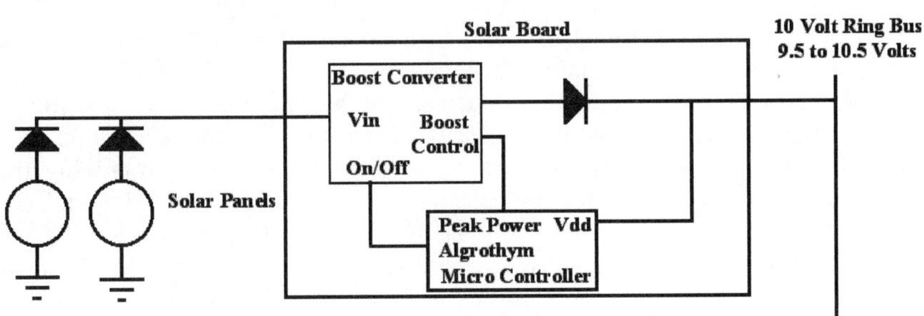

Fig. 13.5 Schematic of PSSC testbed solar-array regulator board.

drops below 10.0 V. When the bus is supported by the batteries, the battery microprocessors turns off all of the battery chargers and goes to a low-power sleep state. The basic building block for the battery subsystem is a battery control element that controls the operation of two individual battery cells and their associated chargers and boost regulators. During the sunlight portions of the orbit, the battery building-block microprocessor turns off one of the battery boost regulators and turns on the associated battery charger. As the bus voltage increases, the microprocessor allows the battery charge current to increase slowly to a maximum charge current for the battery cell. The battery with the lower voltage is charged to maintain a balance between the cells; a lower cutoff of 3.0 V is in place in case of a battery failure. If the bus voltage decreases as a result of an increase in load or decrease in solar illumination, the battery charger immediately decreases the charge current to the maximum amount available after supplying all other loads. If the bus voltage drops below 10.1 V, then the charger turns off, and the boost regulator turns back on.

V. Energy Balance Calculations

Before developing an energy balance model for a specific picosatellite mission, one must define the flight characteristics of the picosatellite and the specifications for the major building blocks that will make up the power system. The bus configuration must also be defined with operating efficiencies of each major component. Thus, to provide the reader with an understanding of what must be included in an energy balance model, a hypothetical spacecraft, mission, and power system will be examined.

The picosatellite power system will have SGES architecture with the ring bus configuration. The picosatellite will have four 1.5 A-h Li-ion rechargeable batteries (Li-ion 18650) and four body-mounted solar arrays (one on each side of the 10 by 5 in. picosatellite body). When fully charged, the battery pack provides 18 W-h at 3.6 V per cell under manufacturer-specified operating temperature. Each of the picosatellite solar arrays consists of four advance triple-junction solar cells, with two cells in a series string and two strings in parallel. The solar array at full sun AM0 and 28°C provides 5.1 V Voc, 0.86 A Isc, and 3.7 W Pmax.

The specific orbit parameters assumed for this example are as follows: low Earth orbit and picosatellite orientation around orbit with solar-cell nadir pointing at all times. The energy balance model consists of two separate elements. The first element is The Aerospace Corporation's Satellite Orbit Analysis Program (SOAP), which calculates the sun angle for the picosatellite in 15-s intervals. The second part of the model is an Excel spreadsheet that uses the sun-angle data from SOAP and the mission power requirement data as inputs. The energy balance model assumes that the spacecraft spins about an axis perpendicular to the sun at 0.25 rpm and calculates the energy available to charge the battery after supplying the 1.08-W mission load and taking into account the losses in each of the dc-dc converters that are utilized in the power system. Figure 13.6 shows plots of power from the solar array into and out of the batteries for one orbit. Figure 13.7 shows the battery state-of-charge for one orbit.

The results of this example show that the spacecraft power system can support a constant load of not more than 1.08 W.

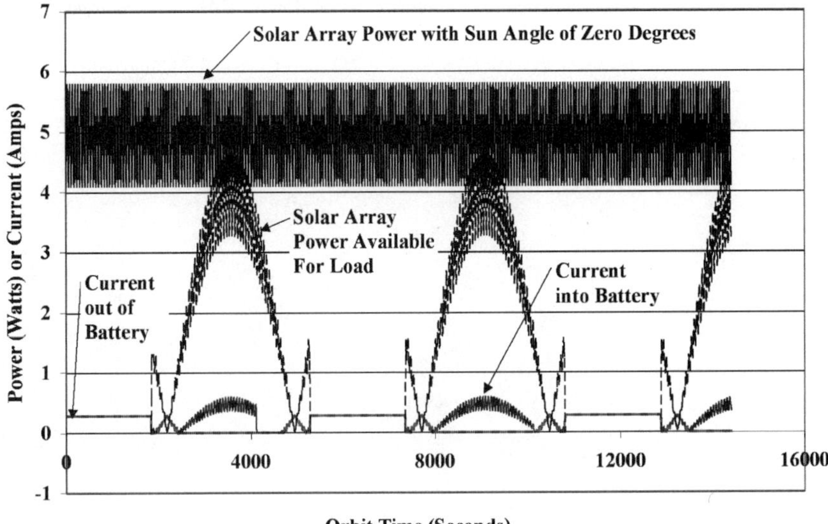

Fig. 13.6 Energy balance model.

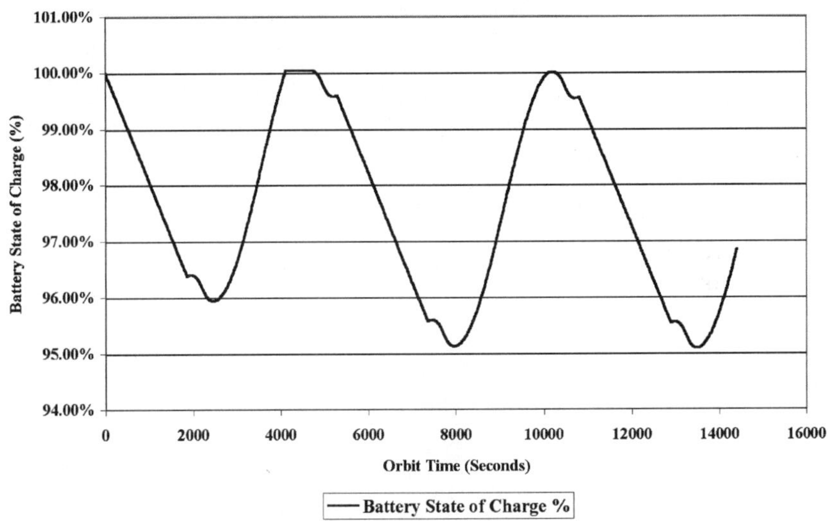

Fig. 13.7 Battery state of charge as function of orbit time.

References

[1] Simburger, E. J., "PowerSphere Concept," *1999 Government Microcircuit Applications Conference Digest of Papers*, Vol. 24, Office of Naval Research, Arlington, VA, March 1999, pp. 426–429.

[2] Prater, A., Simburger, E. J., Smith, D., Carian, P. J., and Matsumoto, J., "Power Management and Distribution Concept for Microsatellites and Nanosatellites," *34th*

Intersociety Energy Conversion Engineering Conference, SAE International, Aug. 1999.

[3] Simburger, E. J., Rumsey, D., Hinkley, D., Liu, S., and Carian, P., "Distributed Power System for Microsatellites," *31st IEEE Photovoltaic Specialists Conference*, Vol. 31, The Aerospace Corp., Jan. 2005, pp. 822–825.

V. Command and Data Handling and Telecommunication Systems

Chapter 14

Evolution of the ION Cubesat Software Architecture

Leon Arber*

University of Illinois at Urbana-Champaign, Urbana, Illinois

I. Background

AS the other sections of this chapter explain, ION's development environment was one of intellectual curiosity and experimentation. The relative lack of involvement from faculty and the freedom provided to ION's developers in the day-to-day technical management and implementation of ION meant that the solutions implemented tended to be simplistic and amateur. Rather than researching how similar problems were solved in other systems, ION's developers chose to go with what they thought would work best and fit within the fiscal, technical, and temporal constraints of the project. Nowhere is this design philosophy more evident than in ION's software architecture.

Rather than being driven by a set of functional requirements, ION's software developers were driven by the capabilities of the payload onboard the satellite. Because of the constantly evolving design and makeup of the students working on the project, it was impossible to enforce a design freeze or even a design specification. As a result, ION's architecture had to be resilient to changes in both hardware and software specifications. Changes were made to wiring configurations, payload sampling frequencies, and power management schemes during the integration phase. If ION's software had to be extensively rewritten each time something onboard changed, the satellite would never have made it past the integration phase. However, because of its modular design and generic scheduler, ION could relatively easily handle major changes to the system's design or even allow these requirements to be worked out after launch and subsequently uploaded to the satellite.

*Graduate Research Assistant, Department of Computer Science (M.S. awarded 2007).

Although ION never made it to orbit, the lessons learned during the implementation of its software proved invaluable to the design of ION2's software architecture. ION2 kept and expanded upon ION's highly modular design and its generic scheduling support, while fixing many of the problems inherent in its monolithic design. The remaining sections will present a detailed overview of how ION's software operates followed by a sketch of ION2's architecture and how it leverages ION's strengths and remedies its weaknesses.

II. ION's Software Architecture, A Bird's Eye View

To provide the needed modularity and generality, ION's software architecture can be decomposed into three "rings" shown in Fig. 14.1. The outermost ring consists of the various device drivers for all of the hardware onboard ION. The middle ring contains of the applications controlling ION's payloads, and at the core of the entire system is the application manager, which is effectively the system scheduler.

The device drivers are the most basic building blocks of ION's software and are responsible for controlling everything from the temperature sensors, to the photometer, the torque coils, and the camera. If there is a physical piece of hardware onboard ION that needs to be controlled or sampled, then a device driver was written for it. The device drivers are also the most cryptic and primitive pieces of code onboard ION because they are design to abstract away the physical interface with a device. They make heavy use of the memory mapped I/O provided by

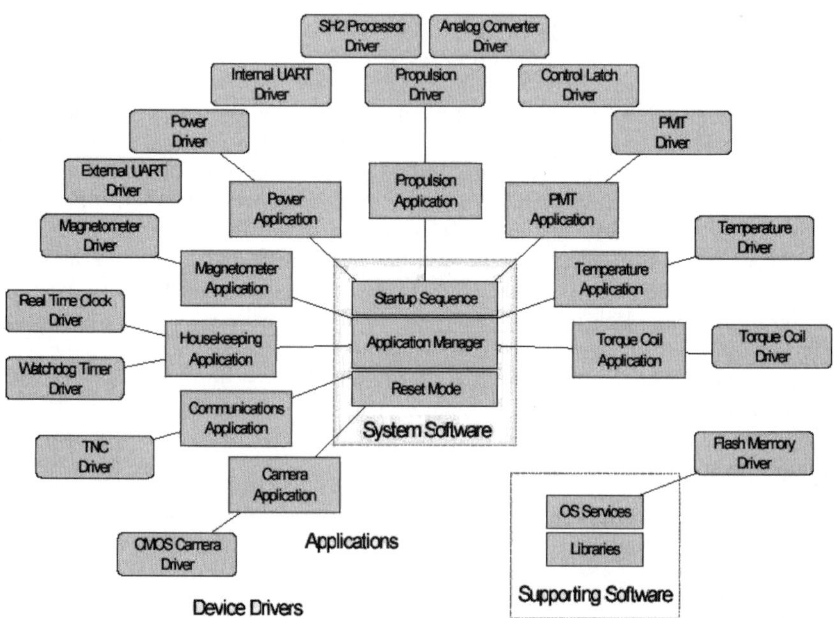

Fig. 14.1 ION's software architecture.

Hitachi SH-2 processor in order to route signals to devices both onboard and off the SID. This results in numerous lines of code that are impossible to understand without consulting the processor manual. The device drivers provide the first line of defense against hardware changes. Most simple changes, such as changes in pin-outs or wiring, can be handled at the device driver level, without needing to touch the rest of the system. Furthermore, if a device has a large amount of functionality, device drivers provide the ability to implement this functionality on an as-needed basis, freeing up the developers to only work on critical features initially, but providing the ability to easily add other functionality later if the time or necessity arises. Lastly, using device drivers provided for some much needed abstraction from the hardware. It was hoped that this abstraction would allow relatively inexperienced developers to contribute to ION's software. Unfortunately, it was quickly realized that any development of ION required an intimate knowledge of all of its constituent parts, and the learning curve proved too formidable for most students. ION2's distributed architecture makes a better attempt at solving this problem and will be discussed later.

The middle ring of components consists of the applications. Whereas there exists one device driver for each hardware component onboard ION, there exists one application for each payload or subsystem onboard ION. Because it often takes more than one device driver to control a particular subsystem, applications further abstract away the hardware by agglomerating the functionality of the various device drivers responsible for controlling a particular payload into a single interface, which is consistent across all of the applications. In addition to merely abstracting away the interaction with the device driver, applications are also responsible for managing a subsystem's settings, its data, and its task schedule. There is one application for each payload onboard ION, in addition to applications that control entire subsystems, such as power and communications. Power and communications, however, can also be thought of as payloads because they both contain device drivers to control their respective devices (such as voltage and current sensors, the radio, and the TNC), and they also maintain log files that must be sent to the ground, as well as configurable settings (such as solar power vs battery power threshold or transmit timeouts).

Lastly, in the middle of everything lies the application manager. This is the central component of ION's software system and is responsible for scheduling all of the applications, ensuring that applications are behaving well (i.e., not hogging CPU or crashing), and giving each application its fair share of CPU time. The flexibility of the application manager is what allows ION to deal with changes in its payload requirements so easily and will be discussed further in the section on system scheduling.

In addition to these three sets of components, there is also a group of miscellaneous libraries that do not neatly fit into the hierarchy defined. These provide general-purpose functionality that is used by many different applications (such as a file system or a time representation). These libraries can also contain very specific code used by only a single application (such as the JPEG implementation). The important thing to note is whereas device drivers, applications, and the application manager either control physical devices or cause things to happen, the libraries never do anything and are only used by other system components to help them achieve their task.

III. System Scheduling Onboard ION

Now that the basics of ION's software architecture have been covered, we can delve into some of the more interesting aspects of this design and what it allows. The most important of these, and the reason this architecture was chosen to begin with, is the flexibility it allows in determining what applications run and when they sample their respective payloads. The control that the application manager exhibits over the individual applications allows for completely arbitrary task schedules, as well as the ability to remotely enable and disable entire payloads.

System scheduling onboard ION centers around "config files." These are small configuration files that reside on the file system, one for each application, and specify both global settings for that application as well as a list of "work units" that this application should execute. Work units define a type of work to do (which is payload specific), settings for that specific execution, as well as an optional duration and period with which to do this work. A single work unit (one without a duration and period) could, for example, specify "take a picture at 5:00 p.m. on Jan 20, 2007, with a shutter speed of 1/10 second." A recursive work unit (one that does specify a duration and period) could, for example, specify "take a picture every 10 minutes from 5:00 p.m. on January 20, 2007 to 7:00 a.m. on January 21, 2007." The settings themselves are payload specific and can control things like integration times, which sensors to sample if more than one is available, or how many samples to take. A default set of config files is placed onboard the file system prior to launch, and new ones can be uploaded as needed to change payload's work schedules and settings.

"Config files" are used to control all aspects of ION, even subsystems that are not explicitly payloads. For example, the communications application is controlled by a communication-specific config file that specifies which beacon to send, how often to send it, when to turn on the radio, and how long a communication window should be. The power application's config file allows ION's power settings to be tweaked from the ground, such as precisely when to switch from solar to battery power, when to switch from charging one battery to the other, and how best to gauge the amount of power left in a battery. Lastly, the application manager's config file specifies which applications are enabled and which are disabled.

Upon completion of ION's boot sequence (to be described in the boot sequence section), the application manager begins to run. The first thing it does is read in its own config file and look at which applications should be started. It builds a list of these applications and then cycles through them, giving each application a slice of the CPU. ION uses cooperative multitasking, that is, although multiple applications can run simultaneously, only one application can use the CPU at any given time and it must relinquish the CPU when it is done with it so that other applications can run. To facilitate this context switching, a GiveUpCPU() function exists. When called it suspends execution of the current application and returns control to the application manager, which then determines what application should run next.

When an application is first given CPU time, it too reads in its own configuration file. This configuration file might contain a list of global settings, which the application then sets, and is guaranteed to contain at least one work unit (either single or recursive). The application then loads any work units it finds in the configuration file into its own local scheduler. Each application maintains its own scheduler, which is really a linked list of work units sorted by execution time. When given CPU time, an application simply pulls the first work unit from the head of the list and checks to see whether the start time on that work unit has approached yet. If yes, it follows the instructions within that work unit and removes it from the list. Otherwise, it leaves the work unit there to be examined again the next time this application gets CPU time. In this manner, each application maintains its own task schedule. The application manager performs a round-robin-type scheduling among all of the applications, and then each application performs a sort of earliest deadline first scheduling among its list of work units.

This sort of scheduling system freed ION's software developers from worrying about when and how often payload data would have to be obtained. The decision could be made after launch, with the comfort of knowing that any settings and schedules could be tweaked from the ground. This also meant that, should anything occur onboard ION, such as a fault in a payload or a dead battery, ION's scheduling system could be set to disable the faulty hardware or work around whatever flaw might develop.

IV. ION's Boot Sequence

Before the application manager can start up, there is an entire sequence of events that is responsible for initializing many of the devices onboard ION, verifying the integrity of the file system, and loading the operating system into memory. This section will detail the workings of ION's low-level implementation, which is only active in the first few minutes after powerup, before the application manager takes control of the system.

ION has 8 MB of nonvolatile flash memory, 1 MB of RAM, and a 256 KB EEPROM. As with any embedded system, the EEPROM is where the bootloader resides, and this is where code first starts executing upon powerup. In small systems, the entire operating system can fit into the EEPROM, greatly simplifying the boot sequence. Unfortunately, because of ION's complexity, the full system load (all of the drivers, applications, the application manager, and the libraries) comes to about 900 KB. Because this is too big to fit into the EEPROM, ION must utilize a smaller bootloader that initializes the flash and reads the system load into RAM from the flash. Although this adds a great deal of complexity to the system, it also provides many opportunities for added fault tolerance. Rather than having *only* a bootloader in the EEPROM (which would take up only a few kilobytes), the developers decided to take a full load of ION and strip off all of the noncritical components, leaving only rudimentary power and communications abilities. The result was something called "reset mode," which can be thought of as similar to safe mode on Microsoft Windows operating systems. Should there be any problems with either the bootloader or the flash, the system will jump into reset mode. Within reset mode, ION has the ability do to rudimentary power management, send beacons, and allow for reading and writing of memory addresses. The hope is that, should a problem occur during system startup, reset mode will allow developers to at least determine what might have gone wrong and possibly fix it.

Assuming that no problems are encountered during the initial boot-up phase, the bootloader proceeds to initialize the flash memory. On the flash, there are actually two partitions, a primary and a backup partition. The primary partition is 7 MB in size, and the backup partition is 1 MB in size. Once the application manager starts running, the backup partition is invisible and cannot be read or written to. The backup partition is effectively read-only, and it contains a copy of a known working system load. If anything every goes wrong or gets corrupted, the bootloader can use the backup partition to boot the satellite into a fully working state, identical to when it was first deployed from the PPOD.

Initially, both the primary partition and backup partition contain an identical system load. The startup sequence always attempts to boot from the primary partition first and will only access the backup partition if it cannot read the data from the flash. Furthermore, whereas the backup partition contains a single, permanent, default system load, the primary partition has room to store several system loads. This gives ION's developers the ability to upgrade the operating system while ION is in orbit. All they have to do is upload a new load. Upon reboot, this new load will be detected and loaded instead of the original load. If the load fails to start for whatever reason, ION will enter reset mode to give the developers a chance to troubleshoot the problem.

Assuming the new load checks out, it is loaded into RAM, and execution begins at this load's entry point. The entry point will then verify the integrity of all of the default configuration files onboard the system and then start the application manager. At this point, the system is now fully operational, and the application manager will read it its configuration file and give CPU time to the first enabled application.

V. Communications

Of the many applications and subsystems onboard ION, perhaps the most interesting is that responsible for enabling communication with the ground. Because of length requirements, it is also the only subsystem that will be described in detail here (although the power and attitude control subsystems are also worthy of further discussion). Communications onboard ION consists of several distinct software components. At the highest level, there is the communications application. The communications application is responsible for executing work units responsible for sending beacons and opening and closing communication windows. ION's communication windows are implemented as two recursive work units, one commanding ION to turn on its radio and listen for commands from the ground and the other, five minutes later, telling ION to turn off its radio. Although the default communication window is only five minutes long, ION's controllers can schedule ION's communication windows to exactly coincide with ION's passes over the ground station in Urbana, Illinois. This means that power usage would be minimized because the radio would never be on at a time when ION was not over the ground station, and it would minimize interference with other cubesats on the same frequency.

Underneath the communications application lies the software responsible for implementing ION's communication protocol, known as SRTP (the semireliable transport protocol). SRTP is designed to maximize throughput and link utilization while minimizing timeouts and unnecessary packets that introduce overhead. It can be thought of as a reliable UDP implementation that does not use sliding windows. At the highest level, it resembles FTP in the sense that it supports many of the classic FTP commands such as download file, upload file, delete file, and perform a directory listing. Because all of ION's configuration and data are stored as files onboard the file system, FTP seemed like a natural choice for a communication protocol with which to interact with ION. To avoid the overhead of timeouts and retransmits, however, SRTP implements a novel download and upload scheme. Rather than confirming that each packet (or a group of packets) was received successfully, ION (or the ground station for uploads) instead sends all of the data needed for the file. After the transmission is complete, the ground station (or ION for uploads) sends a reliable packet (i.e., one that must be acknowledged) containing a list of packets that it missed. Then, only these packets are retransmitted (again, unreliably). This process continues until all of the pieces of the file have been successfully transmitted. The two greatest advantages of this approach are that a large portion of the file is likely to be received on the first attempt (so if a communication attempt is cutoff for whatever reason, most of the file will have been downloaded), and, assuming a clear channel, there is almost no communication overhead.

Below this layer lie the actual drivers responsible for controlling the radio and the TNC. Communication with the TNC occurs via the RS-232 serial protocol. The TNC driver is responsible for sending data from the communication application to the TNC. When it comes to receiving data, the TNC uses interrupts to buffer incoming data until the communication application gets time on the CPU from the application manager. Note that, with regards to scheduling, the communication application is treated no differently than any other application by the application manager. No exceptions or special provisions are made for its operation.

VI. ION2's Architecture

With the demise of ION due to a launch failure, construction began in earnest on ION's replacement. Learning from the mistakes made during ION's implementation, ION2 took a radically different approach towards its system design (Fig. 14.2). Rather than having a centralized processor board (the SID in ION's case) controlling all of the devices onboard, ION2 uses a distributed architecture

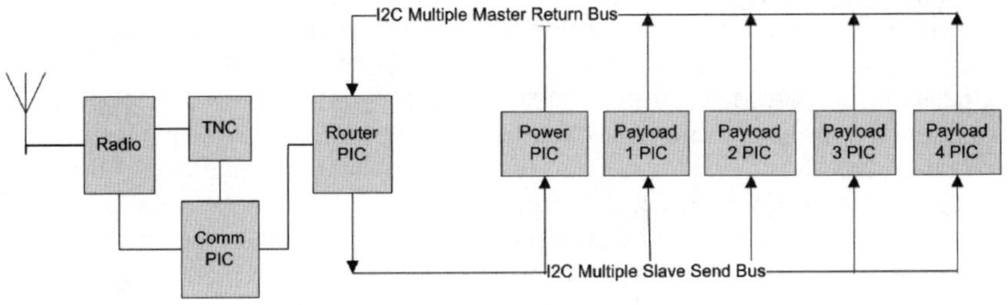

Fig. 14.2 ION2's architecture.

with an I²C bus linking all of the components together. Whereas on ION all of the devices were "dumb" in the sense that they needed to be controlled by the processor, on ION2 each payload is on its own board with a dedicated PIC providing computing resources. In essence, all of the devices onboard ION2 are "smart."

The reason the switch was made to this architecture was because of the difficulties encountered implementing ION, both from a technical and an educational standpoint. Part of what dictated ION's software architecture was the idea that drivers, applications, and even entire subsystems could be black boxed to some extent and handed off for further development. It was thought that, because all of applications only had to interface with the application manager (and this interface was standardized), students should have no problem implementing components on their own. Unfortunately, this turned out to be completely wrong, and ION's two primary developers ended up being responsible for over 90% of ION's software, and the remaining students learned little, if anything. The other problem with a centralized processing scheme is that a failure in any one component could, in theory, affect all other components. Although safeguards in the form of watchdog timers were built in, no one could really predict what would happen to all of the other applications if a single one went haywire because they were all running on the same hardware. Lastly, a centralized architecture made it very difficult to reuse components for future cubesats. Very little of ION's code can be reused for ION2 because of the differences in hardware. A move to a centralized, bus-type architecture allows for much of the software and hardware infrastructure developed for ION2 to be reused for future cubesats.

For these reasons, the move was made to a distributed architecture. With I²C linking everything together, a well-defined protocol could truly free developers to focus solely on their payload and not worry about how power management or communications would be implemented. Furthermore, testing is much easier since there is no competition for hardware (as there is in ION's case) because each team has their own PIC and board to test on. Lastly, in a distributed system, dependencies between components are greatly reduced. If a certain board begins to misbehave, it can simply be powered down without affecting the rest of the system. In fact, because the *only* link between all of the boards is an I²C bus, the amount of a damage that a single board can cause to others is limited.

Furthermore, ION ended up being heavily reliant upon its file system, which was a highly complex piece of software prone to bugs and hardware failures. The file system stored the system load, the configuration files, log files, and the data files from the payloads. A failure of the file system would have been devastating to ION. ION2 has removed this dependency on a file system. With the use of PICs, there is no longer a need for a file system to contain a system load. Instead, each PIC is simply flashed with all of the software necessary to control the payload or devices on its board. Furthermore, instead of a file system to hold data files, the communications board simply has a large outgoing data buffer. As payloads generate data, the PICs time stamp it and send it on to the communications board for storage in the outgoing buffer. When contact is made with the ground, all of the data in the buffer are retrieved and reconstructed on the ground. The only file type to store is the configuration files. For these, either a small amount of nonvolatile memory is attached to each PIC, or the PICs are simply allowed to lose their configurations upon reboot and start in their default state. This is perfectly normal because reboots should occur very rarely, and new settings can always be uploaded from the ground.

Intrasatellite communication on ION2 occurs through the router PIC. This PIC lies at the core of ION2 and is designed to route messages along the I^2C bus between payloads and the communications board. Messages from the ground travel through the communications board to the router PIC, which then forwards the message to the appropriate payload. Data from the payloads are in turn sent to the router PIC, which forwards them along to the communications board for eventual transmission to the ground.

One of the things that ION2 does borrow from ION is its scheduling system. Just like all of the applications onboard ION had their own private scheduler, the PIC on each board of ION2 will also maintain a private scheduler. When work units are sent up from the ground, the router PIC will forward these work units along to the appropriate board. Upon receiving a new work unit, the PIC on this board will insert it into its own scheduler and execute the work unit at the appropriate time. This means that ION2 retains the scheduling control and flexibility that ION's scheduler provided, without having to worry about schedule conflicts of overallocating the CPU (because each payload now has a dedicated PIC and does not need to share computing resources).

Although ION2 is still in its early development stages and its implementation is constantly evolving, the design presented here forms the foundation for the software infrastructure that will hopefully run the next several cubesats from the University of Illinois. It provides the flexibility and modularity that ION lacked and should make designing future cubesats less of a pure technical challenge and more of an educational exercise. By freeing developers to focus on their own payload board and not having to worry about how their software will interface and interact with the rest of the software onboard the satellite, more time can be spent developing interesting payloads rather than solving fundamental systems problems.

Chapter 15

Cubesat Radio Communication Systems

Alex Rein*

University of Illinois at Urbana-Champaign, Urbana, Illinois

I. Introduction

A. Purpose

THIS paper is intended to serve as a general discussion of cubesat radio communication systems. The Illinois Observing Nanosatellite (ION) will be used as a case study alongside the more general discussion. This paper does not present any ground-breaking research or advancements in satellite communication technology. In fact, it does quite the opposite. This paper is an introduction to basic satellite radio communications for student satellite projects where experience might be lacking. It also shows that cubesat communication can be achieved successfully with the use of relatively simple equipment.

Cubesat satellites are often used as test beds for new technology or as inexpensive platforms for science payloads. As such, the communication system hardware simply supports the mission objectives of other hardware. This is not to say that cubesats cannot also be used to test new communication system technology. However a reliable primary communication system is critical to the success of the entire satellite mission. This puts the communication system in a difficult position of both being highly reliable but also meeting the low-cost and fast-development goal of most cubesat programs.

Commercial and military grade satellite equipment is usually too massive and expensive to even be considered for a cubesat system. Commercial-off-the-shelf equipment is generally preferred, but its performance onboard a satellite is unknown or ambiguous. In the following paper the link budget and associated tests are discussed to show that this equipment is indeed viable for a low-Earth-orbit cubesat. Also discussed are various options in the configuration of the communication system with the intention of choosing the simplest and most reliable system.

---

*Graduate Research Assistant, Department of Electrical Engineering; currently HW Engineer, Viasat, Inc.

B. Conventions

The link budget calculated through the following discussion and summarized at the end derives its data from many sources. To differentiate among them, they are grouped into four categories. *Estimates or design goals* are values that are not precisely known or can be controlled by the system designer. *Theoretical or calculated* values come from the underlying theory and are generally not easily directly measurable. *Measured or derived* values come from experimental observation or calculations from experimental data. *Manufacturer or documentation* values are taken from the manufacturer without verification or from other texts on the subject.

It is common practice to report all values in a link budget in decibels because it simplifies computing the total budget outcome. Converting to decibels transforms multiplication of loss and gain factors into addition or subtraction of terms. It also compresses the huge range of magnitudes encountered in communication system calculations into a logarithmic scale. A summary of basic decibel formulas is shown in Eq. (15.1).

$$\begin{aligned}\text{Power in dBW} &= 10\log_{10}(\text{Power in Watts})\\ \text{Power in dBm} &= 10\log_{10}(\text{Power in milliWatts})\\ \text{Power in dB} &= \text{Power in dBm} - 30\\ \text{Gain or Loss} &= 10\log_{10}(\text{output power}/\text{input power})\\ \log(x) + \log(y) &= \log(x \bullet y)\end{aligned} \tag{15.1}$$

C. System Overview and Conceptualization

The block diagrams showing the actual ION satellite and ground station communication system hardware are shown in Figs. 15.1 and 15.2, respectively. The satellite system is relatively simple and consists of three pieces of hardware. The TNC and transceiver are both commercial-off-the-shelf (COTS) equipment. The antenna system was custom built by students and consists of a phase inverter chip and deployable dipole antenna.

The ground station is composed of three subsystems that function together to achieve radio communication with the satellite. The radio subsystem is similar to the satellite radio system except there is also a lightning arrestor and pre-amp inline

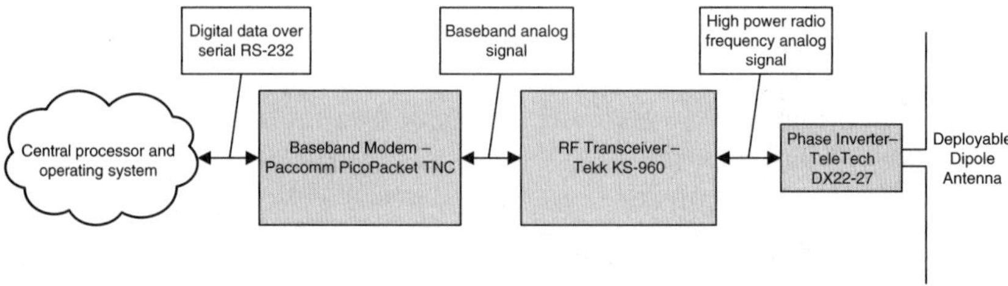

Fig. 15.1 ION communication system.

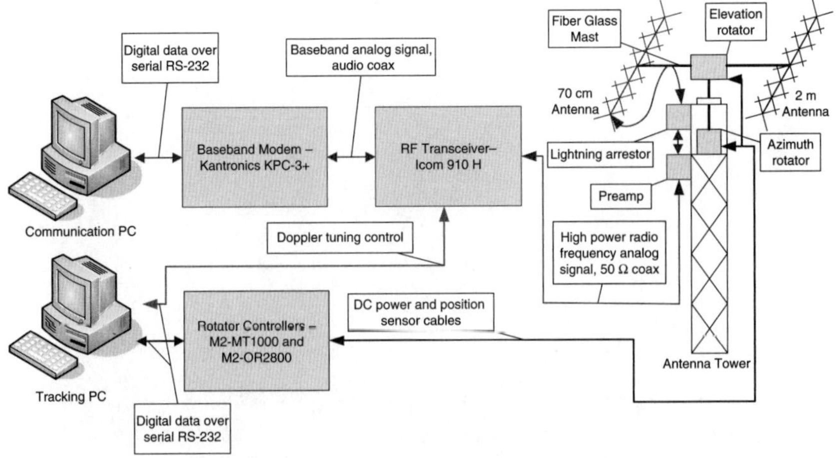

Fig. 15.2 Ground station system.

between the transceiver and antenna. The tracking system uses the Nova orbital propagator to calculate satellite's position very accurately. Nova interfaces with the rotator controllers to position the directional high-gain antenna to point at ION as it passes. The last subsystem compensates for Doppler shift. A custom PERL script takes relative velocity information from Nova, then calculates the Doppler shift, and retunes the radio over a serial interface.

In the following sections it will be more helpful to use the system conceptualization in Fig. 15.3. It is a generic one-directional block diagram of a communication system. It should be sufficiently generic to apply to most cubesat communication systems in both directions (uplink or downlink). Each block, except for the propagation medium, has a parallel block on the Rx or Tx side and will be discussed together in the following sections. The mapping between the ION hardware and generic block diagram is straightforward. On uplink (ground to satellite) the Icom transceiver is the "RF transmitter," and the Tekk KS-960 is the "RF receiver" and vice versa on downlink (satellite to ground). On uplink the Kantronics TNC is the "baseband modulator," and the Paccomm Pico Packet TNC is the "baseband demodulator" and vice versa and the downlink. The same applies to the transmission lines and antennas. However, the propagation medium is the same in both directions.

II. RF Spectrum and Regulation

There are literally an infinite number of frequencies to choose from for a cubesat communication system. However both regulatory and design consideration will determine the best frequency to use. The current trend with cubesats is to use one of the many available licensed amateur bands. The amateur bands have several benefits for cubesats. COTS equipment is readily available for use in many of these bands. The regulations regarding bandwidth, modulation, power, and other factors are relatively lax and allow for experimentation with communication

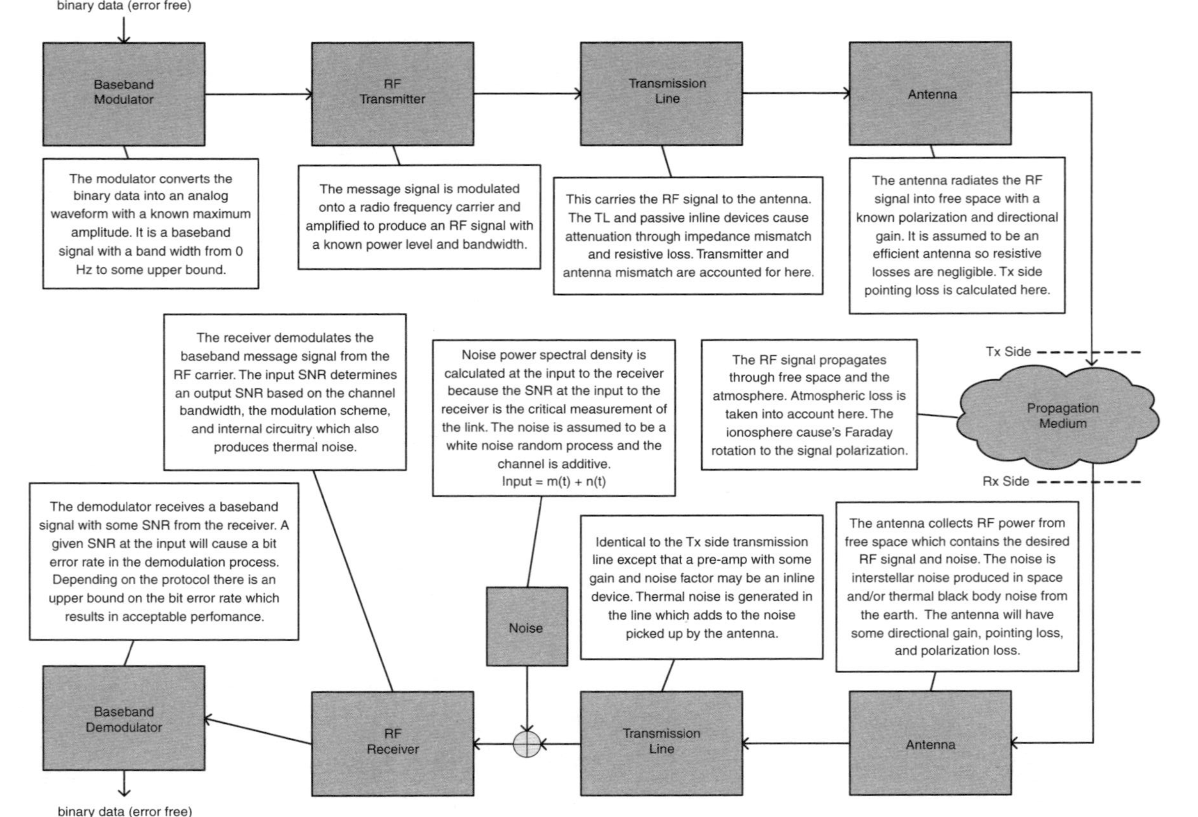

Fig. 15.3 Conceptual communication system.

Table 15.1 Available Amateur Bands

Frequency range	Common name
21.00–21.45 MHz	15 m
24.89–24.99 MHz	12 m
28.00–29.70 MHz	10 m
50–54 MHz	6 m
144–148 MHz	2 m
220–225 MHz	1.25 m
420–450 MHz	70 cm
902–928 MHz	33 cm
1.24–1.30 GHz	23 cm
2.3–2.305 GHz	12 cm
2.40–2.45 GHz	12 cm
3.3–3.5 GHz	NA
5.650–5.925 GHz	NA
10.0–10.5 GHz	NA
24.00–24.25 GHz	NA
47.0–47.2 GHz	NA
75.5–81.0 GHz	NA
119.98–120.02 GHz	NA
142–149 GHz	NA
241–250 GHZ	NA

systems. The licensing process for amateur bands involves much less red tape than commercial bands. There are many cubesats and amateur satellites already in orbit using these bands, which allows testing and verification of the ground station system prior to launch. Last, there are amateur bands spread throughout the spectrum giving sufficient freedom to the system designer to choose the best frequency of operation for communication.

The discussion in following sections should make clear some of the effects that different frequencies will have on the system. Table 15.1 provides amateur bands for frequencies above 20 MHz. There are many bands below 20 MHz, but they are impractical because of ionosphere reflection and the physical size of the antennas. The amateur bands are commonly referred to by their wavelength and not the actual frequency range. The most common amateur bands for cubesats are the 2-m, 70-cm, 33-cm, and 23-cm bands, and the discussion in the following sections can be assumed to apply to these four bands. Amateur bands at significantly higher or lower frequencies might have different design and physical considerations. For more information see the actual regulations defined by the regulatory bodies in the following corporations: International Telecommunication Union, Federal Communications Commission, and International Amateur Radio Union.[†]

[†]International Telecommunication Union, data available online at http://www.itu.int/ITU-R/; Federal Communications Commission, data available online at http://wireless.fcc.gov/rules.html, "Code of Federal Regulations, Title 47" (In Particular Part 15 & 97); International Amateur Radio Union, data available online at http://www.iaru.org/satellite/.

III. Orbit and Propagation Medium

A. Slant Range and Free Space Loss [1]

Cubesats are typically launched into circular (low eccentricity) low Earth orbits. The distance between the ground station and satellite is very important to the communication system design because this determines the free space signal loss. This is the loss caused by the radio signal power density spreading out as it radiates away from its source. It follows an inverse square law, and it is the dominant loss mechanism in the entire system. Because the orbit can be approximated as purely circular, only the law of cosines is needed to calculate the distance between the satellite and ground station, often referred to as slant range. Given the slant range, the free space loss can be calculated. The free space loss is dependent on frequency. The geometry, relevant equations, and results are presented in Fig. 15.4 for the projected orbit of ION. ION was predicted to achieve an apogee altitude of 518.4 km, and this is used in the calculations because the apogee is the maximum altitude of the satellite. The following calculations do not take into account the altitude of the ground station.

For ion we chose a target goal of 10-deg minimum usable elevation angle to ensure that we can use the majority of the line-of-sight pass time to communicate with the satellite. At this elevation the slant range is 1739.9 km, the free space loss is –150.1 dB, and the pointing loss angle is 65.61 deg. Also, at 10-deg terrestrial sources of radio noise are eliminated as a result of the radiation pattern of our ground station antenna, which is discussed in Section VIII.D.

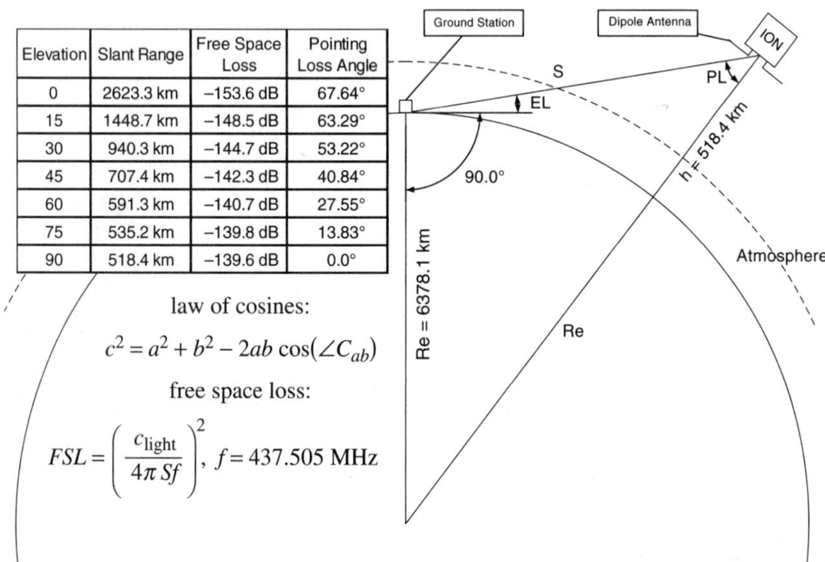

Elevation	Slant Range	Free Space Loss	Pointing Loss Angle
0	2623.3 km	–153.6 dB	67.64°
15	1448.7 km	–148.5 dB	63.29°
30	940.3 km	–144.7 dB	53.22°
45	707.4 km	–142.3 dB	40.84°
60	591.3 km	–140.7 dB	27.55°
75	535.2 km	–139.8 dB	13.83°
90	518.4 km	–139.6 dB	0.0°

Fig. 15.4 Simple circular orbit and free space loss.

B. Communication Time and Data Throughput

Another very important consideration of the communication system is the available data bandwidth. In this context data bandwidth is the amount of data that can be sent through the system (both uplink and downlink) in a given amount of time. The data bandwidth requirements of the payloads must be satisfied by the communication system, or the payloads will generate more data than can possibly be downloaded from the satellite. The orbit altitude, minimum usable elevation, and bit transmission rate all affect the data bandwidth. Table 15.2 presents daily average available communication time and data bandwidths for several minimum usable elevation angles and data transmission rates. It was generated using the STK satellite simulation software for the projected orbit of ION and the UIUC ground station. The ground station is at an altitude of 715 m and has a clear view of the horizon. It shows the average daily data bandwidth assuming an 100% link efficiency (no dropped packets and no protocol overhead). The table shows the total data through put for both uplink and downlink in one day for a half-duplex system.

The simulations show that beyond 20-deg minimum elevation using 1200 bps transmission the data bandwidth for ION drops below the 75-kB/day requirement for payload and telemetry data. Taking 10 deg as the minimum elevation angle yields a 58% margin in the daily bandwidth, which allows for dropped packets and communication protocol overhead. The data bandwidth drops off rapidly with angle at roughly 62% between 0 and 10 deg and at roughly 50% per 10 deg beyond that. This might seem surprising, but it is because a LEO satellite as seen from the ground station spends most of a pass at a low elevation angle. Furthermore, some passes might not even exceed the minimum elevation angle and are lost completely. Thus, there is a balancing act between minimum usable elevation angle and data transmission rate to achieve the required data bandwidth per day. In Fig. 15.5 the elevation vs time plot shows that ION spends 89% of a zenith pass (directly overhead) below 60 deg and 69% below 30 deg. A zenith pass also has the longest possible communication window, lasting a total of 716 s for ION's orbit.

C. Atmospheric Attenuation

In addition to free space loss, the gaseous medium of the atmosphere also attenuates the radio signal. In Fig. 15.4 it is clear that the amount of atmosphere

Table 15.2 Throughput vs Elevation

Minimum elevation angle, deg	Avg daily pass time, s	Data BW 1200 bps, kB	Data BW 9600 bps, kB
0	3119	468	3743
10	1182	177	1418
20	531	80	637
30	262	39	314
40	136	20	163
50	71	11	85
60	35	5	42

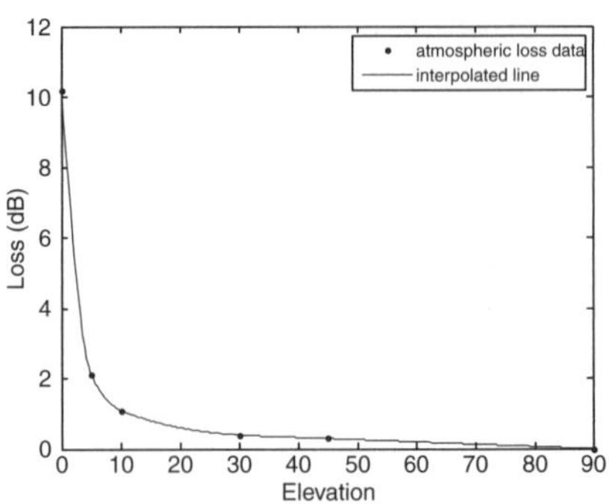

Fig. 15.5 Atmospheric losses (<2 GHz).

the signal must pass through increases with lower minimum elevation angle. The amount of attenuation that a signal experiences will vary with frequency, temperature, clouds, and other atmospheric effects. However for frequencies below 2 GHz, the amount of attenuation can be approximated as only dependent on the elevation angle. This relationship is taken from a communications textbook [2] and interpolated between points in Fig. 15.5. At 10-deg minimum elevation the attenuation is 1.1 dB.

The ionosphere also attenuates signals and causes Farraday rotation to the signal's polarization. The ionosphere can induce significant attenuation of 10 dB or more on signals below 100 MHz, and below 20 MHz signals are fully reflected or absorbed by the ionosphere [1]. This places a limit on the lower bound of feasible cubesat radio frequencies. The amount of attenuation for a range of frequencies is given in Table 15.3. Farraday rotation is a phenomenon that will rotate the polarization orientation of the electric field of a signal passing through the ionosphere by some amount. It does not cause the signal to become circularly polarized. Quantifying and predicting the amount of rotation is difficult and depends on frequency, solar activity, and other atmospheric conditions. For reasons explained in Section VIII.B, this phenomenon can be mitigated.

IV. Doppler Shift

A. Physics of the Doppler Shift Phenomenon

The Doppler shift phenomenon occurs when a wave propagates between two objects that are moving relative to each other. In radio communications the shift causes the frequency of the received signal to be shifted up or down when the transmitter is approaching or receding respectively. In terrestrial radio

Table 15.3 Ionosphere Losses

Frequency, MHz	Attenuation, dB
146	0.7
438	0.4
2410	0.1

communications this is usually not important because the relative velocities are so small that the shift is negligible. In the case of a LEO satellite such as ION, which is traveling with a linear velocity of roughly 7.6 km/s, the Doppler shift must be taken into account. The formula for calculating the relativistic Doppler shift is given in Eq. (15.2)

$$f_{Rx} = f_{Tx}\sqrt{1 - \frac{v}{c}\Big/1 + \frac{v}{c}} \tag{15.2}$$

where f_s is the transmitted frequency, f_r is the received frequency, c is the speed of light, and v is the relative velocity. The velocity v is negative if the transmitter is approaching the receiver and causes the received frequency to be shifted up. The opposite is true when the transmitter is receding from the transmitter.

Doppler shift phenomenon also causes dilation of the signal bandwidth. Consider a signal centered at 100 MHz, which is experiencing an upward shift of 1% as a result of a high relative velocity. Say the signal has a bandwidth of 200 kHz. The upper and lower band edges of the signal are shifted to 101.101 MHz and 100.899 MHz, respectively. Taking the difference between these band edges show that the Doppler shifted signal bandwidth is now 202 kHz. This could cause filtering problems and demodulation distortion in the receiver system. However, for a 1% shift the relative velocity must be roughly 3000 km/s. Fortunately both the relative velocity and bandwidth of cubesat systems are small enough to make bandwidth dilation negligible in practice. For ION the maximum dilation is calculated to be only 0.47 Hz.

B. Doppler Shift and Satellite Communication

In any radio communication system the receiver will have a channel selection filter that has the narrowest filter bandwidth in the receiver (but not necessarily the demodulator). It defines the maximum bandwidth of the received signal and the maximum permissible tuning error. The channel selection filter and received signal are depicted in the frequency domain in Fig. 15.6.

To maximize the signal-to-noise ratio (SNR) in the receiver, the channel selection filter bandwidth is chosen to closely match the transmitted signal bandwidth. (Noise is discussed in Sec. IX.) Thus, the filter gap is typically a small fraction of the transmitted signal bandwidth. The tuning error cannot exceed the filter gap, or some of the signal will be filtered out. This will cause the demodulated signal to

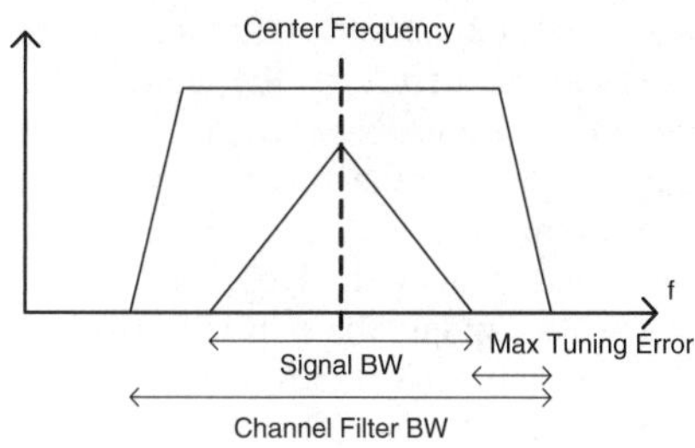

Fig. 15.6 Tuning error.

become distorted and have a degraded SNR. The absolute Doppler shift in Hertz is proportional to the transmitted frequency and the relative velocity. For ION that attains a maximum relative velocity to ground station of 7.08 km/s (at 0-deg elevation during a zenith pass) and transmits at 437.505 MHz, the maximum Doppler shift will be 10.32 kHz. The signals between ION and the ground station will have a bandwidth of 12 kHz, and the narrowest channel selection filter is 20 kHz for the satellite radio. Thus, the received signal will drift outside the satellite's channel selection filter.

C. Overcoming Doppler Shift

Obviously, either the transmitter or receiver must be retuned in order to keep the received signal within the filter bandwidth. It is common practice to leave the satellite at a fixed frequency and retune the transmit *and* receive frequencies in the ground station. This design eliminates unnecessary hardware and software complexity in the satellite. Otherwise, the satellite must be "aware" of its orbit and the ground station position in order for it to correctly retune the satellite transceiver. This requires the satellite to have a GPS receiver onboard or that accurate ephemeris data are loaded into the satellite prior to launch. The ephemeris data will not be known precisely until after orbital insertion. Retuning only the ground station also allows multiple ground stations to communicate with the satellite at the same time given the satellite transmitter and receiver center frequencies. This is necessary to allow secondary ground stations to listen for beacons.

The consequence of retuning the ground station transceiver is that it must be tuned to a different frequency for transmit (uplink) and receive (downlink). This can be seen from considering Eq. (15.2) and letting either f_{Rx} or f_{Tx} be fixed as it would be on the satellite. Using STK simulation software, the relative velocity between ION and the ground station was calculated for a pass that crossed through zenith (directly overhead). This type of pass produces the most extreme relative velocity and hence the most extreme Doppler shift. In Figs 15.7a and b the Doppler shift as a fraction of center frequency is plotted as a function of time and elevation. These plots define the parameters of the Doppler tuning system in the ground station. For ION the ground station must accommodate a shift of ±10.3 kHz. Furthermore in Fig. 15.7c the maximum rate of change in the Doppler shift occurs at zenith and is 147 Hz/s. Given the maximum permissible tuning error for ION is 4 kHz, the transceivers must be retuned at least every 27 s. Figure 15.7d shows pass time as a function of elevation.

The ION ground station adjusts for Doppler shift using the Nova satellite tracking program and custom radio interface software. Nova is essentially an orbital propagator with lots of extra features. It takes as input ephemeris data in the form of two line elements (TLEs). TLEs are made available first from the launch provider and later the NORAD radar satellite tracking facilities. Nova can then propagate the satellite's orbit into the future and calculate the relative velocity (range rate) between it and the ground station at any point in time. The radio interface software interacts with Nova to poll the range rate, calculate the Doppler shift for uplink and downlink given any center frequency, and update the radio to maintain no more than a 1-kHz tuning error that is well within the 4-kHz limit. The ground station transceiver, an Icom 910 h, provides a serial RS-232 connection for

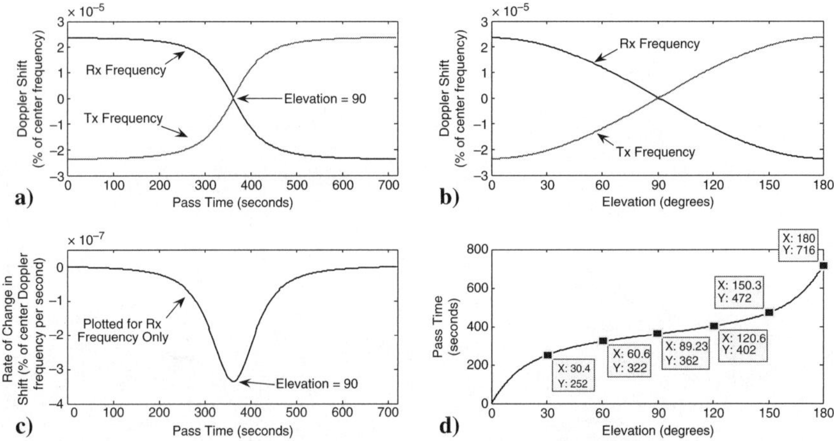

Fig. 15.7 Doppler shift profile for zenith pass: a) Dopper shift as a fraction of center frequency plotted as a function of time, b) Doppler shift as a fraction of center frequency plotted as a function of elevation, c) maximum rate of change in Doppler shift, and d) pass time as a function of elevation.

two-way communication with a PC. The radio software is implemented using PERL, which is a high-level scripting language.

One major drawback to using a half-duplex system like ION's is that the ground station oscillator must be retuned between the Doppler adjusted receive and transmit frequencies whenever the transceiver changes between receive and transmit modes. The oscillator will take a small amount of time ("key-up time") to stabilize at the correct frequency every time this is done. This wastes scarce communication time and increases the likelihood of packet collisions. Fortunately, most communication time is spent in downlink to retrieve sensor data, and switching between modes does not happen so frequently that it becomes a major problem.

The internal implementation of the ground station radio also makes adjusting for Doppler shift difficult. There is a single digitally synthesized PLL oscillator controlled by one of two registers that contains the transmit or receive frequency. However, only the active register can be written to over the serial connection. Attempting to update both registers while the transceiver is actively transmitting or receiving data will cause that packet to be dropped. Also the radio can change states while a register is being updated causing the wrong register to be overwritten. When this occurs, the software must recover immediately, or no communication will succeed because the transmit and receive frequencies have been swapped. This complicates the software significantly. A basic flowchart of the tuning software is shown in Fig. 15.8.

D. Alternative Method to Compute Doppler Shift

Satellites in low Earth orbits attain the maximum relative velocity with a ground station during a zenith pass at 0-deg elevation. Given this fact and simple geometry of circular low Earth orbits (Fig. 15.4), the relative velocity can be estimated

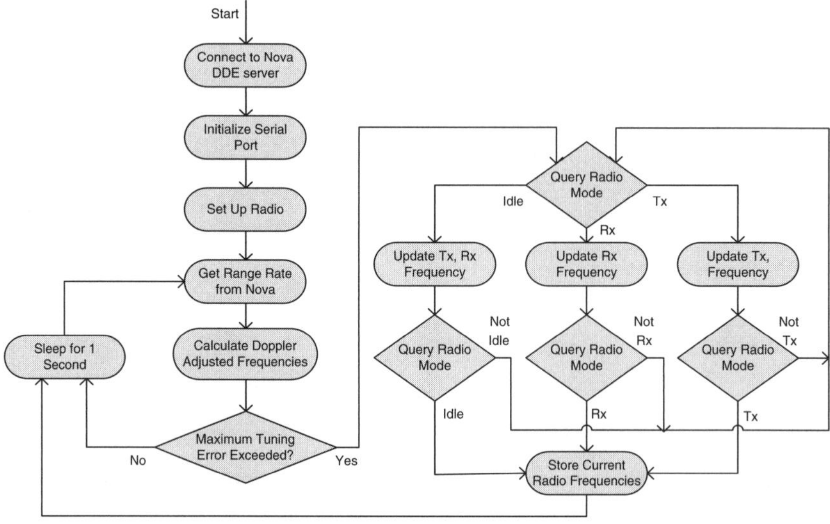

Fig. 15.8 Doppler tuning software.

using the formula in Eq. (15.3) for any elevation during a zenith pass. The maximum relative velocity and hence maximum Doppler shift occur at 0-deg elevation. The maximum Doppler shift for the communication system can influence its design and should be calculated. The formula was derived by approximating the orbit as purely circular and equating centripetal acceleration of the satellite with the acceleration caused by the gravity of the Earth. The error is less than 1% compared to simulations. To compute the relative velocity during any arbitrary pass, the mathematics becomes much more complicated, and a simulator like STK is preferred.

$$\begin{aligned} V_{\text{linear}} &= \sqrt{\frac{3.9873 \times 10^{14}}{R_e + h}} \\ V_{\text{relative}} &= \cos\left(90 - \angle PL\right) V_{\text{linear}} \end{aligned} \tag{15.3}$$

V. Baseband Modulation

A. Common Baseband Modulation Techniques [3]

In the world of HAM radio and cubesat technology, there are two modulation standards used far more frequently than anything else. They are the Bell 202 standard, which is a 1200-bps audio frequency shift keying (AFSK) scheme, and the 9600-bps G3RUH standard, which is a Nyquist pulse scheme (often misleadingly referred to as FSK). The simple fact that these schemes are so prevalent makes them appealing for cubesats. There is a large amount of documentation about both schemes, and widely available commercial equipment supports them. Furthermore, many cubesat ground stations and amateur radio operators use modems of these types, which enable them to receive beacons from compatible cubesats.

The modulation of binary data into an analog waveform is accomplished by mapping m bits into 2^m unique analog waveform symbols. A serial stream of binary data produces a continuous-time analog wave form referred to as the message signal. The message signal is formed by summing the time-shifted symbols as described in Eq. (15.4),

$$m(t) = \sum_{n} S_n\left(t - nT\right) \tag{15.4}$$

where S_n is one of 2^m symbols, T is the signaling interval, and n is the index of the symbol in the analog waveform. In general, the symbol S_n is not constrained in time to only occupy one interval T and can overlap onto other symbols. The signaling interval specifies the rate at which the symbols are sent. The total data transmission rate in bits per second is given by m/T.

The Bell 202 standard uses two symbols with a signaling interval of 833 μs resulting in a data transmission rate of 1200 bits per second. It was developed by Bell Telephone as an early data modulation scheme over telephone lines. Binary bits are mapped directly to audio frequency tones of 1200 and 2200 Hz. A segment of the waveform generated from the ground station Kantronics KPC 3+ modem for random binary data is shown in Fig. 15.9 along with the spectrum of the signal in Fig. 15.10. In the time domain the symbols do not overlap, and

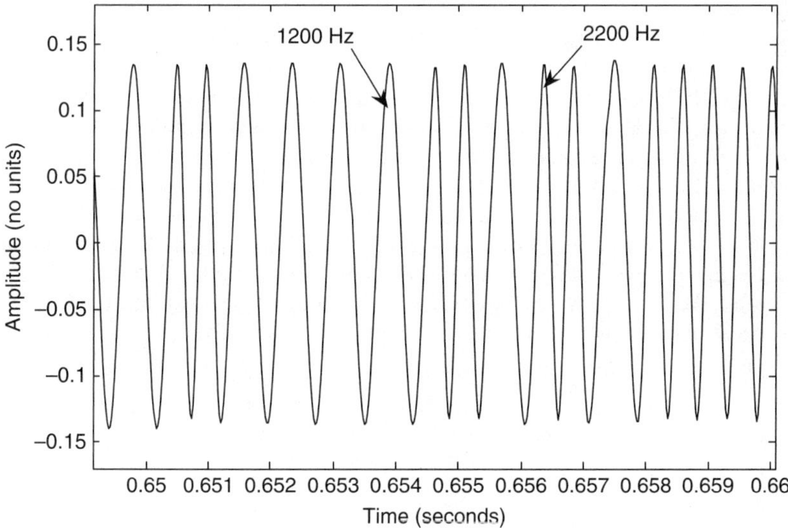

Fig. 15.9 Bell 202 waveform.

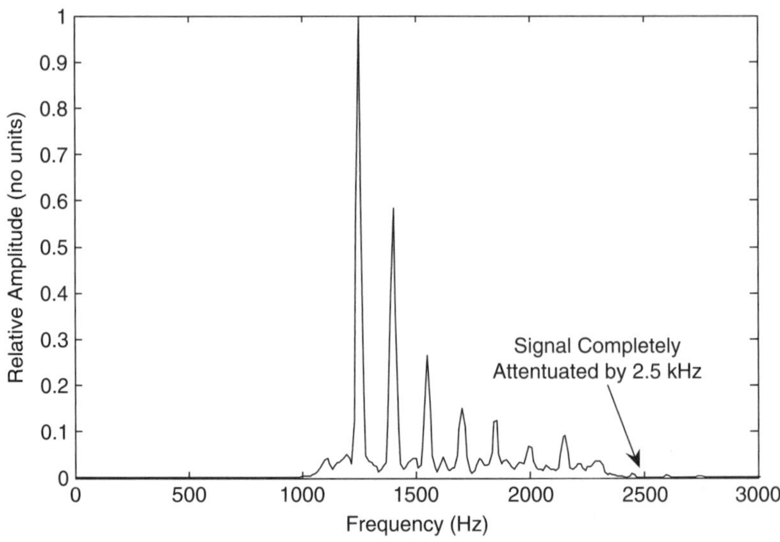

Fig. 15.10 Bell 202 PSD.

adjacent symbols are connected at zero crossings to minimize truncation distortion. In the frequency domain the expected two impulses are not observed. This is because the sinusoidal tones are multiplied by a binary pulse train resulting in convolution in the frequency domain. The signal is completely attenuated by 2.5 kHz, which effectively defines the bandwidth of the message signal.

$$m(t) = \sum_n a_n p_n \left(t - nT\right) \tag{15.5}$$

The G3RUH standard is a Nyquist pulse-based modulation scheme. It was named after the HAM call sign of James R Miller who popularized its use. A Nyquist pulse is an analog waveform that has a zero value at nonzero integer multiples of the symbol period. In the G3RUH scheme a "raised cosine pulse" is used, and data are encoded by scaling it with a factor a_n. Again a_n must have 2^m unique values to represent m bits. This lends itself well to DSP implementation because the received waveform only needs to be sampled at intervals of nT to determine the value of a_n. The message signal formula is shown in Eq. (15.5) and is a special case of the general message signal equation (15.6).

$$p(t) = \frac{\sin\left(\pi t/T\right)}{\pi t/T} \frac{\cos\left(\beta \pi t/T\right)}{1 - \left(2\beta t/T\right)^2} \tag{15.6}$$

For the G3RUH scheme, a_n has two symmetric values c and $-c$, where c is a scalar and its exact value is not important yet (see FM modulation section). The formula for a raised cosine pulse in the time domain is given in Eq. (15.6). The frequency-domain representation is given in Eq. (15.7). The primary benefit of this type of pulse is the "brick wall" shape of the spectrum.

$$P(f) = \begin{cases} T, & |fT| \leq \frac{1}{2}(1-\beta) \\ \frac{T}{2}\left(1 + \cos\left(\frac{\pi}{\beta}\left(|fT| - \frac{1}{2}(1-\beta)\right)\right)\right), & \frac{1}{2}(1-\beta) < |fT| \leq \frac{1}{2}(1+\beta) \\ 0, & |fT| > \frac{1}{2}(1+\beta) \end{cases} \quad (15.7)$$

Unlike many other Nyquist pulse shapes, the raised cosine spectrum is band limited to a finite bandwidth and controlled by a single parameter β. The bandwidth also depends on the signaling interval T, but this is required to be 104 μs for 9600-bps transmission. In Fig. 15.11 the analog waveforms of two time-shifted pulses are plotted, and in Fig. 15.12 the frequency-domain representation is plotted. The parameters were $\beta = 0.458$ and $T = 104$ μs for both plots that correspond to the parameters for the PacComm UP9600 modem that will be used on ION2.

In Figs. 15.13 and 15.14 the actual waveform and power spectral density (PSD) of the signal produced by the Paccomm UP9600 TNC for random binary data are shown. In the analog waveform we can see the summation of many pulses multiplied by symmetric scaling factors $+c$ and $-c$. The power spectral density of the signal is again distorted by convolution in the frequency domain with a binary pulse train. There is a small impulse at 9600 Hz. However this is not part of the

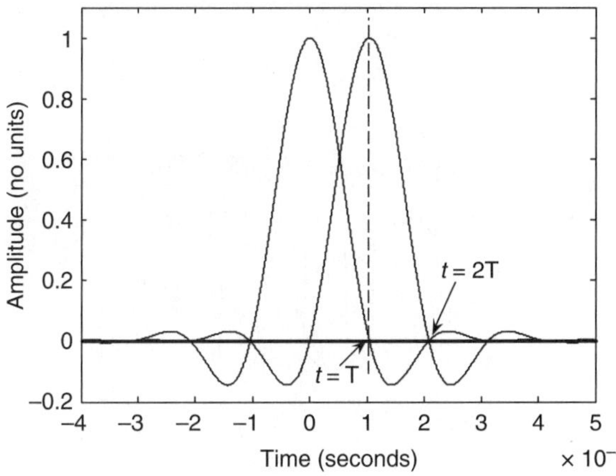

Fig. 15.11 Superimposed raised cosine pulses.

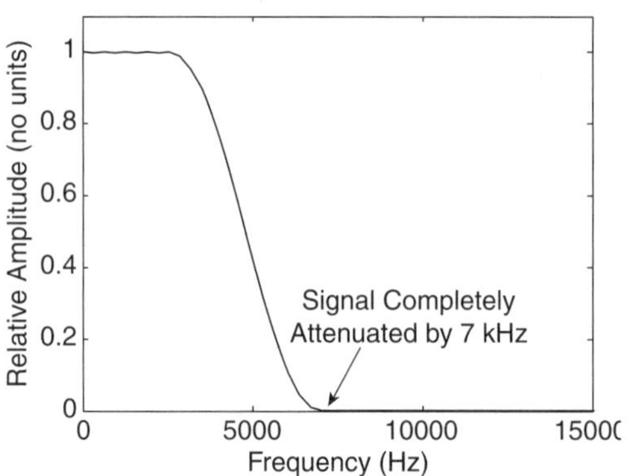

Fig. 15.12 Raised cosine pulse Fourier transform.

message signal because it is only an artifact of truncating the raised cosine pulses so they may be stored in a lookup table.

B. Demodulator Testing

In the conceptual link budget (Fig. 15.3) the modulator and demodulator (or modems) form the bridge between the digital and analog domain. They are critical to the communication link, and their performance must be quantified in order to document an accurate link budget. The critical measure of a modem is the bit error rate (BER) for a given SNR at the analog input to the demodulator. The unfortunate drawback to using inexpensive COTS components for the communication system is that data on BER to SNR performance are often not documented or made publicly available. There are theoretical models for the BER of different modulation schemes in the presence of noise. However, these ignore these necessarily ignore implementation loss. For the purposes of the link budget, it is more practical to directly test the demodulator performance in the presence of noise.

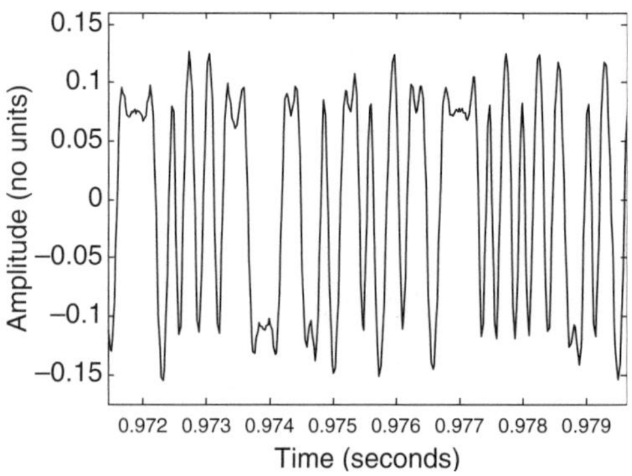

Fig. 15.13 PacComm UP9600 waveform.

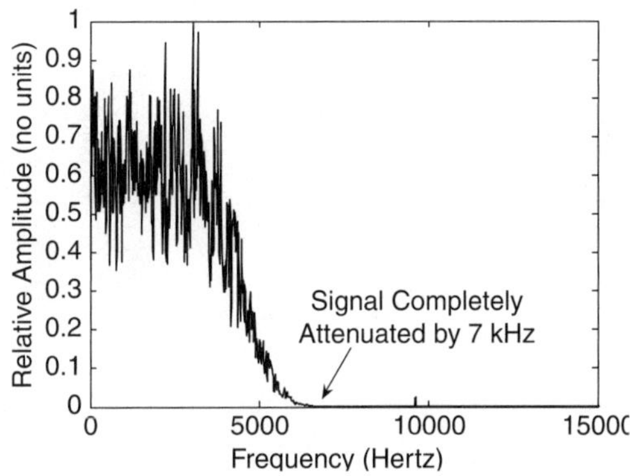

Fig. 15.14 PacComm UP9600 PSD.

The AX.25 protocol uses a noncorrective checksum so that any corrupted packets are dropped. In this test only the dropped packet rate is recorded, and the BER can be calculated from this using statistical analysis. The dropped packet rate is still the more appropriate measure of link performance. The testing setup for the demodulators is relatively simple. All that is required in the test setup is a PC with a sound card and serial port. Digital data are entered into the modem, and the resulting analog output is recorded by the computer. Then the recording is played back to the modem after adding some amount of noise to achieve a given SNR. MATLAB® was used to record, add noise, and playback the signal. White Gaussian noise was added using the MATLAB® *awgn* function. White Gaussian noise is typically how noise in communication channels is modeled [3]. The noisy output is played back 100 times for increasing SNR values, and the percentage of dropped packets is recorded. The experimental setup is shown in Fig. 15.15.

The results are graphed for both the PacComm UP9600 and Kantronics KPC 3 in Figs. 15.16 and 15.17, respectively. For the PacComm modem, the performance degrades rapidly regardless of packet length for an SNR below 6 dB. For the Kantronics modem, the performance degrades rapidly regardless of packet length

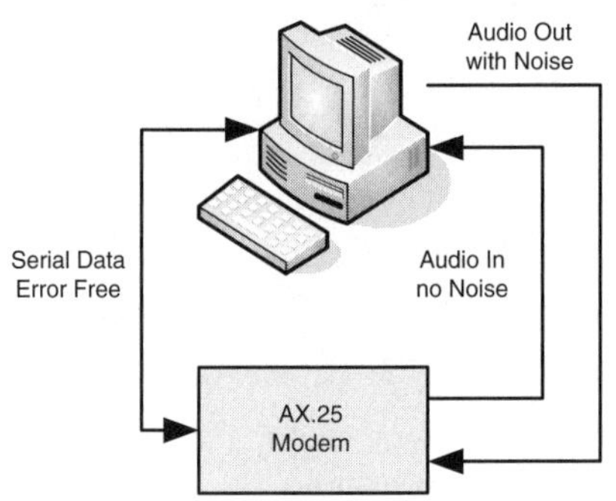

Fig. 15.15 Modem test setup.

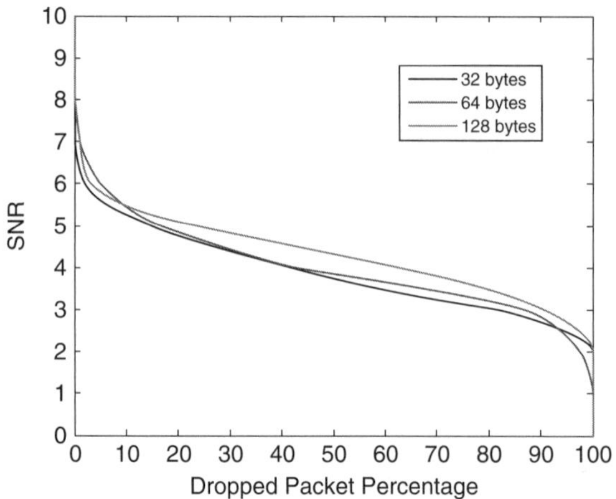

Fig. 15.16 PacComm 9600 bps performance.

for an SNR below 3 dB. Thus, we take these SNR levels as the minimum required SNR at the output of the RF receiver for a functional link. The next step is to determine the SNR at the output of the receiver given the link parameters.

The bandwidth of the noise for the preceding tests is 22.05 kHz. This bandwidth is dictated by the sound card, which had a maximum sample rate of 44.1 kHz and an analog filter stage to prevent aliasing. The analog filter stage could not be adjusted, and so the bandwidth of the sound card was defined as the bandwidth of the test. Modems typically have an analog filter to remove noise outside the message signal bandwidth, but this is of no consequence because the modem test is

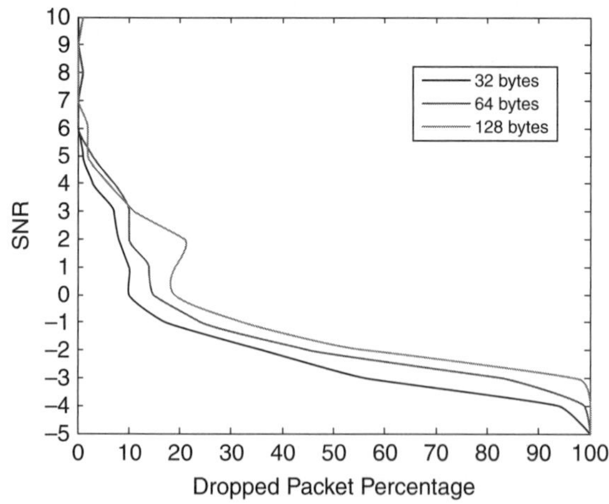

Fig. 15.17 Kantronics 1200 bps performance.

paired with the radio receiver test. The baseband SNR output of the receiver is tested in Section VI.D with the same noise bandwidth. Even if the receiver or modem filter out noise at some cutoff frequency below 22.05 kHz, the test remains valid assuming the noise is white (constant across frequency). The actual SNR into the demodulator might vary, but the total performance that is actually being tested is the SNR input to the receiver vs dropped packet rate. Thus, the SNR output from the receiver and input into the demodulator is an intermediate step where only the SNR across a specified bandwidth is important. The only requirement of the tests is that the demodulator input and receiver output is tested with the same noise bandwidth and that the noise output from the receiver is approximately white.

VI. FM Modulation

A. FM Modulation Theory

The next stage in the conceptual link (Fig. 15.3) is the radio frequency transmitter and receiver. Here the baseband signal is modulated onto a radio carrier for transmission, and then it is received and demodulated back to a baseband signal. Just like the baseband modems, there are many available modulation schemes to accomplish this. This section will begin with a discussion of the basic theory behind frequency modulation (FM). Then in the following section the benefits of frequency modulation will become apparent.

The general form of an FM signal is shown in Eq. (15.8):

$$s(t) = A\cos\left[\omega_c t + k_f \int_{-\infty}^{t} m(t')\,\mathrm{d}t'\right] \tag{15.8}$$

where $m(t)$ is the message signal produced by the baseband modulator. There are three variables that completely characterize an FM modulator: the amplitude A independently controls the transmitted signal power; the carrier frequency ω_c (rad/s), which is the constant transmitter center frequency; and the frequency deviation constant k_f controls the transmitted signal bandwidth.

Several import results are derived from considering a message signal $m(t)$ that is a sinusoidal tone [3,4]. The results are derived in several steps in Eqs. (15.9–15.13):

Sinusoidal message signal:

$$m(t) = A_m \sin(\omega_m t) \tag{15.9}$$

FM signal with sinsoidal message signal:

$$s(t) = A\cos\left[\omega_c t + \frac{k_f A_m}{\omega_m}\cos(\omega_m t)\right] \tag{15.10}$$

Fourier transform of Eq. (15.10):

$$S(\omega) = \pi A \sum_{n=-\infty}^{\infty} J_n\left(\frac{k_f A_m}{\omega_m}\right)\delta[\omega \pm (\omega_c + n\omega_m)] \tag{15.11}$$

Instantaneous frequency and maximum frequency deviation:

$$\omega_{\text{inst}} = \omega_c - k_f A_m \sin(\omega_m t), \qquad \Delta\omega_{\text{inst max}} = A_m k_f \tag{15.12}$$

Carson's Rule for FM signal bandwidth:

$$BW_{FM} \cong 2\left(\frac{\Delta\omega_{\text{inst max}} + \omega_m}{2\pi}\right) = 2\left[\frac{A_m k_f}{2\pi} + BW_{m(t)}(\text{Hz})\right] \tag{15.13}$$

The most important are Eqs. (15.11) and (15.13). In the Fourier transform equation $J_n(x)$ is the Bessel function of the first kind of order n. Carson's rule defines an effective bandwidth for the FM signal, which contains 98% of the signal power. Carson's rule extends to nonsinusoidal message signals by taking ω_m to be the highest frequency-domain component in a band-limited nonsinusoidal message signal. The frequency deviation is also controlled by $A_m k_f$ for nonsinusoidal signals where A_m is the maximum amplitude of the message signal.

B. Benefits of FM for Cubesats

Frequency modulation has several important features that make it well suited to cubesat communication systems. First, the power of an FM signal is controlled only by the signal amplitude A. The message signal and other parameters have no effect on the power. Even in the absence of a message signal, the FM signal is transmitted at constant power. This allows the transmitter power to be adjusted to an optimum value independently of other considerations. This is not true for amplitude modulated signals where the message signal directly affects the transmitted signal power.

Second, FM demodulators get an SNR performance boost if more bandwidth than necessary is used to transmit the signal. The bandwidth of the transmitted signal is given by Carson's rule. The bandwidth is controlled by $A_m k_f$ and the message signal bandwidth $BW_{m(t)}$. The message signal bandwidth is predetermined by the baseband modulation scheme. Thus, the signal bandwidth and SNR boost can be controlled by adjusting either k_f or A_m. A rule of thumb for FM systems is that for each doubling of the maximum frequency deviation $A_m k_f$, the demodulator output SNR increases by a factor of 6 dB when other variables are held constant [3].

Third, and perhaps most importantly, FM receivers are insensitive to small tuning errors caused by Doppler shift. In Fig. 15.18 a block diagram for an ideal FM receiver is shown. So long as the signal remains within the channel selection filter as discussed in Section IV.B, the output $m'(t)$ will only be a scaled version of the original $m(t)$. This can be seen from considering the effects that the

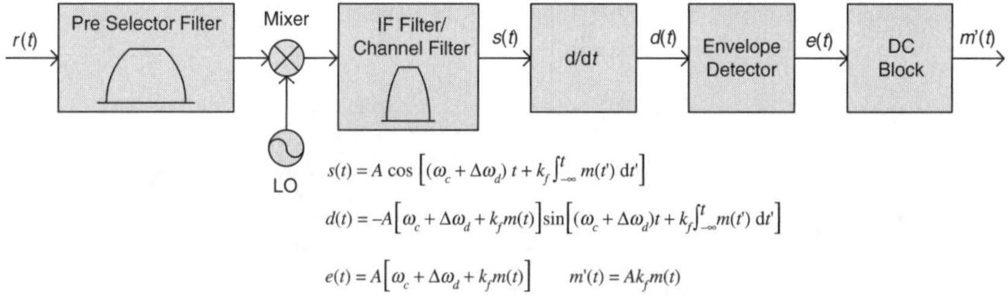

Fig. 15.18 FM ideal demodulator.

differentiator, envelope detector, and dc block would have on an FM signal with some tuning error. The signal after each stage in the block diagram is given in the figure to show this. The only restriction is that the Doppler shift varies with time much more slowly than the message signal so that the $\Delta\omega_d$ term is removed by the dc block. The maximum rate of change in the Doppler shift for ION is only 147 Hz/s, which satisfies this assumption.

C. Modulator Alignment [3,4]

Again, the drawback to using commercial-off-the-shelf (COTS) equipment is that basic parameters of the system are not supplied with the manufacturer documentation. In the case of FM radio transmitters, k_f must be known to calculate the signal bandwidth. To calculate or adjust this value, a procedure referred to as "alignment" is performed. Consider the Eq. (15.11) for the frequency-domain representation of the FM signal with a sinusoid message signal. The RF spectrum consists of impulses spaced at intervals of the input frequency ω_m about the center frequency ω_c and scaled by the Bessel function of the first kind of order n. An example of this is shown in Fig. 15.19 taken on a signal analyzer with a sinusoid message signal of 7 kHz. The Bessel function takes as an argument $A_m k_f/\omega_m$, which can be adjusted to cause the Bessel function to become zero. The arguments corresponding to zeros for several orders of the Bessel function are in Table 15.4.

To calculate k_f, the transmitter baseband input is driven with a function generator, and the RF output is fed into a signal analyzer. Then the amplitude A_m and

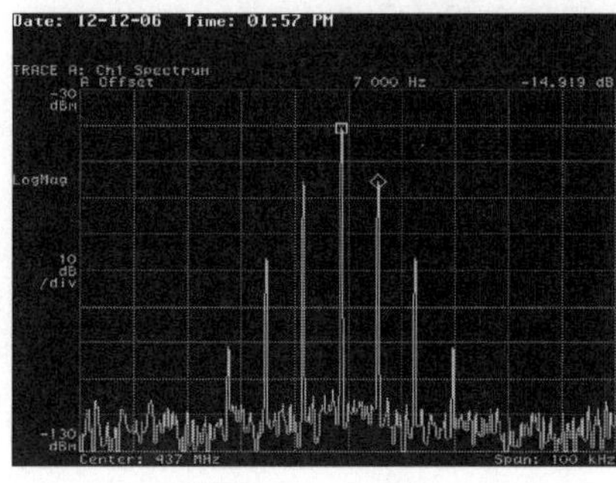

Fig. 15.19 FM spectrum.

Table 15.4 Bessel Zeros

		Order n					
		0	1	2	3	4	5
Zero	1	2.4048	3.8317	5.1356	6.3802	7.5883	8.7715
	2	5.5201	7.0156	8.4172	9.761	11.0647	12.3386
	3	8.6537	10.1735	11.6198	13.0152	14.3725	15.7002
	4	11.7915	13.3237	14.796	16.2235	17.616	18.9801
	5	14.9309	16.4706	17.9598	19.4094	20.8269	22.2178

frequency ω_m are adjusted until the nth impulse in the frequency domain goes to zero. At that point k_f can be solved for by equating the values in Table 15.4 with $A_m k_f / \omega_m$ for the appropriate impulse and zero crossing. Several data points should be collected across frequency and amplitude because in practice k_f is actually weakly dependent on input frequency.

For the ground station radio and satellite radio k_f was determined to be 78.19 Hz/mV and 76.16 Hz/mV, respectively. Values around this range are commonly implemented by commercially available transceivers. For these radios k_f is not adjustable, but the amplitude of the output from the modems is adjustable. Using Eq. (15.12), the amplitude was adjusted to give a maximum deviation of 3.5 kHz on each radio. This corresponds to a transmitted signal bandwidth of 12 kHz using Carson's rule, Eq. (15.13). For ION2 the deviation will be adjusted to 2.5-kHz maximum deviation and a transmitted signal bandwidth of 19 kHz. These signal bandwidths are typical for the respective data transmission rate in the amateur bands.

D. Demodulator Testing

The RF demodulator will have some relationship between input SNR and output SNR. The output SNR determines the performance of the baseband demodulator (Sec. V.B). There are theoretical models for the SNR input–output relationship, but again the real-world implementation can introduce significant degradation. The manufacturer documentation often gives a minimum sensitivity measurement of the radio, but this is a one-point sample out of a multidimensional performance. A method for direct testing was used to determine this relationship.

Using an RF signal generator with frequency modulation capabilities to drive the receiver under test the relationship between input SNR to output SNR was measured using a computer to capture and analyze the receiver baseband output. The experimental setup is shown in Fig. 15.20. There are three dimensions that can be varied in this test: the power level from the signal generator, which maps to an input SNR; the maximum deviation (product of k_f and A_m); and the frequency of the sinusoidal message signal. The maximum deviation was determined in Section VI.C to be 3.5 kHz, and so this is held constant throughout the test.

The computer runs a MATLAB® script that records five seconds of the output from the receiver and takes user input to specify the sinusoid message signal frequency. The script then computes a fast Fourier transform (FFT) of many

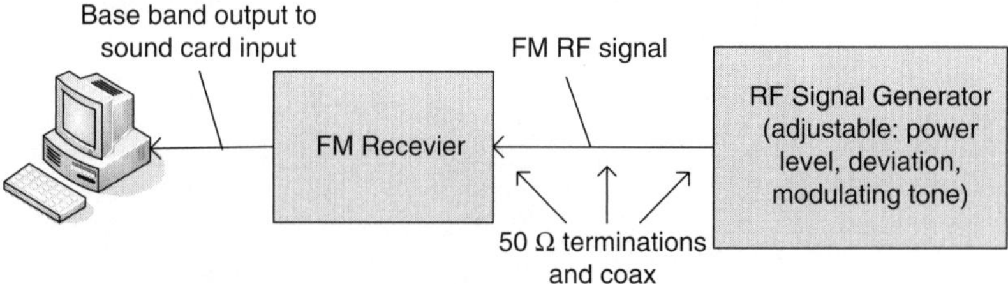

Fig. 15.20 FM receiver test setup.

overlapping windowed segments and computes the average PSD of the receiver output. Based on the user input frequency, the script sums the FFT bins for the signal and noise power separately and computes the SNR. A basic flowchart describing this process is given in Fig. 15.21. A 4096-point Chebyshev window with side-lobe attenuation set to 120 dB was used.

The RF signal generator is used to simulate a noisy signal, but the generator itself has no additive noise function. Instead, the power output is decreased to near the minimum, and the termination impedance of the source is used as a thermal noise source. The connecting coaxial cable is very short and assumed to be

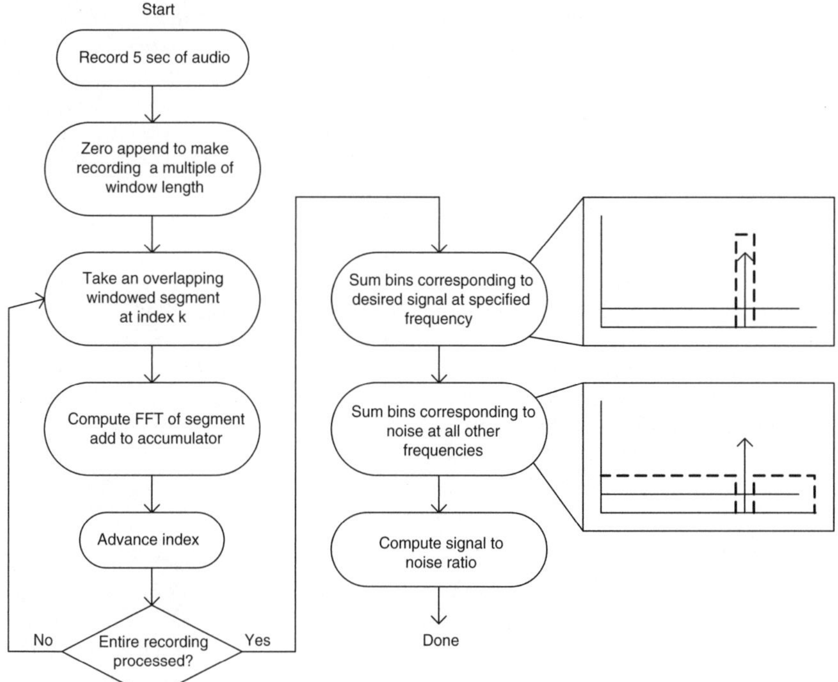

Fig. 15.21 SNR receiver test script.

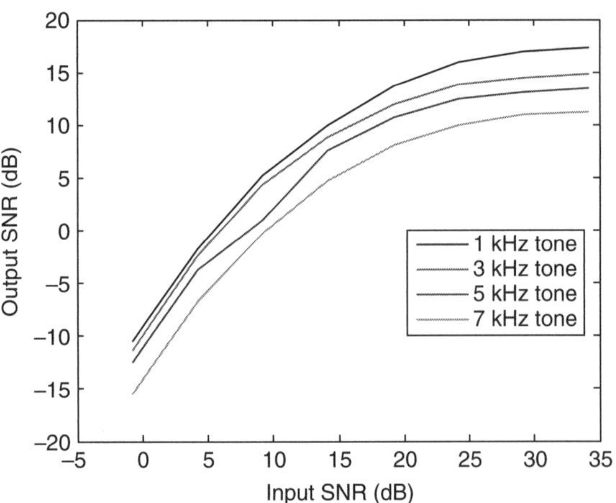

Fig. 15.22 FM receiver results.

lossless. Using a signal generator as a thermal noise source is a very conservative approximation for two reasons. First, the signal generator is really a "hot" noise source with noise contributed by active components so that the noise output is higher than the thermal noise floor. Two, the output signal power must be decreased to near the thermal noise floor so that thermal noise generated in the receiver will also degrade performance. The input noise power is computed using Eq. (15.15) with noise bandwidth taken to be the channel bandwidth of the receiver and temperature at 290 K. The results for the ground station Icom 910H transceiver (30 kHz BW) are presented in Fig. 15.22. The SNR tends to decrease with increasing frequency. This is because a frequency-modulated message signal is more susceptible to noise at higher frequency [3]. However, the output SNR at the highest frequency in the message signal will be used as the SNR for the receiver output across all frequencies, which is another conservative approximation. The results show that the input SNR to the receiver must be at least 9 dB to get at least a 3-dB output SNR, which results in acceptable demodulator performance for 1200-bps AFSK modulation (Sec. V.B). The satellite side transceiver for ION was tested in the same manner and had very similar performance. The results are listed in the link budget spreadsheet (Table 15.5).

VII. Transmission Line

A. Transmission Line Theory [5]

The transmission line (TL) portion of the communication link carries RF signals between the transceiver and antenna. In the context of the generic communication system (Fig. 15.3), the TL also includes passive inline devices and on the *Rx* side sometimes a preamplifier. Between the input and output of the TL, the signal experiences attenuation through resistive loss and impedance mismatch reflection. If there is a preamplifier, the signal power experiences a gain but might

Table 15.5 TL Loss Test and Results

Condition	SWR	Loss, dB
1.TL short load	1.592	−6.41
2.TL ant. load	1.173	−0.35
3.Transceiver input	1.97	−0.49
Total TL loss	NA	−7.25

also have impedance mismatch loss at the input and output of the preamplifier. The preamplifier is only active during reception; thus, during transmission it acts like another passive inline device. In Fig. 15.23 the formulas for calculating impedance mismatch loss and resistive TL loss are given for a source, load, and segment of transmission line. The source, load, and TL each have a complex impedance Z at the frequency of operation. The TL has a loss tangent α at the frequency of operation, which determines the resistive loss per unit length.

The formulas in Fig. 15.23 can be extended to a more complex system containing many segments and numerous inline devices by chaining segments together as shown by the dotted lines. If the inline device were a preamplifier in receive mode, then the loss factor (less than one) is actually a gain (greater than one).

B. Transmission Line Testing

The ION ground station was upgraded shortly before the anticipated launch with new coaxial cable (the transmission line) because the old cable had been destroyed by water damage. The new cable used was Belden 9914 RG8/U 50 Ω coaxial cable. The cable and inline devices were installed without any testing because the launch date was quickly approaching. Once the cables and inline devices were installed, testing was performed to ensure the transmission line was delivering the expected amount of power into the antenna. However, because the

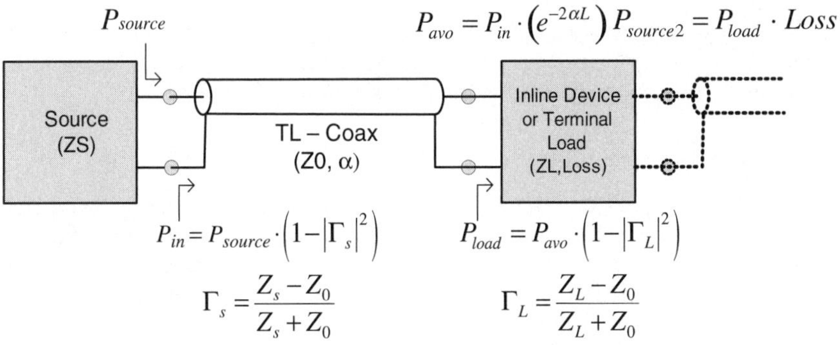

Fig. 15.23 TL theoretical diagram.

cable was in a permanent installation, the loss and impedance mismatch of the cable and inline devices could not be directly measured. Instead standing wave ratio (SWR) measurements were used to calculate the total loss and reflection at the antenna input.

The SWR is the ratio of maximum to minimum voltage or current magnitudes present on the transmission line. It can be used to compute the loss tangent α and reflection coefficient Γ for a TL terminated by a load with some arbitrary impedance. It can also be used to determine the reflection loss at the output of the transceiver. The SWR was measured inside the ground station at the junction between the transceiver and coaxial TL. Measurements were taken under three conditions: 1) with the TL shorted at the junction to the antenna, 2) with the TL connected to the antenna, and 3) at the input to the transceiver. The first two measurements were used to solve for two unknowns, TL resistive loss (the product of α and d) and antenna reflection loss (Γ_{ant}). The results are given in Table 15.5, along with Eq. (15.14). The distance d is taken to be the distance from the load, which is L for the first two conditions and zero for the third condition. The reflection coefficient for a short circuit load is known to be –1 from TL theory [5]. The total loss agrees well with the manufacturer documentation for the coaxial cable and antenna. A power meter was later used to directly measure the power loss throughout the cable. It showed the SWR measurements to be accurate within 1%. A total loss of 7.25 dB means that 81% of the power is lost end to end in the transmission line. This might seem unacceptably high, but it is unavoidable for long cable runs even with high quality coaxial cable.

$$\begin{aligned} \text{SWR}(d) &= \frac{1+|\Gamma_L|e^{-2\alpha d}}{1-|\Gamma_L|e^{-2\alpha d}}, \quad \Gamma_{\text{Short}} = -1 \\ \text{Loss (dB)} &= 10\log\left(1-|\Gamma_L|^2\right) = 10\log\left(e^{-2\alpha d}\right) \end{aligned} \tag{15.14}$$

The only assumption made when using the SWR method to calculate line loss is that the passive inline devices are well matched to the cable impedance. If this is true, then the reflection at each of the inline devices is small and contributes a negligible amount to the SWR reading under the short and loaded conditions. Then the reflection and resistive losses of the inline devices are added to the cable loss and make it appear longer and add to the total loss. If the devices are not well matched, then the reflection at a particular inline device will dominate, and the SWR measurement is only valid up to the input of that device.

C. Preamplifier

On the receive path the preamplifier boosts the received signal to overcome noise generated in the TL. With respect to total signal gain, the preamplifier can be placed anywhere in the line because the gain and loss factors are commutative. However, noise considerations dictate that the preamplifier be placed as close as possible to the antenna feed point to maximize the SNR at the input to the radio. This is discussed more in Section IX.C. The preamplifier for the ION ground station is an Icom AG-35 70-cm band preamplifier. It was tested using an Agilent

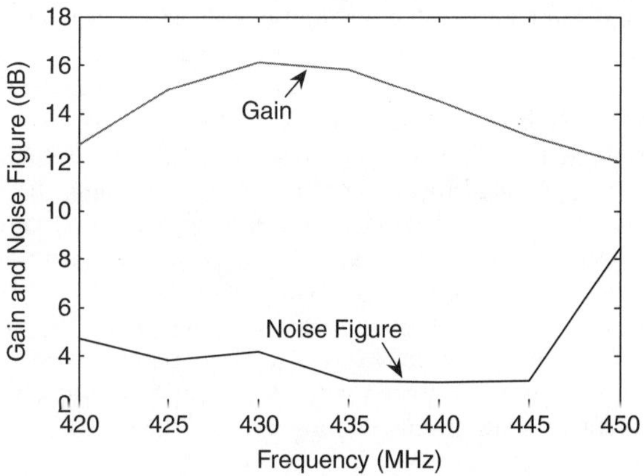

Fig. 15.24 Gain and noise.

N8973A noise figure analyzer, and the results are shown in Fig. 15.24 over the range of the 70-cm band. The AG-35 has a 15.3-dB gain and a 2.9-dB noise factor at 437.505 MHz.

VIII. Antennas [5]

A. Directional Gain

The most common measure of an antenna is the gain. This indicates the signal gain in the direction(s) of maximum gain. A common misconception is that antennas somehow add signal power as if it were an active device. Antennas are passive devices. If an ideal antenna were enclosed in a sphere, the power radiating through the sphere would be equal to the power delivered into the antenna feed point. The directional gain arises from the power being focused towards a certain area and away from others. The total gain in all directions averages out to 0 dB. The formula for directional gain is given by Eq. (15.15). P_{ant} is the power delivered into the antenna, and $P_R(\theta_{\max}, \phi_{\max})$ is the maximum power radiated towards any single point. An important concept in antenna theory is reciprocity, which states that the receiving and transmitting pattern of an antenna are identical. This allows the antenna to be completely characterized by one radiation pattern for both transmission and reception.

$$G_{\max} = \frac{4\pi \left\langle P_R\left(\theta_{\max}, \phi_{\max}\right)\right\rangle}{P_{\mathrm{Ant}}} \tag{15.15}$$

Another point of confusion comes from the reference level of the gain. Often an antenna gain is specified as dBd, which is referenced to the directional gain of a dipole. The direction gain of dipole antenna is 2.15 dBi, where dBi is referenced to an isotropic radiator. An isotropic radiator is a fictional antenna that radiates power equally in all directions. Thus, the total gain of the antenna in dBi is the gain in dBd plus 2.15 dB.

In documenting a link budget, the maximum gain is usually tabulated, and then a pointing error loss is subtracted from the maximum gain. Radiation pattern plots are usually drawn in this manner too with the pattern normalized to the maximum gain such that angles away from the maximum gain appear as a loss. There will be a pointing loss for both the ground station and satellite antennas because they will not always be optimally oriented towards each other. Determining the exact pointing loss can be difficult, and often one must resort to using the worst-case scenario.

Upon ejection from the launch vehicle or after an attitude control system (ACS) failure, a cubesat will be tumbling randomly. Any radiation pattern when spun about randomly will average out over time to an isotropic pattern. This is a useful approximation when documenting a cubesat link budget. The link should be designed so that it is successful even when there is no attitude control by taking the satellite directional gain minus pointing loss to total 0 dBi.

B. Polarization

The next major parameter of an antenna is its polarization. The polarization refers to the orientation and relative phase of the electric field components, which make up a propagating EM wave. There are three flavors of polarization: circular, elliptical, and linear. Upon closer inspection linear and circular polarization are just special cases of elliptical polarization. Consider Fig. 15.25, which shows two orthogonal E fields out of phase by 90 deg. If E_1 and E_2 are equal, then the EM wave is circularly polarized (CP). If E_1 and E_2 are not equal, then the EM wave is elliptically polarized (EP). If the wave is propagating into the page, the wave has a right-hand polarization (RHP), and if the wave is propagating out of the page, then it has a left-hand polarization (LHP). If either E_1 or E_2 is zero, then the EM wave is linearly polarized (LP).

Communication and broadcast satellites often use polarization for channel isolation on the same frequency to effectively double the data throughput per unit of bandwidth. This can be accomplished by using vertically and horizontally oriented LP waves or using circular LHP and RHP polarized waves. This would seem like an attractive option for cubesats to isolate the uplink and downlink channels for a full duplex link on one frequency.

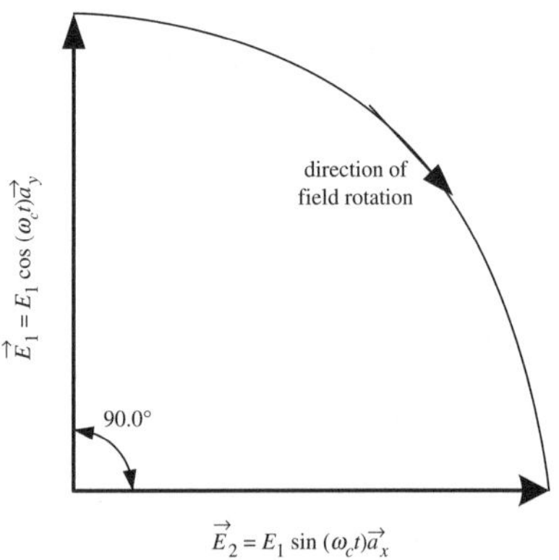

Fig. 15.25 E field polarization.

However, there are several major problems with this. A circularly polarized antenna will produce a different direction of field rotation when the transmitting antenna is viewed from the opposite side. This results in the propagating wave changing its handedness (LHP vs RHP) when the antenna is flipped 180 deg. For example, view Fig. 15.25 from the back side of the page, and note that the direction of field rotation is now CCW. Thus, the ground station must be able to switch the polarization of the transmitting and receiving antennas on the fly if the satellite is tumbling. For the linear scheme, the receiving and transmitting antennas must have their orthogonal linear elements aligned nearly perfectly for adequate isolation. At just 20-deg misalignment the isolation is decreased to only 8.8 dB [1]. This requires the satellite to have an excellent attitude control system or the ground station to have linearly polarized antennas, which can be reoriented to match the signal from the satellite. Further complicating the linear scheme is the phenomenon of Farraday rotation.

To avoid these unnecessary complications, a linear-circular scheme is the simplest and most effective. The satellite has a linearly polarized antenna, and the ground station uses a circularly polarized antenna. Choosing a LHP or RHP ground antenna is inconsequential. This allows the satellite to be in any orientation without polarization effects breaking the link. Furthermore, only one antenna needs to be used on the satellite and ground station. This system was implemented for the ION antenna system. The ground station CP antenna will be equally sensitive to a linear signal at any orientation. The satellite LP antenna can be at any orientation, yet it will be equally sensitive to a CP wave. This is another manifestation of reciprocity. The only drawback is there is a 3-dB polarization loss between linear and circularly polarized antennas (see Fig. 15.26 for derivation). This will be incurred on both uplink and downlink transmission. This loss is worth the simplifications it buys. The radiation pattern might be too weak for certain orientations of the satellite, but that is addressed with pointing loss calculations in the link budget.

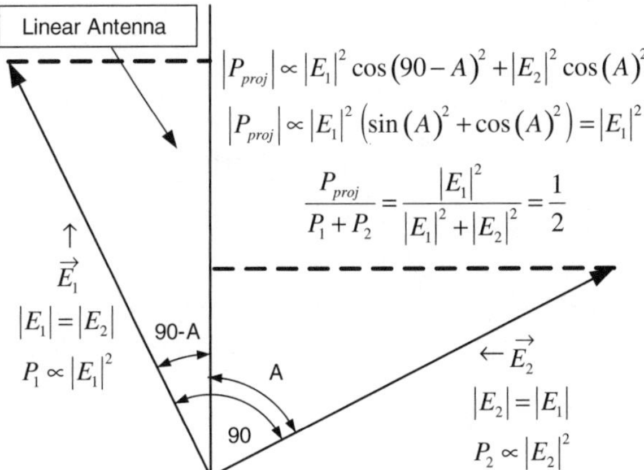

Fig. 15.26 Polarization loss CP to LP.

Ellipticity is a parameter often specified in the documentation for a CP antenna. It refers to the deviation away from an ideal circular pattern to an elliptical pattern. It specifies the amount of power gain along the main elliptical axis and loss along the minor elliptical axis usually in decibels or a ratio. When the receiving antenna is a linear antenna, the ellipticity is taken as a worst-case loss when the linear antenna is parallel to the minor elliptical axis.

C. ION Satellite Antenna

Because of the considerations in the preceding section, a linearly polarized antenna design was chosen for ION. To choose among the many types of linear antennas, primary consideration was given to the radiation pattern. A highly omnidirectional pattern is desirable because of the uncertainty regarding the ACS and to enable communications without stabilization. One of the simplest approximately omnidirectional designs is a half wave center fed linear dipole. At 437 MHz the antenna elements are shorter than 20 cm, the length of the double cubesat like ION, which simplifies the deployment mechanism. The normalized ideal radiation pattern for a half-wave dipole is shown in Fig. 15.27 along with the formula that describes it. Note that the plot is only a two-dimensional slice, and the three-dimensional pattern is radially symmetric around the dipole. The current and voltage distributions along the dipole are shown in Fig. 15.28. The current is at a maximum and moving in the same direction at the feed point, which requires that the phase of the feed point current be shifted 180 deg for one antenna element.

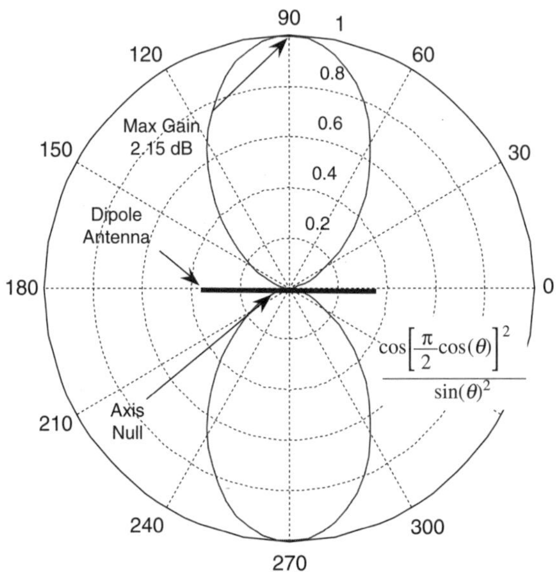

Fig. 15.27 Half-wave linear dipole radiation pattern.

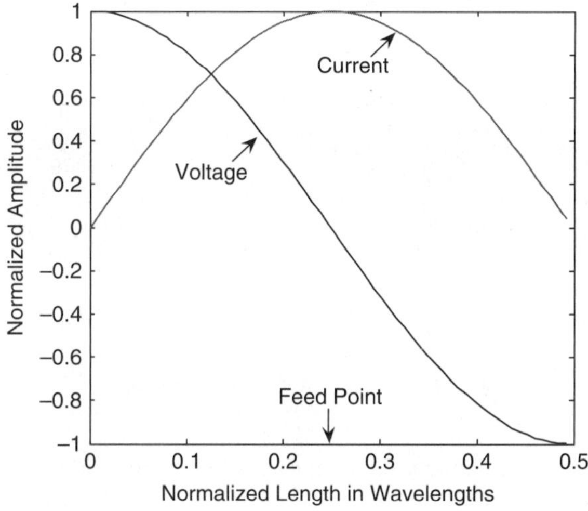

Fig. 15.28 Half-wave dipole curren and voltage distribution.

The theoretical maximum gain for a dipole antenna is 2.15 dBi and is constant in the plane perpendicular to the dipole axis, which passes through its midpoint. ION's dipole antenna will be parallel to the nadir face. To calculate the pointing loss, there are two scenarios. If the satellite is tumbling, assume that over time the radiation pattern becomes isotropic. Thus, the pointing loss is –2.15 dB for a 0-dB total directional gain. If the satellite is stabilized in the nadir direction (towards the center of the Earth), then the worst-case scenario is that the antenna is oriented in the plane that contains the triangle in Fig. 15.4. The pointing loss angle away from maximum gain in this orientation is tabulated in Fig. 15.4 as well. To determine the pointing loss, evaluate the formula for the radiation pattern at the corresponding angle away from the angle of maximum gain maximum gain (i.e., 90-deg-PL; note θ is in radians in Fig. 15.27). The calculated maximum pointing loss is –9.41 dB at 10-deg elevation, and the pointing loss decreases with elevation if the satellite is stabilized in the nadir direction.

The satellite antenna system consists of the power divider phase inverter IC (TeleTech DX22-27), a two-layer circuit board mount for the IC with microstrip lines, and rolled steel antenna elements. The circuit board dielectric is made of Rogers Corporation RT/Duroid 6002 PTFE composite ($\varepsilon_r = 2.94$) with a copper-clad ground plane. The copper cladding was recommended by the manufacturer for the thermal environment of space. CAD software was used to calculate the microstrip width given the dielectric thickness of 0.762 mm and the frequency of operation at 437.505 MHz. The resulting width was 1.9 mm. The dimensions of the antenna circuit board were determined by its placement in the satellite. The phase inverter IC was placed midway between the connection points to the antenna elements, and the feed point to the IC was placed to minimize the distance to the transceiver. The antenna elements were cut to quarter-wavelength pieces at 437.505 MHz. They were connected using only solder and epoxy for reinforcement.

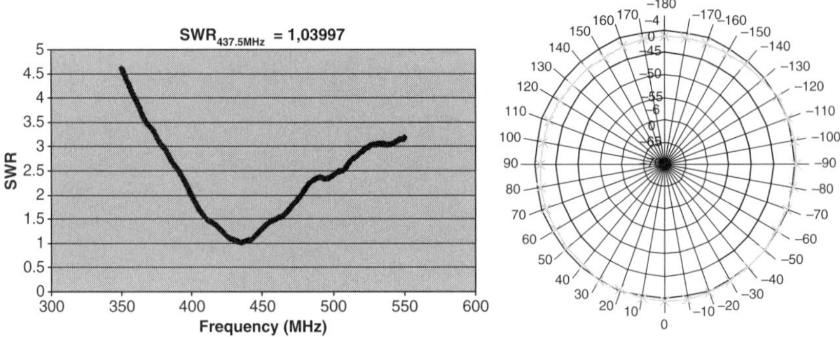

Fig. 15.29 ION antenna test results.

The last step was to attach the antenna feed point to a network analyzer and trim down the antenna elements until the SWR indicated a good match. Using exactly quarter-wavelength elements results in a radiation impedance of 73 Ω[5]; this would have reflection loss of 0.15 dB. This reflection loss is acceptable but can be improved upon by shortening the antenna elements slightly. The final VSR for the ION antenna was 1.04, which indicates an excellent match and a negligible reflection loss. The SWR vs frequency and the results of testing the antenna pattern are shown in Fig. 15.29. The antenna pattern was tested in the plane of maximum gain to test the effects the satellite body would have on the radiation pattern. The body of the satellite slightly distorted the otherwise uniform pattern.

D. ION Ground Station Antenna

The ground station antenna is a circularly polarized high-directional-gain Yagi antenna from M2 Antenna Systems (436CP42 U/G). It has a total gain of 18.95 dBi and a narrow –3-dB beam width of 21 deg. The radiation pattern from the manufacturer documentation is given in Fig. 15.30. Again this is a one-dimensional slice, and it is radially symmetric around the axis through 0 and 180 deg. This pattern is typical of high-directional-gain Yagi antennas and is well suited for a cubesat ground station. A high-gain ground station antenna allows the cubesat to transmit at very low power and use a low-gain antenna. Typical cubesat transmitter power is between 2–0.5 W (roughly the same power output of a single Christmas tree light bulb). This is critical because the power generation of a cubesat is very limited. The reason for choosing a circularly polarized antenna was already discussed.

The pointing loss on the ground station side is a total of three errors: 1) the antenna rotator alignment error estimated to be roughly 2 deg; 2) the mechanical accuracy of the rotators, which is given in documentation for the M2 rotators as 1 deg; 3) the tracking software error, which should have a negligibly small error with accurate ephemeris data. Summing and rounding up the pointing errors for the ground station gives an estimated 5-deg maximum pointing error. Based on the radiation pattern plot, this pointing error results in a –1-dB pointing loss.

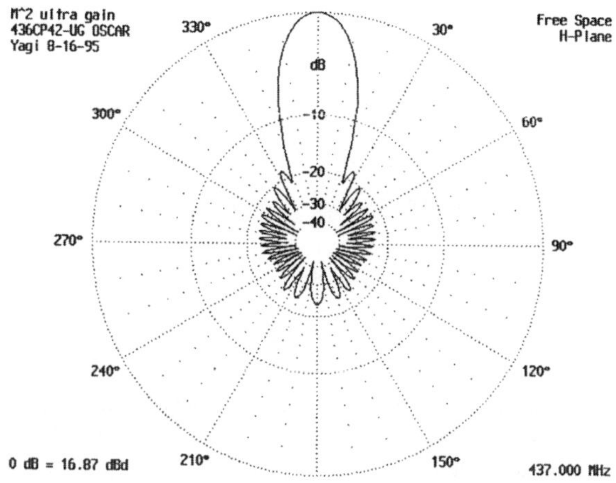

Fig. 15.30 Ground station Yagi antenna pattern.

E. Antenna Rotator and Calibration

Because the beam width is extremely narrow, the ground station antenna must be positioned to point at the satellite while it passes overhead. This requires the use of mechanical rotators. The rotators should be accurate and reliable. The ION ground station rotators were also supplied by M2 and have thus far proven to be very reliable and accurate. There are generally two methods for determining the rotator position in commercially available rotator equipment: a switch with a pulse counter or an analog potentiometer coupled with an analog-to-digital converter. The potentiometer-based scheme, in our experience with other equipment, is very problematic. The resistance can change with environmental conditions and must be recalibrated often. A pulse counting scheme is much more accurate and reliable. Furthermore, rotators with worm gear drive mechanisms tend to be more mechanically robust.

To calibrate the antenna rotator position, it is easiest to use celestial bodies in the night sky. Most satellite tracking software is also capable of tracking the position of the moon, planets, and stars with high accuracy. Select a celestial body that is easy to see at your geographic position. Point the antenna boom at the body by sighting along the boom. Then take the azimuth and elevation position of the body calculated by the tracking software for the reference point for the antenna rotators. For the ION ground station we use Nova tracking software and the moon to align our antennas because there is too much light pollution to see much else. We estimate the error in this process to be no more than 2 deg based on the viewing angle of the moon, which is approximately 0.5 deg.

An alternative calibration method is to use the North Star (Polaris), which does not require any tracking software. Of course, this only works in the northern hemisphere. Polaris is relatively bright and can easily be identified as the last star in the tail of the Little Dipper constellation. The rotational axis of the Earth points almost directly (currently 0.7 deg away) at the star, and so it appears to be directly over the North Pole or at 0-deg azimuth. The elevation of Polaris above the horizon at your geographic position is exactly the latitude (±0.7 deg) of your position. Thus, the elevation and azimuth can be calibrated very accurately just using Polaris.

IX. Noise

A. Noise Theory

Noise is the limiting factor in any communication link and must be quantified in order to document the link budget properly. The SNR performance tests documented in Sections V.C and VI.D are steps in a chain leading back to the sources of noise that will determine the SNR at the input of the receiver. The SNR at the input of the receiver determines the SNR at the input to the baseband demodulator, which in turn determines the BER or dropped packet percent. Now the SNR at the input to the receiver must be calculated.

Noise in communication systems is typically specified as an effective input temperature. This is done in order to treat all noise sources as a noisy resistor and to refer all noise sources to the input of the system. With noise referred to the input, the system can be considered noiseless with a noisy resistor at the input. There are just four formulas that are needed to quantify noise in the system when all components are treated as two port networks. Each two-port network has an effective input temperature and gain (or loss) associated with it. If it is a passive attenuator two port, such as a piece of transmission line, then the effective input temperature is described by Eqs. (15.16–15.19), where T_0 is the physical temperature of the attenuator and L is the loss.

Thermal noise available from resistor:

$$N_{\text{avo}} = kT_e B \tag{15.16}$$

Effective temperature of passive attenuator:

$$T_e = T_0 \left(L - 1\right) \tag{15.17}$$

Noise figure–effective temperature conversion:

$$NF = 10\log_{10}\left(1 + T_e/T_0\right) \tag{15.18}$$

Friis' formula for cascaded:

$$T_e = T_1 + \frac{T_2}{G_1} + \frac{T_3}{G_1 G_2} + \cdots + \frac{T_n}{G_1 G_2 \cdots G_{n-1}} \tag{15.19}$$

If it is an active device, such as a preamplifier, then the noise figure must be known and then converted to an effective input temperature by Eq. (15.18), where T_0 is again the physical temperature. When two-port networks are connected in a series cascade, the overall effective input temperature is given by Eq. (15.19), or Friis' formula. The effective input noise temperature can be converted to a noise power using Eq. (15.16). This formula describes the noise power available from a noisy resistor with temperature T_e over some bandwidth B, which is typically the receiver channel selection filter.

The antenna is also treated as a noisy resistor except that its effective temperature is determined by the amount of noise it receives. When an antenna is connected to the input of the system, it is modeled as a noisy resistor in series with the fictional noisy resistor that models the system noise (referred to the input). The effective temperatures of the series resistors are summed for the total effective input temperature. The antenna also receives the desired signal, and the SNR can be determined at the output of the antenna. The SNR at the input to the receiver will be the same because both the signal and noise will experience the same gain and attenuation throughout the transmission line.

B. Antenna Noise

The receiving antennas can receive three sources of noise: 1) sky noise from space, which in the frequency range of 50 MHz to 4 GHz is of interstellar origin [4]; 2) thermal blackbody radiation noise that is generated by anything having a nonzero temperature; and 3) human-made interference that is generated by a wide variety of sources.

The ground station antenna will primarily only receive sky noise. The minimum elevation angle was chosen as 10 deg, and the antenna beam width is 21 deg, so that only a small 0.5-deg slice of the antenna pattern will be receiving terrestrial noise sources at 10-deg elevation. Therefore, terrestrial noise sources are neglected for the ground station antenna. The sky noise that the ground station antenna receives is interstellar noise generated within our galaxy. Figure 15.31 shows the effective antenna temperatures over a wide range of frequency. In the range of 50 MHz to 4 GHz, it specifies a daily minimum and daily maximum because twice daily the Earth rotates through the galactic plane where interstellar noise is generated. In keeping with taking the worst-case scenario for the link budget, the daily maximum will be the antenna temperature used, which at 430 MHz corresponds to a temperature of 200 K.

The satellite antenna will receive both terrestrial noise sources and interstellar noise. The terrestrial noise sources consist of thermal blackbody radiation from the surface of the Earth and human-made noise sources. The blackbody radiation from the Earth is also described by Eq. (15.15), and T_0 is taken as the average temperature of the surface of the Earth. Human-made sources of noise and interference will raise the apparent temperature of the surface of the Earth to $T_0 + T_h$ and an estimated value of 300 K is used (+13 K over the average surface temperature of 287 K). To determine the effective noise temperature of the satellite antenna, one must take a weighted average of all of the noise sources that the antenna "sees." The weighting for each source is determined by the antenna radiation pattern. For a linear dipole this is simple because it has a radially symmetric pattern, and it will "see" an equal amount of sky and Earth in a LEO orbit. The satellite antenna temperature is then 1/2(300 K + 200 K) = 250 K.

C. Noise Calculations

The ground station system is modeled as a cascade of two-port networks in Fig. 15.32. The results of the calculations described in Section IX.A are also given in the figure. The physical temperature of the ground station components is

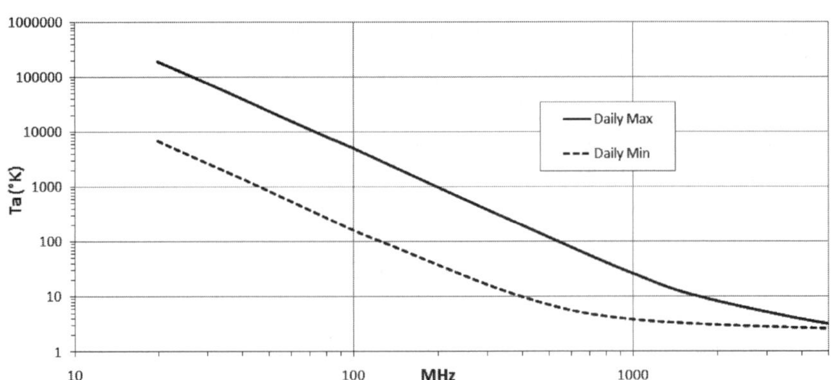

Fig. 15.31 Antenna temperature.

estimated at standard temperature, 290 K. The main result is the total effective temperature because it describes the noise power available at the output of the antenna.

The satellite system is modeled as a single two-port network in Fig. 15.33. The results of the noise calculations are also given. Here the single two-port network is the lossy transmission line between the transceiver and antenna. The length of this TL is extremely short, and the loss is not more than 1 dB. The TL temperature is estimated to reach at a maximum 273 K because it is not a heat-generating device nor is it directly exposed to sunlight.

Nowhere in the calculations does the directional gain or actual impedance of the antenna matter. We have also assumed that sky noise is spatially uniform. This is of course not true, but a safe approximation when the daily maximum sky noise temperature is used as the antenna temperature. The noise power available from a resistor is independent of its impedance. However, this implicitly requires the system to be perfectly matched for all the available noise power to actually be transmitted to the load. The noise formulas in Eqs. (15.15–15.18) assume the system are perfectly matched. Under the condition that the system is "well" matched, the calculations are still valid.

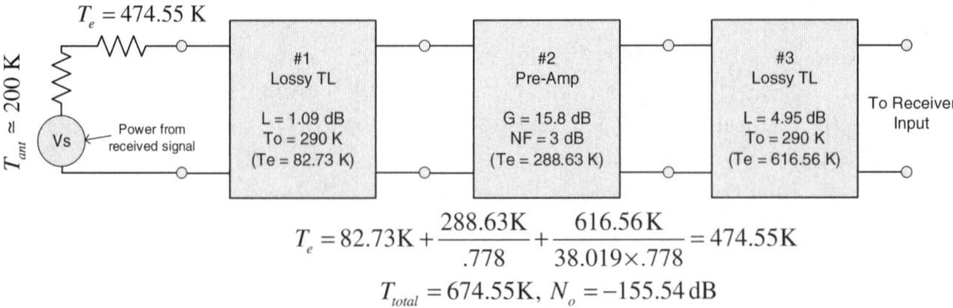

$$T_e = 82.73\text{K} + \frac{288.63\text{K}}{.778} + \frac{616.56\text{K}}{38.019 \times .778} = 474.55\text{K}$$

$$T_{total} = 674.55\text{K},\ N_o = -155.54\,\text{dB}$$

Fig. 15.32 Ground station two-port noise calculations.

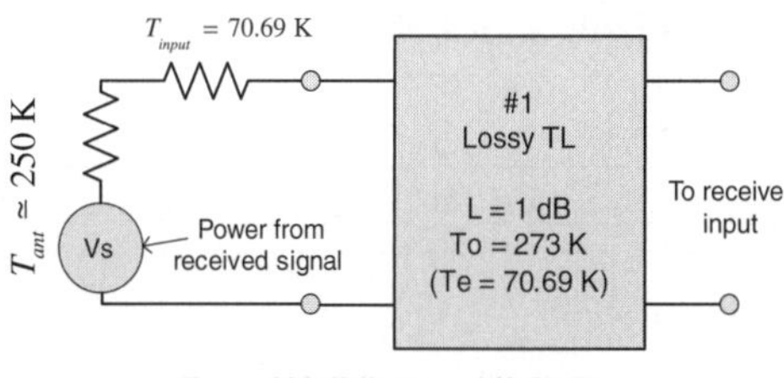

Fig. 15.33 Satellite two-port noise.

X. Conclusions

A. Link Budget

At this point a formal link budget can be documented with the parameters presented throughout this paper. The link budget spreadsheet for ION is presented in Table 15.6. The worst-case and average-case scenarios are tabulated to show the difference between antenna loss and power considerations. Of interest is that the worst case is calculated with only a 1-W transmit power instead of the 2-W power actually implemented on ION. Output power of the ground station radio is also taken to its minimum value of 5 W for the worst case and 50 W for the average. Pointing loss plus directional gain is taken to be zero for the average case, corresponding to a tumbling satellite. Ellipticity is taken as a zero for the average case because over time the random orientation of the satellite will average out the elliptical gain or loss.

The link budget is broken in sections corresponding to the conceptual link in Fig. 15.3. When the final SNR into the radio is calculated, the SNR output is interpolated from the receiver test data in Section VI.D. Then the packet error rate is interpolated from the demodulator performance data in Section V.B. If the dropped packet rate is acceptable, then the link is considered functional or "closed." The results show that for both the uplink and downlink worst-case scenarios the link is closed at 10-deg minimum elevation. A roughly 10% packet drop rate is rather large, but acceptable so long as the communication protocol is able to request retransmission of dropped packets. Of course, the link will only improve as the elevation angle increases.

The analysis presented here was also performed for the ION2 system, which will use the same ground station hardware, the PacComm modems, and a different satellite radio. The analysis shows that at 10-deg minimum elevation angle the link will be marginal. By 15 deg the link will be closed because the higher data rate modems require a better SNR at the demodulator input to close the link. This is not a problem with respect to the available data bandwidth because the transmission rate is eight times faster for the ION2 modems.

B. Recommendations

The link budget presented here shows two important things with respect to the communication system. First, the satellite can transmit at only 1 W and the ground station at only 5 W if a high-directional-gain antenna is used on the ground station. ION was built with a 2-W transmitter because in the initial design phase the

Table 15.6 ION Link Budget Analysis

Parameter	Ground -> Sat		Sat -> Ground	
	Worst case	Average case	Worst case	Average case
1. Radio power output Tx	7[a]	17[a]	0[a]	3.01[b]
2. Transmission line Tx				
Radio TL mismatch	−0.49	−0.49	−0.5	−0.5
Coax + inline device loss	−6.41	−6.41	−0.5	−0.5
Antenna mismatch loss	−0.35	−0.35	−0.01	−0.01
3. Propagation Loss				
Antenna gain Tx	18.95[c]	18.95[c]	2.15[d]	2.15[d]
Pointing loss Tx	−1	−1	−9.41	−2.15
Free space loss	−150.1	−150.1	−150.1	−150.1
Atmos loss	−1.5	−1.5	−1.5	−1.5
Antenna gain Rx	2.15[d]	2.15[d]	18.95[c]	18.95[c]
Pointing loss Rx	−9.41	−2.15	−1	−1
Polarization loss Rx	−3	−3	−3	−3
Ellipicicity loss Rx	−1.5	0[c]	−1.5	0[c]
4. Transmission line Rx				
Antenna mismatch loss	−0.01	−0.01	−0.35	−0.35
Pre-amp gain	0[a]	0[a]	15.3[b]	15.3[b]
Coax + inline device loss	−0.5	−0.5	−6.41	−6.41
Radio TL mismatch	−0.5	−0.5	−0.49	−0.49
5. Rx Parameter				
Power into radio	−174.77	−166.01	−174.77	−166.01
Noise into radio	−161.54	−161.54	−147.49	−147.49
Noise at antenna output	−160.53	−160.53	−155.54	−155.54
Effective Noise Temp (K)	320.69[d]	320.69[d]	674.55[d]	674.55[d]
Channel BW Rx (kHz)	20[c]	20[c]	30[c]	30[c]
Input SNR	14.87[d]	33.71[d]	9.12[d]	20.89[d]
6. Rx radio SNR performance				
SNR in –> out profile	9.3[b]	14.8[b]	4.21[b]	12.7[b]
7. Rx TNC Performance				
SNR –> dropped packet (%)	~0%[b]	~0%[b]	<10%[b]	~0%[b]
Minimum tolerable drop rate	10%[a]	10%[a]	10%[a]	10%[a]
Results				
Link closed?	yes	yes	yes	yes

[a]Estimate or design goal.
[b]Measured or derived.
[c]Manufacturer or documentation.
[d]Theoretical or calculated.

link budget had not been carefully documented. Commercial 70-cm transmitters are roughly 25% efficient, and so a 2-W transmitter consumes about 8 W, which is a significant power drain on the satellite. One watt or less should be sufficient for a LEO cubesat.

Second, a high-gain antenna is necessary for the ground station. The antenna should have a gain on the order of 15–20 dBi. However, a high-gain antenna is not needed or desirable on the cubesat. In fact, in the link budget worst-case scenario the satellite antenna has a total directional loss. In the average case the antenna is approximated as isotropic radiator and the link still closes. Furthermore, a highly directional antenna requires the satellite to be oriented correctly and makes the communication system dependent on the attitude control system. The one significant problem with a linear dipole antenna is the on-axis nulls in the radiation pattern. If the satellite is tumbling and a null points towards the ground station, then the link will fail. However, if the satellite is stabilized, the on-axis null will never point at the ground station.

Another important design consideration is polarization. For the reasons outlined in Section VIII.B, the linear-circular antenna scheme is the simplest and most reliable. It only introduces a 3-dB polarization loss, which does not break the link. Channel isolation via polarization multiplexing is not feasible for a primary cubesat communication system.

A preamplifier is usually necessary for the ground station if the coaxial cable connecting the receiver to the antenna is of appreciable length (contributing more than 1–2 dB of attenuation). The preamplifier is used to overcome noise in the coaxial TL not the resistive attenuation of the TL. Equation (15.18) can be manipulated to show that the preamp should be located as close as possible to the antenna feed point for the lowest effective input temperature [4]. A preamplifier will not improve the SNR for the satellite receiver, which is also shown by Eq. (15.18).

The effects of Doppler shift can be mitigated by using a lower frequency. For the 2-m band the maximum shift for ION would have been only 3.4 kHz. The transmitted signal would not drift outside the channel selection filter, and thus the receiver would not need any active correction.

References

[1] King, J. A.. "Amsat IARU Link Budget Spread Sheet V2.2.1," Amateur Satellite Corp., Salt Lake City, UT, 5 May 2005.

[2] Ippolito, L. J., *Radiowave Propagation in Satellite Communications*, Reinhold, New York, 1986.

[3] Lathi, B. P., *Modern Digital and Analog Communication Systems*, 3rd ed., Oxford Univ. Press, New York, 1998, pp. 208–250, 541–557.

[4] Franke, S. J., *ECE 453 Radio Communication Circuits and Systems Course Notes,* Dept. of Electrical and Computer Engineering, Univ. of Illinois, Urbana, IL, 2006, pp. 11–60, 61–86, 311–341.

[5] Rao, N. N., *Elements of Engineering Electromagnetics*, 6th ed., Prentice-Hall, Upper Saddle River, NJ, 2004, pp. 439–526, 675–725.

[6] Jordan, E., and Balmain, K., *Electromagnetic Waves and Radiating Systems*, Prentice-Hall, Upper Saddle River, NJ, 1968, Fig. 11–44.

Chapter 16

Ground Station Design: Mobile Approach

Dylan J. Ichikawa,* Reece T. Iwami,† Rick J. Rodrick,‡
Justin M. Akagi,§ and Wayne A. Shiroma¶

Hawaii Space Flight Laboratory, University of Hawaii at Manoa, Honolulu, Hawaii

WHETHER you're part of a well-established small-satellite program or just beginning to develop one, it's important to realize that no mission is complete without first considering the ground station.

The ground station is of primary importance for mission success as it is the first and final piece in the communication link. Its main purpose is to track and receive data from cubesats for data analysis. The role of the ground station in satellite communications is analogous to that of the base station in cellular phone communications. As long as one is within range of a base station, calls can be made and received; outside that range, there is no reception, and one is cut off from communication. Similarly, without a ground station serving as the vital access point on Earth, the cubesat is of minimal value.

In this chapter we discuss the following topics related to ground stations, focusing on an innovative mobile ground station developed by the University of Hawaii Small-Satellite Program:

- Organizational considerations and planning
- Technical considerations and system design
- Stationary vs mobile ground stations
- Mobile ground station design

---

*Undergraduate Research Assistant; currently Systems Engineer, Lasers and Sensors Systems Engineering, Northrop Grumman Aerospace Systems, Redondo Beach, CA.

†Undergraduate Research Assistant; currently Circuit Designer, RF Products Center, Northrop Grumman Aerospace Systems, Redondo Beach, CA.

‡Graduate Research Assistant; currently Industry/Government Outreach Coordinator, College of Engineering.

§Professor, Department of Electrical Engineering, and Co-Director, Hawaii Space Flight Laboratory.

¶Professor, Department of Electrical Engineering, and Co-Director, Hawaii Space Flight Laboratory.

A basic ground station includes transmit and receive antennas, orbital-prediction software compatible with the hardware for autotracking, and a transceiver to transmit and receive data and commands. Before discussing these various components, we first offer some suggestions on organizing the process.

I. Organizational Considerations and Planning

Building a ground station is a complex task that should not be underestimated. Forming the project framework will make everything clear to the members involved, and so the work can be delegated and accomplished. This section provides an outline of the steps in planning a ground station.

A. Form a Team

The first step is to form a team, with the tasks delegated among the team members. The team leader is responsible for overall coordination, with the other team members providing support. Each member should become familiar with certain aspects of the ground station. Split up the work as much as possible, and have a couple of team members working on the same task, but don't keep the same members paired together. This will provide a broad base of knowledge because one person will be working with different members in different areas of the ground station. Also, redundant knowledge coverage will be achieved because two people know the same information. Specializing in certain areas can benefit the team because the person who knows a lot can teach others. Everyone should be learning and helping out, making the working experience that much more fun and exciting for everyone involved.

B. Task Assessment

Many factors dictate the ground station design. A particular satellite might have a set of specific requirements, and a particular organization might have different priorities and ideas. So, the first step is to understand what is important and necessary by outlining the tasks and goals of the team. This assessment should include what is feasible with the resources available to the team. The primary design goal should be to establish basic functionality. Additional capabilities can be added at later times. Here are factors to consider:

- *Timeframe*: Knowing the available timeframe dictates what is possible to get done for the ground station. If the deadline to have a ground station is near, it might be necessary to invest in a smaller, simpler setup like a handheld Yagi antenna and a handheld transceiver. However, if a longer time period is available, then it is possible to order larger components that take longer to be shipped out. Identify the deadline and how much time can be spent on the project: this will help to gauge what can be done.
- *Location*: When designing a ground station, location must be considered. Some locations do not allow for any tracking or communication. Both the natural and man-made surrounding geographic features dictate the ease and availability of a communication link between the satellite and the ground station. Choosing an area with many obstructions will reduce the amount of

satellites that can be tracked and also reduce the duration that communication can occur. An analogy is watching a fireworks show. When choosing an area to watch the fireworks, a reasonable person will choose a site where there is the least amount of obstructions, allowing the largest viewing area and less of a chance of missing some of the show behind an obstruction. Because of this dependency on open area without "blind" spots, the ground station should be designed with a good tracking location in mind. To begin design of the ground station's antenna support structure, the type of ground station should be determined, be it mobile or stationary. The tradeoffs of choosing either of these options will be discussed later in the chapter.

- *Current knowledge*: The knowledge that is present within the team can play a large role in building the ground station. If someone already knows a lot about a certain aspect of ground station development, it makes everything much easier. This is one of the main reasons why we are writing this: having as much knowledge as possible before diving into a project will make the project flow much smoother.
- *Budget*: Everyone is on a budget. This might be one of the limiting factors for many ground station teams. Some of the equipment is very expensive, and if everything needs to be purchased, it can be a project manager's nightmare. Find out what kind of funding is available for your project. Knowing what kind of money is available will help in determining what components will be able to be implemented.
- *Available parts*: The available parts will also determine what can be used and done with the ground station. Amateur radio parts are readily available through many distributors. Other frequency bands might not have as many commercial-off-the-shelf (COTS) parts. You might be stuck with some frequency because it is being used on your satellite. This can cause major problems if special parts are needed. Special-ordered components might also take lots of time and money. This is why the ground station should be considered when initially designing the satellite.
- *Minimum communication requirements*: Different satellites might have different communication requirements/protocols. If vital information can be pulled off the satellite's beacon, it might not be necessary to have a very complicated setup. If a lot of data must be retrieved, a more complicated ground station might be required, and thus more time and money must be spent.
- *End goal*: The end goal is what you want the ground station to accomplish. It can be as simple as hearing a beacon from your satellite or as complicated as communicating with your satellite for an entire pass. All of the other areas must be accounted for when deciding on a final goal. Some goals are feasible whereas others are not.
- *Final assessment*: The final assessment takes into account all of these factors, providing an overall look at the logistical requirements of the ground station. The final outcome in this assessment should include a deadline, budget, and some sort of final goal statement. With this, the next step in ground station design can be taken.

All of these concerns should be addressed in a brainstorming meeting to get the team thinking. Another meeting should be held soon after, to give enough time for some research and compile all of the information. At this second meeting, the team should review the logistical concerns and then come up with a complete assessment. Frequent meetings are also important throughout the entire project timeline. Getting (and keeping) everyone on the same page will help the project run much more smoothly.

C. Make a Plan

Armed with the knowledge of what needs to be accomplished, make a plan to accomplish the goals. Allow a reasonable amount of time in the schedule. Stressing out in this type of project takes the fun out of it. Keep in mind all of the information from the task assessment. Start with the necessary components, and then work in the extras.

Set deadlines for completion of individual parts of the system. Make sure that the schedule allows ample time because finishing early is better than getting behind schedule. Internal deadlines should be set prior to any external deadlines, such as the all-important launch.

Planning is important, and having a good plan will translate into a good ground station. Changes in the plan can occur over time, but just keep the end goal in mind. This starting strategy section is very important because it provides a base, by which everything else will be supported.

II. Technical Considerations and System Design

Once the organizational phase of the ground station is complete, work can begin on the design and construction of the ground station.

Because the main objective for the ground station is to provide a communication link for the satellite, the ground station should be designed around the characteristics of the satellite. A large emphasis is put into the design of the satellite, but sometimes it can hurt to think only about the satellite. Ideally, the ground station and satellite should be designed in parallel, so that all of the pieces fit together perfectly in the end. For example, if the satellite was designed for a 10-GHz link and there is no money for a 10-GHz ground station dish, using 10 GHz is pointless. Important satellite parameters for ground station design include frequency, power, downlink, uplink, and satellite antenna.

Once these characteristics are determined, ground station equipment can be researched, and an overall design can be drafted. Figure 16.1 provides a flowchart of the basic design.

The following sections explain the various parts of the flowchart in further detail. As Fig. 16.1 shows, there are two main sections that compose the ground station: software and hardware.

A. Software

There are many functions of the different software including tracking of the satellite (azimuth, elevation, duration, etc.) with respect to a given time (date, time, and duration) and mapping to give a general overview of the satellite path as well as compensation for the Doppler shift.

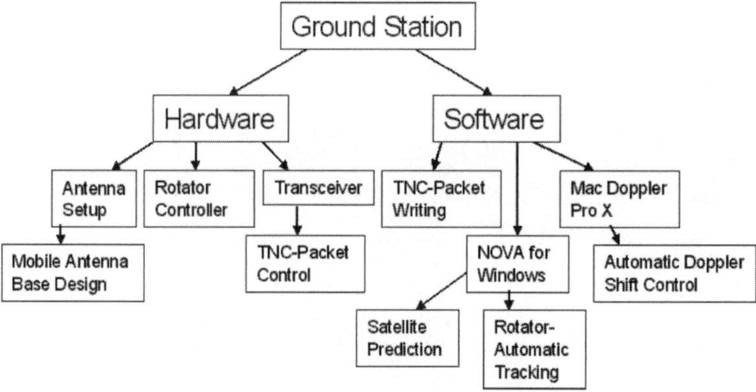

Fig. 16.1 Ground station design flowchart.

1. Tracking

Tracking software is used to determine the location of the satellite at any point in time. Information about satellite orbital patterns is needed to communicate with the satellite. Communication takes place by pointing an antenna at the satellite to establish a communication link. The antenna has a finite main beamwidth, which means that there is a direction in which the antenna works best. The antenna main beam should point toward the satellite for the best possible link. If it is not pointing the right way, the signal might be very weak and quite possibly undetectable. Therefore, we need something to give us a pretty good idea of where the satellite is in its orbit for us to point our antennas in the right direction. This is what the tracking software does.

Tracking software takes data that are known about the satellite orbit parameters and uses it to predict where and when the satellite will be overhead. Although tracking software is not directly associated with the satellite-to-ground communication link, it is very important because it provides information that is required to obtain the link. One of the most popular software programs used by cubesat enthusiasts is NOVA. A screenshot of this software is shown in Fig.16. 2.

NOVA uses Keplerian elements to calculate the satellite position. Keplerian elements are the inputs to a standard mathematical model of spacecraft orbits. With the "keps," the correct time, and station location, NOVA calculates when the satellite will be in view and where to point the antenna.

2. Communication Link

Once the satellite's position is known, a communication link must be made. The tracking software controls the positioning of the antennas to point at the satellite. The link software is what creates and maintains a link from the ground station to the satellite. As long as the antenna points in the correct direction, it is the job of the link software to communicate with the satellite. Link software includes Doppler shift correction and the control of transceiver by the computer.

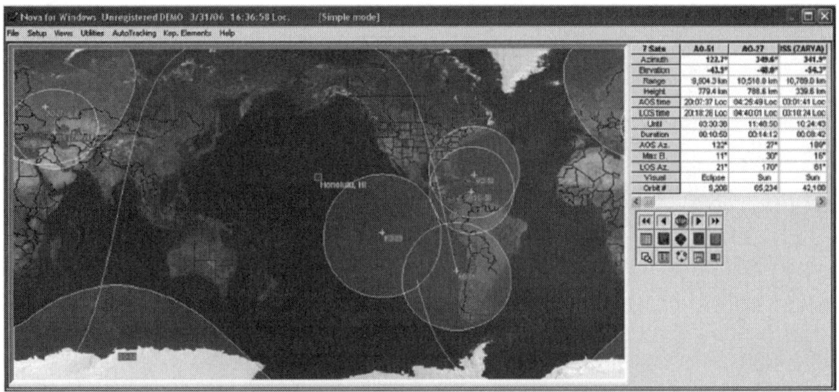

Fig. 16.2 Screenshot of NOVA software.

The Doppler shift is a change in the observed frequency of a wave occurring when the source and observer are in motion relative to each other, with the frequency increasing when the source and observer approach each other and decreasing when they move apart. The motion of the source causes a real shift in frequency of the wave, whereas the motion of the observer produces only an apparent shift in frequency. The Doppler shift is encountered when tracking satellites.

At the time of acquisition of signal (AOS), the apparent frequency is higher than what is listed as its operation frequency. As the satellite starts its pass, the rate of change of frequency, due to Doppler shift, is small, and as the satellite makes its way overhead the rate of change of Doppler shift increases. At the time of loss of signal (LOS), the apparent frequency is lower than the listed operation frequency. The rate of change of frequency is shown in Fig.16.3.

NOVA includes a useful tool in calculating Doppler shift. This utility takes the original uplink and downlink frequencies, compensates for Doppler shift, and adjusts the uplink and downlink frequencies accordingly to communicate properly with the satellite. These values change as the satellite moves closer and then farther away from the ground station; therefore, the values need to be updated constantly on the transceiver while tracking. This can be done either manually or automatically using additional software.

B. Hardware

Each hardware component has a distinct function and may or may not be needed, depending on the ground station application. This section provides

Fig. 16.3 Diagram of Doppler shift.

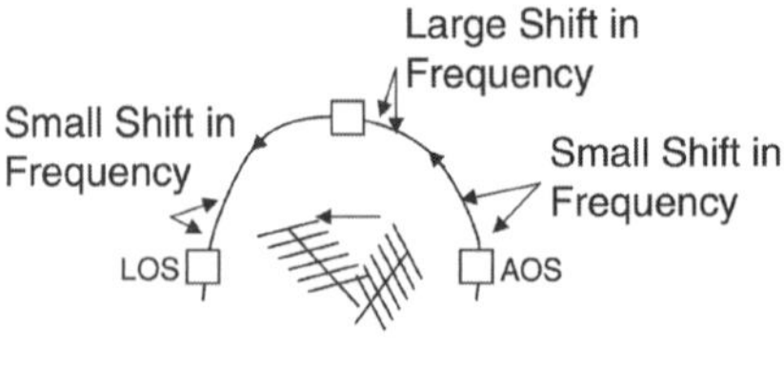

guidelines for a basic ground station; it is by no means a complete or comprehensive list of ground station components.

1. Antenna

The antenna enables transmission and reception of electromagnetic waves, which are the signals used to communicate with the satellite. Ground station antennas should be very directive and efficient. The downlink signal power levels are very weak and must be received at the highest possible level. A small-satellite antenna is not very effective because of its small size and minimum complexity. Therefore, the antenna at the ground station must compensate. The received signal can be amplified with a low noise amplifier (LNA) and later decoded using a terminal node control (TNC).

Yagi antennas are good because of their high gain and simple design. A drawback to Yagi antennas is that they are narrowband, meaning they work for a narrow range of frequencies. A diagram of a basic Yagi antenna is shown in Fig. 16.4.

Antenna polarization is another factor that must be considered. When dealing with satellite-to-ground station communications, it is important to employ a circularly polarized antenna. If both the ground station and satellite antennas are linearly polarized, there is a chance of cross-polarization, which will result in a poor communication link. This is because the satellite has a random physical orientation when deployed in space. Having a circularly polarized ground station antenna helps to alleviate this issue.

2. Low-Noise Amplifier

The signal strength of an electromagnetic wave decreases as the square of the distance traveled. Even for LEO orbits, the resulting attenuation can be quite considerable. To account for some of the losses, a low-noise amplifier (LNA) can be connected to the ground station antenna to strengthen the received signal. The signal will also weaken as it travels through a coaxial cable so that it is most beneficial to place the LNA closest to the received signal with longer connecting cords following the amplifier. An LNA is shown in Fig. 16.5.

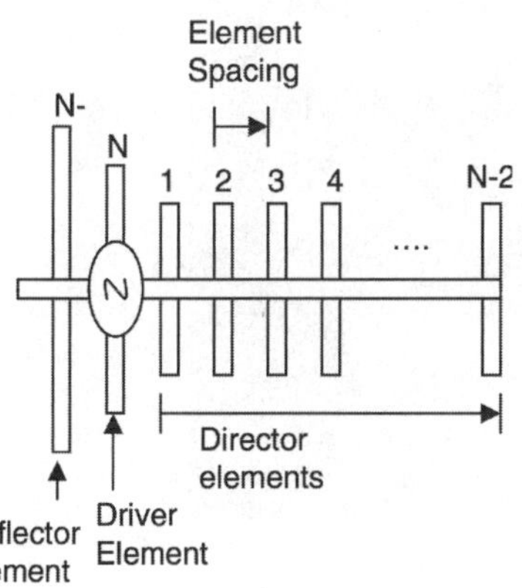

Fig. 16.4 Diagram of a basic Yagi antenna.

Fig. 16.5 LNA (pins from left to right: antenna, 13.8 VDC, transceiver).

3. Transceiver

The ground station transceiver transmits and receives the signal to and from the satellite. The word "*transceiver*" is a combination of *transmitter* and *receiver*. All ground stations and satellites need a transceiver, without which communication is impossible.

There are many different types of applications where transceivers are used. For example, a cell phone has a transceiver. It transmits and receives voice, text, and other types of media. The type of transceiver used in a ground station depends on many different factors. One of the most important factors is the frequency of operation. Other factors such as link structure and data formats should be considered. Figure 16.6 shows a UHF/VHF transceiver.

4. Terminal Node Controller

The terminal node controller (TNC) is the link between the transceiver that is receiving the information and computer that displays the information. Without the

Fig. 16.6 Yaesu FT-847 transceiver.

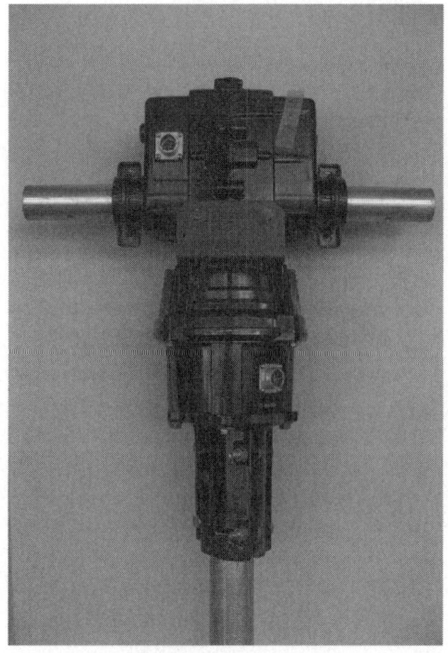

Fig. 16.7 Example rotator —Yaesu G-5500.

TNC, the incoming signal would not be able to be decoded and would be just a bunch of beeps and tones.

The TNC takes the digital data and encodes it onto a packet to be transmitted on to a radio-frequency (RF) wave. This process is called modulation. Modulation refers to the manipulation of a carrier signal to transmit information. The RF wave is then sent down to the ground station where the TNC on the ground decodes the RF data into digital data. This step is the opposite of modulation and is called demodulation. The computer connected to the TNC can then display the data on its screen.

Common among cubesat communication schemes is the AX.25 protocol. If your satellite does not use the AX.25 packet radio protocol, then a TNC might not be needed. However, if digital data must be sent, some type of protocol must be used to send data wirelessly. This idea is analogous to the IEEE 802.11a/b/g protocol for wireless networks/internet. Satellites could follow existing protocols or the user could write their own protocol.

5. Rotator

A rotator is used to point the antenna in the direction of the passing satellite. Some models use a control that is interfaced with a computer for automatic tracking of satellites. A reliable handheld antenna is able to track satellites at lower cost, but without the automatic tracking capability. However, because the antennas used for ground stations are directional, using a rotator for automatic tracking proves very beneficial. An example rotator and rotator controller can be found in Figs. 16.7 and 16.8, respectively.

Fig. 16.8 Yaesu G-5500 azimuth and elevation controller.

III. Stationary vs Mobile Ground Station

Aside from the hardware and software components of the ground station, there is one last characteristic that must be considered before design can proceed. This is the physical support structure of the ground station and whether or not it will be mobile or stationary.

To begin designing the ground station's antenna support structure, the type of ground station had to be determined, be it mobile or stationary. Traditionally, ground stations have been primarily stationary in nature. The decision for selecting between the two types can be influenced by a number of factors. In many cases, geography determines the type of ground station. City areas or areas surrounded by mountains can pose communication problems. Oftentimes it is difficult to gain clearance for ground station construction on building rooftops. When this is the case, a mobile ground station can be constructed instead. A well-designed mobile ground station allows for ease of mobility and a quick setup. This is highly desirable in situations where there is short notice of satellite pass periods. By using a hand truck and a pushcart containing the necessary hardware and tools, the entire ground station can be moved to the desired location that will offer an unobstructed link and line of sight to a given satellite.

The task of transporting all of the components of a ground station might seem like a daunting task; however, the idea might seem more attractive after considering the following characteristics of mobile and stationary ground stations. See Table 16.1 for a summary of the benefits and drawbacks of each type of ground station.

The following section illustrates the steps involved in constructing a mobile ground station. The motivation behind the development of this particular design was a lack of space for a stationary structure. This ground station design has been successful in tracking a number of cubesats. Additions and enhancement can further improve the performance of this design.

IV. Mobile Ground Station Design

The University of Hawaii at Manoa has several physical obstructions that can hinder wireless communication. They include wide-tall buildings made from concrete, glass, and other types of materials, and Manoa Valley to the north, which consists of high mountains and thick cloud cover. This eliminates the options of

Table 16.1 Advantages and Disadvantages of Both Stationary and Mobile Ground Stations

Type	Pros	Cons
Stationary	Stationary, no need to transport and store ground station parts	Difficult to obtain permission (from buildings) to mount stationary ground stations
	Stable, little variation in physical orientation and calibration	Mounting structure, building modification, and cable routing can be expensive
	Always ready for use	Constant exposure to the elements. Neglect can lead to increase spending on repair
Mobile	Can be shielded from the elements, prolonging equipment life	Must be setup and broken down whenever it needs to be used
	Does not require authorization to setup	Tracking can be interrupted by sudden adverse weather conditions
	Can be transported to unobstructed areas	

placing a permanent ground station at ground level because the window for communication would be too small. Therefore, a higher elevation for a ground station was needed. A permanent on-campus ground station placed on the roof of the engineering building was initially proposed, but the University's rules and regulations prevented the construction of student-designed structures on rooftops, eliminating the possibility of a permanent ground station at a suitable height. Therefore, a mobile modular ground station was designed to achieve the option of possibly using a higher location for tracking.

The mobile modular ground station hardware consists of a collapsible quad-pod connected to a foldable square base, as shown in Figs. 16.9 and 16.10, respectively.

The quad-pod and square base were chosen to be constructed out of wood because it is heavy enough to support the rotator, fiberglass boom, and antennas in

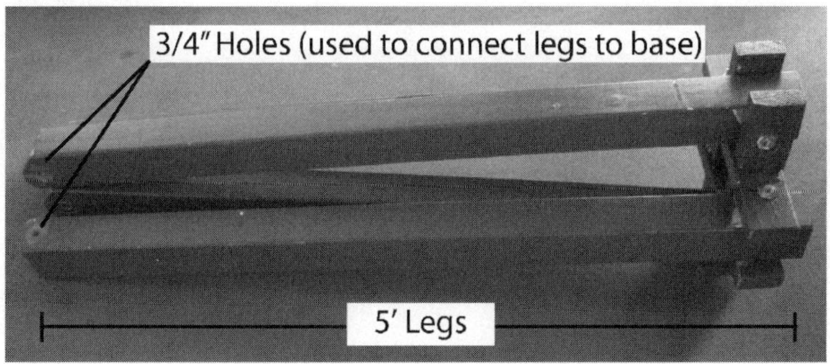

Fig. 16.9 Quad-pod of antenna base.

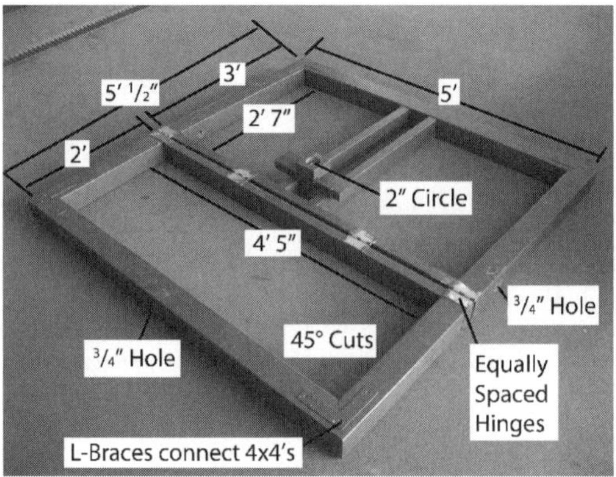

Fig. 16.10 Square bottom of antenna base.

light winds, but not so heavy that it would be too hard to move. The telescoping mast is provided to obtain unobstructed rotation of the antenna-boom assembly. Once the rotator is mounted, it can be controlled autonomously using computer software. A block diagram of an example computer-to-rotator system is shown in Fig. 16.11.

To provide unobstructed rotation of the antenna/boom assembly, a ground station base was designed to support a telescoping mast, which could be extended to elevate the rotator. Though it was important that the base be structurally sound, it was necessary that it also be light enough to be transported. After considering a number of designs, a two-section modular base was chosen. The design of each of the two base sections can be found next.

Figure 16.10 shows the bottom portion of the antenna/rotator base. Construction of this module began by cutting 4×4's into four 5-ft-long pieces. These pieces were brought to a local hardware store, which made 45-deg cuts at each end. The 5-ft pieces were then connected to form a square using L-shaped braces at all four corners (top and bottom). These braces along with the 45-deg cuts provided a sturdy frame. Next, the 5×5-ft square was cut into two sections; one, 2×5 ft, another, 3×5 ft. This was done to allow folding of the frame. The next step was to connect the cut ends of each section with a 2×4. Afterwards, the two

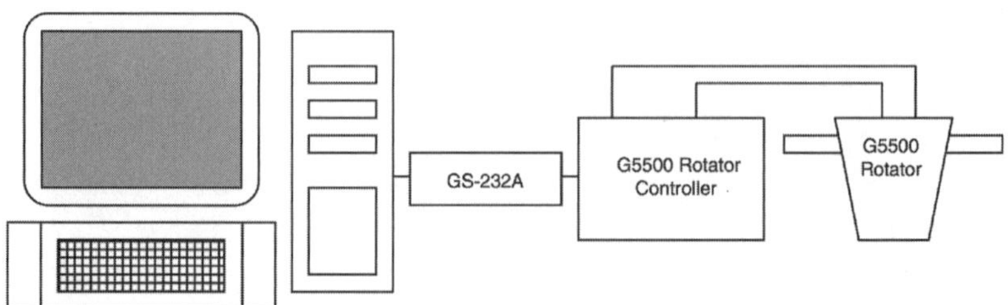

Fig. 16.11 System block diagram.

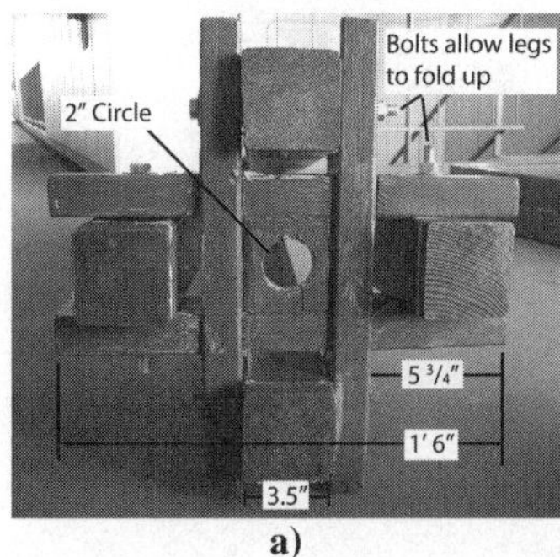

a)

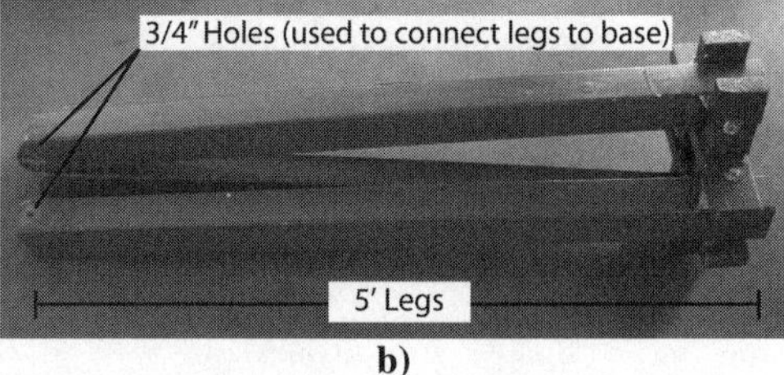

b)

Fig. 16.12 Quad-pod: a) top view and b) side view.

rectangular-shaped sections were connected using four equally spaced door hinges. Next, two 2 × 4's were cut to fit into the 3 × 5-ft section of the frame then fastened using brackets (Fig. 16.10). On top of these 2 × 4's were stacked two small pieces of 2 × 4's. A 2-in. hole was cut into the top piece. This hole would eventually help support and stabilize the rotator mast. Finally, ¾-in. holes were drilled into center of the outward-facing side of the frame. These holes were used to bolt down brackets connecting the two sections of the base.

Figure 16.12 shows the quad-pod. This section was also made by starting with four 5-ft lengths of 4 × 4's. One side of each of these pieces was rounded using sandpaper. The rounded edges allowed for an easier connection to the bottom section described earlier. Holes were also drilled in the bottoms of the legs in order to allow bolts to connect them to the bottom section. The "X"-shaped device pictured in Fig. 16.12a connected the tops of the legs. The device was made up of four interlocking pieces 1.5-ft 2 × 4 stock. The tops of the quad-pod legs were placed between the 2 × 4's (see Fig. 16.12b). Once in place, holes were drilled through both the 2 × 4's and 4 × 4 legs. Bolts were then inserted in the hole, holding the whole unit together while allowing the legs to fold. A square piece of 2 × 4 with a 2-in. hole drilled into it was placed at the center of the "X." This hole allowed for insertion of the mast, which was fed down to the bottom section of the base. Figure 16.13 shows the fully assembled ground station base.

Fig. 16.13 Fully assembled ground station base.

Fig. 16.14 Complete raised VHF/UHF mobile-modular ground station.

The nature of this modular ground station base allows it to be folded up and stored under a table or in a closet. Its relatively simple design also makes setup and breakdown quick and painless; a six-person team can assemble the ground station in as little as five minutes and break it down in even less time. Figure 16.14 shows the entire ground station setup, complete with computer workstation.

After reading this chapter, you should now be familiar with the following ground station related concepts:

- Importance and application of a ground stations
- Organizational and managerial aspects of ground station design
- Technical aspects of ground station design
- Characteristics of stationary and mobile ground stations
- Mobile ground station construction

Overall, we've learned that ground stations are an essential part of any small-satellite program, big or small. They provide the means by which a satellite is able to communicate with Earth and thus serve its purpose. Once a goal is established, a ground station team can distribute tasks and begin the process of creating a design and selecting components. Ultimately, through creativity and research, the result of the team's hard work will be a fully functional ground station.

Chapter 17

Retrodirective Antenna Systems for Cubesats

Monte K. Watanabe,* Justin M. Akagi,† and Wayne A. Shiroma‡
Hawaii Space Flight Laboratory, University of Hawaii at Manoa, Honolulu, Hawaii

RECENTLY, some countries have demonstrated the ability to intercept and destroy an orbiting satellite using a ground-launched ballistic missile, exposing the vulnerability of military communications, global positioning system (GPS) tracking, and defense satellites. If a vital satellite were destroyed in such an attack, the security of a country, its citizens, and soldiers could be compromised. This new military tactic of disabling a nation's intelligence network has warranted much concern among officials throughout the world. This vulnerability inherent in large satellite systems could be exploited in times of conflict, leaving a country defenseless.

Small-satellite technology offers a proposed solution to the vulnerability of conventional satellite systems. A number of networked small satellites could replace a large satellite, decreasing the system's vulnerability to ballistic missiles. Small satellites are less susceptible to ballistic missile attack because of their small size. But even if a few satellites were destroyed, remaining satellites in an autonomously reconfigurable network would be capable of continuing with the mission. The combination of a smaller target size and a high level of redundancy make small-satellite networks an attractive alternative to conventional satellites. However, developing a reconfigurable distributed network creates many design challenges in autonomous control and satellite-to-satellite communications.

This chapter reviews the state of the art in retrodirective antenna technology for satellite-to-satellite communication. Retrodirective antennas have the unique ability to maintain autonomous, self-steering crosslinks, making them attractive for randomly orbiting small satellites in a reconfigurable network.

---

*Undergraduate Research Assistant.

†Graduate Research Assistant; currently Industry/Government Outreach Coordinator, College of Engineering.

‡Professor, Department of Electrical Engineering, and Co-Director, Hawaii Space Flight Laboratory.

I. Motivation

In addition to defense applications, there has been considerable interest in small satellites for disaster detection, environmental sensing, technology demonstrations, and other scientific experiments. The small size and minimal complexity of nanosatellites (10 kg) and picosatellites (1 kg) make them economical to develop and launch into orbit. However, the very features that make small satellites attractive at the same time limit their functionality and capabilities. Autonomous networks of small satellites have therefore been proposed to distribute the tasks and subsystems typical of a single large satellite among numerous networked small satellites. The distribution of tasks and subsystems reduces the possibility of catastrophic single-point failures. The idea of replacing a single large satellite with a distributed network of small satellites, such as cubesats, offers several advantages. A small-satellite network is less susceptible to single-point failures because a fraction of the satellites can malfunction before the entire network is rendered inoperable; even if a few satellites fail, others can take up the workload until replacements are launched. Needless to say, replacement satellites are also more economical than their big-brother counterparts.

One of the limitations of cubesats is that it is difficult to accommodate attitude-control systems. This presents a problem when trying to establish crosslinks for network operation if no prior knowledge of the satellite's position or orientation is available. Omnidirectional antennas such as monopoles and dipoles have been the usual choice for crosslinking these networked satellites that are subject to constant unknown repositioning, but this leaves the network susceptible to eavesdropping by unauthorized ground stations as well as by satellites outside the network but still within range of the constellation. In most situations, the small-satellite network must demonstrate adequate security because of the sensitive nature of the information being passed through the network. A complicated signal-encoding scheme could solve the security problem, but would require a significant amount of the small satellite's limited processing power. In addition, omnidirectional antennas are also inefficient, as power is radiated in all directions, not just in the direction of the receiver. It is well known that small-satellite subsystems need to use power efficiently because of their limited power-generating capability.

In covert or security-sensitive networks, signal interception can be prevented by employing highly directive antennas. Directive antenna arrays help maintain security and efficiency of the network through the use of narrower beams, which direct more of the signal in the direction of interest and less in unwanted directions where eavesdroppers could be. An alternative to omnidirectional broadcasting is to use a conventional phased-array antenna, whose narrower beamwidth eliminates the security and efficiency problems described earlier. However, a phased-array antenna still requires a priori knowledge of the target direction to point the antenna correctly. A small satellite with a conventional phased array must have some input telling the satellite where to point its beam, obviating the desired autonomous nature of the satellite-to-satellite link.

A smart antenna can overcome this security issue, but it involves digital signal processing algorithms. For nanosatellites and picosatellites, processing power is a valuable resource and dynamic beam steering using smart antennas would add another layer of complexity to the system, negating the advantages of the simple, low-cost features of these small satellites. Smart antennas can also incur an antenna pointing delay as a result of the processing time associated with the smart antenna's decision-making algorithm. A pointing delay could significantly hurt link performance especially if the satellites are moving rapidly or erratically.

For small-satellite applications, an attractive alternative to dynamic beam steering is a *self-steering* retrodirective antenna array that permits secure cross-link communications between satellites moving randomly in space. Retrodirective antennas are able to sense the direction of an incoming radio transmission and send a reply in that same direction. This incoming radio transmission is usually called an interrogator signal. The retrodirective system gets all necessary information to steer the beam from this interrogator signal. The automatic, self-steering function of retrodirective systems is performed autonomously, without user-controlled phase shifters, digital circuitry, or computational software. In retrodirective antenna arrays, steering is accomplished through purely analog circuitry, which does not require any processing. There is no need for complex algorithms or target orientation/position information to enable beam steering. In addition, the analog circuitry allows the beam to be steered quickly with minimal pointing delay because digital computations are not needed. The high-directivity associated with retrodirective arrays not only improves network security, but also improves the communication link efficiency by minimizing power consumption; this reduces the burden on transmitting and receiving amplifiers while still maintaining a reasonable link margin. The lower complexity associated with retrodirective antenna arrays makes them more attractive than smart antennas for small satellites. The simplicity of the retrodirective approach is important for small satellites in which energy and processing power are at a premium.

Figures 17.1–17.4 help to illustrate the effectiveness of retrodirective antenna arrays when compared with other types of antennas. When $t = 0^-$, the corresponding figure represents the link status just before a sudden satellite orientation change, which occurs at $t = 0$. At $t = 0^+$, the satellite orientation has just occurred, and the crosslink status is shown.

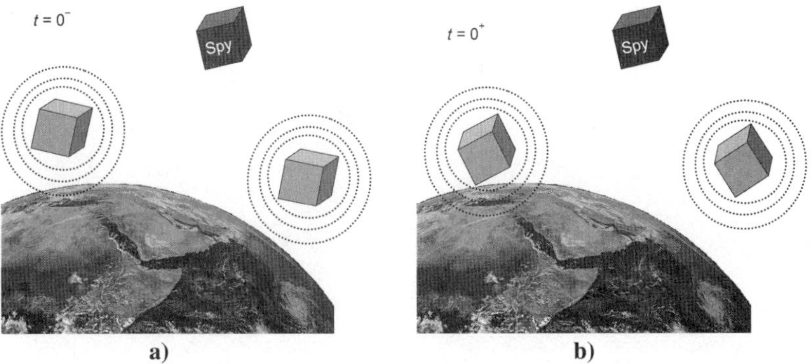

Fig. 17.1 Omnidirectional antenna: a) before repositioning and b) after repositioning.

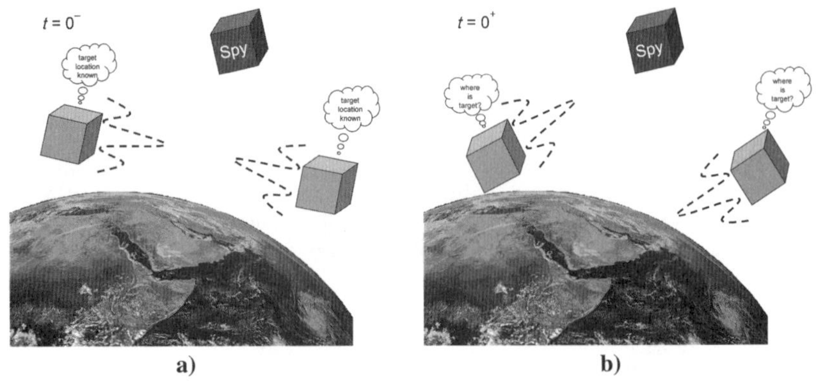

Fig. 17.2 Phased-array antenna: a) before repositioning and b) after repositioning.

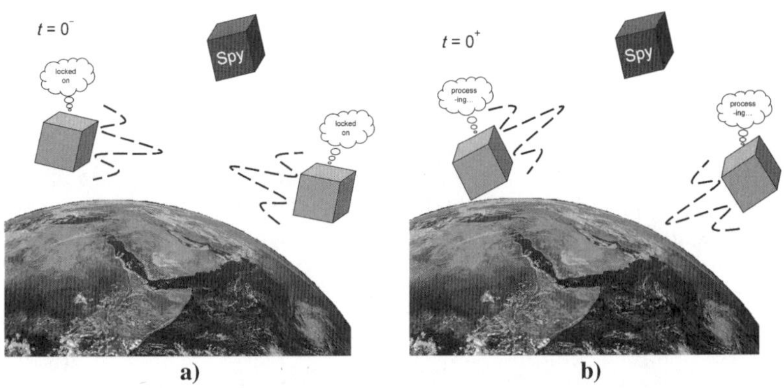

Fig. 17.3 Smart antenna: a) before repositioning and b) after repositioning.

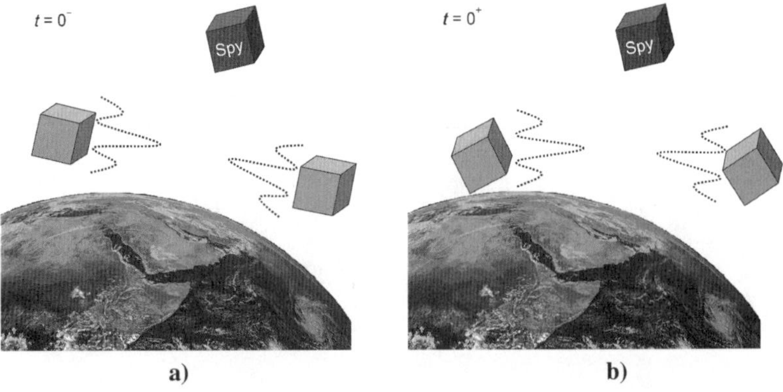

Fig. 17.4 Retrodirective antenna: a) before repositioning and b) after repositioning.

In recent years, the University of Hawaii (UH) Small-Satellite Program has been developing technologies for distributed, dynamically reconfigurable, autonomous small-satellite networks for crisis management and disaster mitigation. The planned network requires a new type of architecture along with sufficient retrodirective technology to provide adequate crosslinks. UH has been doing research in retrodirective array systems since 1999 [1–15]. For a network containing multiple satellite nodes, each satellite would have both an onboard omnidirectional interrogator and retrodirective transponder. The link is initiated by one satellite's omnidirectional interrogating signal, which is coded for an intended receiver. That receiver then sends a retrodirective signal back to the interrogator, whereupon it switches to retrodirective mode and the link is established. The use of coded signals prevents unauthorized third-party interrogation and helps in setting up multiple links between nodes. Research is currently underway in realizing this small-satellite network.

II. Fundamentals of Retrodirectivity

Retrodirective systems have the unique property in that when the system is interrogated, the system automatically directs electromagnetic waves in the direction of the interrogating source. Consider an electromagnetic wave in free space impinging on a flat metal sheet as shown in Fig. 17.5. In this example, the electromagnetic wave serves as the interrogator. If the direction of propagation is normal (i.e., perpendicular) to the metal surface, the wave reflects off the metal sheet in the same direction from which it originated. This is the simplest case of retrodirective behavior, or simply, retrodirectivity. However, this does not imply that the metal sheet is a retrodirective system because this behavior is observed only at normal incidence, when $\theta = 0$ deg. If the incoming wave is incident at an arbitrary, nonnormal angle θ, the reflected wave does not bounce off the metal sheet at θ, as required for a retrodirective system, but instead at the specular angle $-\theta$ predicted by Snell's law. Moreover, if the direction of propagation is parallel to the metal surface ($\theta = 90$ deg), the wave only grazes the sheet, and there is no reflection at all. Retrodirectivity requires that the incoming wave bounce back in the same direction from which it originated, regardless of what that direction is.

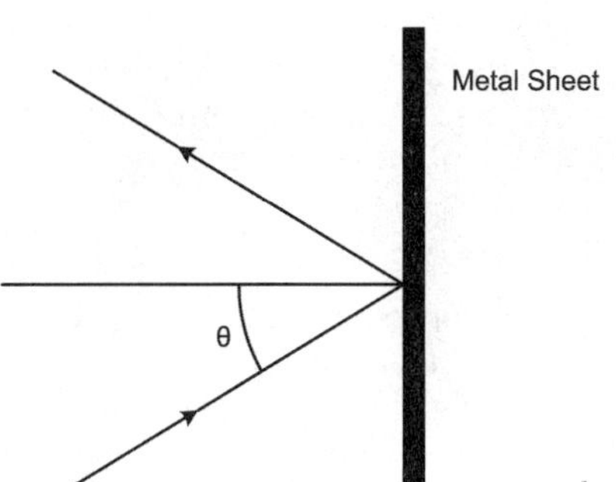

Fig. 17.5 Metal sheet reflector, retrodirectivity observed only when ($\theta = 0$ deg). Other angles of incidence do not exhibit retrodirective behavior.

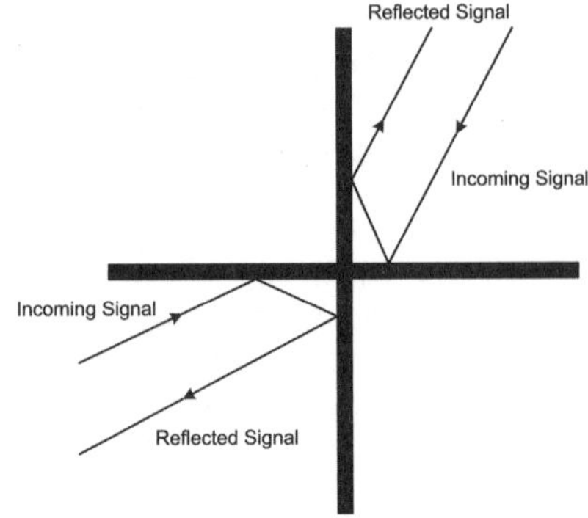

Fig. 17.6 Two-dimensional corner reflector; retrodirectivity is observed for waves incident from any angle in the *x–y* plane.

Although one metal sheet cannot function as a retrodirective system, two or more metal sheets can. Figure 17.6 illustrates a retrodirective system formed by intersecting two mutually perpendicular, flat metal sheets. Because each quadrant forms a corner shape, the resulting structure is called a corner reflector [16] and provides retrodirectivity for any angle of incidence in the *x–y* plane. The system is easily extended to three dimensions by intersecting the two metal sheets in Fig. 17.6 with a third metal sheet that is mutually perpendicular to both, creating a three-dimensional corner in all eight octants. Multiple reflections off of any one corner cause incoming waves from all possible angles in three-dimensional space to be reflected back in the same direction of the originating wave, giving rise to true retrodirectivity. We define a retrodirective system, then, as one that reacts to an incoming interrogator signal from an unspecified direction by transmitting a response to that same direction, regardless of what that direction is. The response is completely automatic and is performed without any prior knowledge of the location of the source.

The two- and three-dimensional corner reflectors are examples of retrodirective systems, but they are not true retrodirective antenna systems. Antenna systems are usually able to reply with some type of signal, which is a response to the received interrogator signal; this is true whether or not the antenna system is retrodirective. Corner reflectors can only reflect the interrogator signal, retransmitting a duplicate interrogator signal only at a fraction of the power because of the losses in the metal. Thus, there is not much practical use for corner reflectors as antenna systems, for they are not able to transmit a modulated retrodirective signal, which is usually a requirement of the antenna system. Corner reflectors are usually used as radar targets because of their large radar cross sections and not much more. However, corner reflectors are a useful tool in describing retrodirectivity.

III. Basic Retrodirective Antenna Array Architectures

For wireless communication applications, it is more common to use retrodirective antenna arrays, which lend themselves to integration with electronic circuits, are smaller and lighter than corner reflectors, and can be made planar and low profile. Unlike corner reflectors, retrodirective antenna systems are useful for communication applications. As mentioned earlier, retrodirective antenna systems excel in applications where communication links are not fixed, as in base-to-mobile or mobile-to-mobile communications. The autonomous beam steering is achieved at a purely hardware level, providing a means for creating secure, efficient, directive systems for satellite communications, wireless local-area networks, terrestrial peer-to-peer communications, or covert battlefield communications, without adding additional processing or system design strain.

An antenna array [17] is incorporated into a retrodirective antenna system because a single antenna element cannot provide retrodirectivity. An antenna array is composed of individual radiating elements that interact with each other to create a new radiation pattern different from each individual radiation pattern. The radiation pattern can be manipulated by changing element spacing and phase difference between elements. Steering of the beam in an antenna array is accomplished through creating a phase gradient across the individual elements in the array. By combining individual antennas in an array, one can achieve higher directionality and the ability to steer the beam by changing the progressive phase shift across the elements in the array.

For a retrodirective system to be useful, it must be able to send a response with actual information back to the interrogator. If the response is simply a duplicate of the interrogating signal, no information is actually conveyed. A retrodirective antenna system only uses the interrogating signal to determine the direction of the interrogator. Once the beam is pointed correctly, it should ideally transmit modulated information back in target direction. In all cases where retrodirective antenna systems will be used, the antenna system must be able to return information in the form of a modulated signal as the retrodirective response.

A. Van Atta

Another simple way of achieving retrodirectivity is through the use of the Van Atta array [18], consisting of pairs of antenna elements equally spaced from the center with equal-length transmission lines. As shown in Fig. 17.7, the progressive phase shift associated with the incoming signal is phase-lagged going right to left across the array. The signal received by one antenna is reradiated by its pair, causing the order of reradiation to be flipped with respect to the center of the array. The phase gradient of the array is therefore reversed (positive to negative, negative to positive). It is this reversal of phase progression for the outgoing signal, which causes it to retroreflect back in the same direction.

Because the lengths of the connecting transmission lines are equal, the only frequency-dependent component in the Van Atta array is the antenna element. The use of broadband antennas and nondispersive transmission lines allows the array to perform over a wide bandwidth. However, this array configuration is restricted to planar wavefronts and planar topologies. Two-dimensional Van Atta arrays have been demonstrated for two-dimensional steering, but the geometrical arrangement of the Van Atta array makes it spatially inefficient because of the numerous numbers of transmission lines needed to connect antenna elements required for a directive antenna beam. Also, integration of Van Atta arrays with transceivers can be difficult because of its manner of operation.

Van Atta arrays are useful in introducing how reversing phase gradients can create retrodirectivity. Van Atta arrays do not have very many practical applications as antenna systems because of complicated issues with electronics integration, but

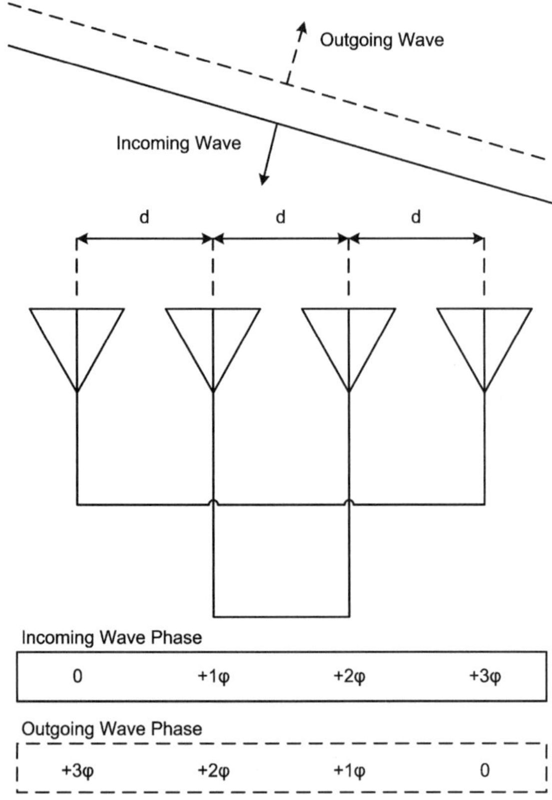

Fig. 17.7 Van Atta array—the two transmission lines connecting antenna pairs are of equal length.

are still important to illustrate the simple concept of retrodirective antenna technology.

B. Phase Conjugation

Because both corner reflector and Van Atta methods for achieving retrodirectivity are infeasible for many applications, including satellite crosslinks, a more popular technique is the phase-conjugating array. This approach uses the same idea of the reversal of a phase gradient in the Van Atta array, but phase reversal is achieved at each antenna element instead of relying on antenna pairs.

Both passive and active methods for achieving phase conjugation have been demonstrated [16]. Passive, diode-based phase conjugators generally provide better RF-intermediate-frequency (IF) isolation than active, transistor-based phase conjugators, especially when no bias is required. The major shortcoming of passive phase conjugators is the conversion loss, which limits the distance between the interrogator and the retrodirective array. The use of active devices such as

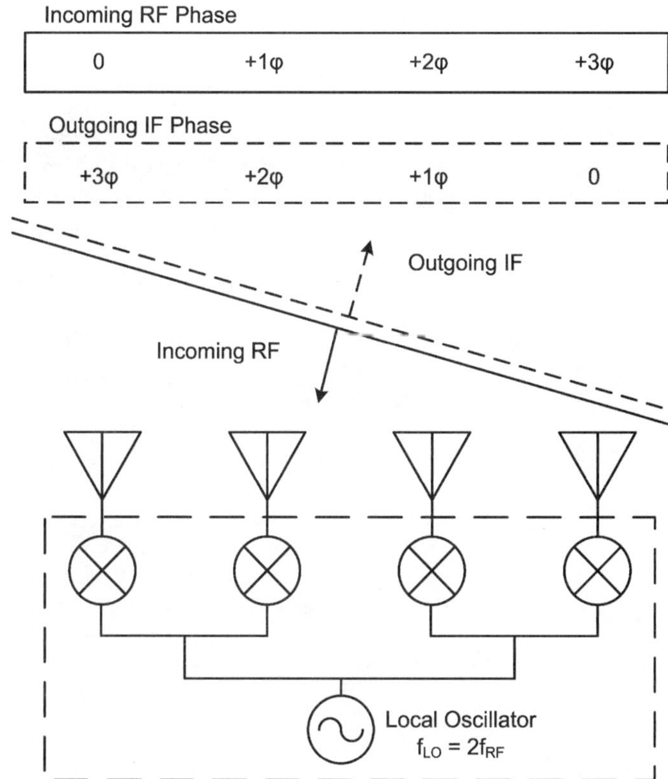

Fig. 17.8 Schematic of a heterodyne phase-conjugating retrodirective antenna.

metal semiconductor field effect transistors is attractive as they can provide conversion gain during the mixing process. Active phase conjugators make it possible for the retrodirective array to transpond an amplified signal back to the source location, without the need for additional amplifiers.

Usually the heterodyne technique [19], an active form of phase conjugation, has been used for retrodirective antenna systems. Figure 17.8 illustrates the typical heterodyne phase-conjugating array. In the heterodyne technique the incoming RF signal at each element is mixed with a local-oscillator (LO) signal at twice the frequency. The mixing process results in the following IF signal:

$$
\begin{aligned}
V_{IF} &= V_{RF}\cos(\omega_{RF}t + \varphi) \times V_{LO}\cos(\omega_{LO}t) \\
&= \frac{1}{2}V_{RF}V_{LO}\left[\cos((\omega_{LO} - \omega_{RF})t - \varphi) + \cos((\omega_{LO} + \omega_{RF})t + \varphi)\right]
\end{aligned}
$$

If $\omega_{LO} = 2\omega_{RF}$, then

$$
V_{IF} \propto \cos(\omega_{RF}t - \varphi) + \cos(3\omega_{RF}t + \varphi) \tag{17.1}
$$

Note that the first term in Eq. (17.1) has the same frequency as the RF signal, but with a conjugate phase. The resulting phase conjugation across the entire array results in retroreflection of the IF signal back towards the RF source, just as in the Van Atta array. The second term in Eq. (17.1) is an undesired, non-phase-conjugated signal that radiates in accordance with Snell's Law. Fortunately, this signal is easily filtered and suppressed because of the large difference between this frequency ($3f_{RF}$) and the RF (f_{RF}). For the same reason, any LO leakage ($2f_{RF}$) can also be easily filtered. A narrow-bandwidth antenna can also contribute to the filtering process.

Most of these recently demonstrated retrodirective arrays [20,21] incorporate heterodyne mixing to self-phase conjugate a retrodirective response. This method achieves phase conjugation through hardware with minimal added circuit complexity. Modulation of the retrodirected response can also be easily incorporated to make the retrodirective antenna system practical for many wireless communication systems.

The strength of the phase-conjugating technique is the recreation of a wavefront, which allows nonplanar wave fronts and surfaces to be handled by the retrodirective antenna array. Unlike the Van Atta architecture, this technique allows an array to retrodirect a signal to the source through obstacles or in the near field, as shown in Fig. 17.9. Phase-conjugating antennas have also been used to reduce the effect of multipath fading [6], which is a problem in many urban wireless communication systems.

With the advancement of the heterodyne technique for phase-conjugating retrodirective systems, the use of retrodirective antennas for wireless communication systems has finally become practical. Many retrodirective arrays have been

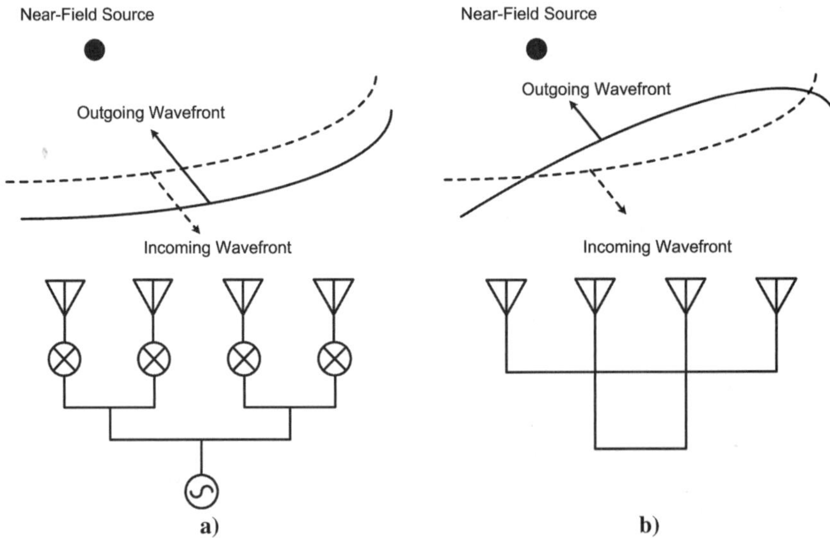

Fig. 17.9 Reconstruction of the wavefront: a) in a phase-conjugating array and b) in Van Atta array adapted from [16].

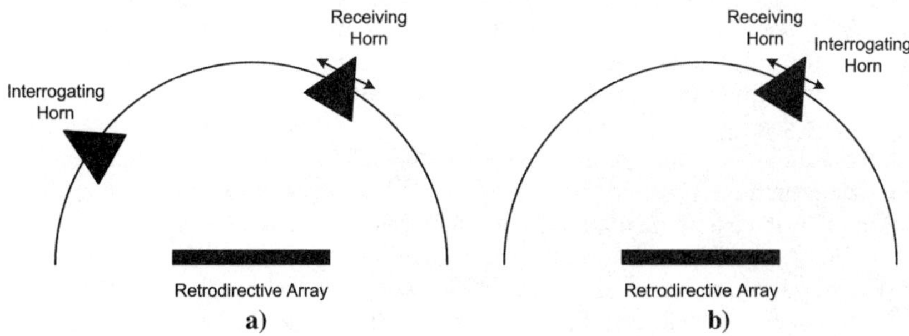

Fig. 17.10 Measurement setup for a) bistatic RCS and b) monostatic RCS.

demonstrated using the heterodyne technique, but the transition to small satellites in the space environment introduces many new challenges.

IV. Array Characterization

Before actual detailed retrodirective array architectures can be described, the method of characterizing retrodirective arrays must be explained. Retrodirective antenna arrays require an interrogating signal to operate, and so a different type of radiation pattern must be taken to characterize the array. Figure 17.10 illustrates the two standard measurement setups for characterizing the self-steering performance of a retrodirective antenna array: bistatic and monostatic radar cross sections (RCS). In both cases, an interrogating signal provided by a horn antenna impinges on the array under test, whereupon the retrodirected signal is transmitted back (ideally in the same direction from which it originated) to a second receiving horn antenna.

In the bistatic case, the interrogating antenna remains stationary while the receiving antenna is scanned. The resulting measurement shows a main lobe in the direction of the source, with nulls or sidelobes in other directions as shown in Fig. 17.11.

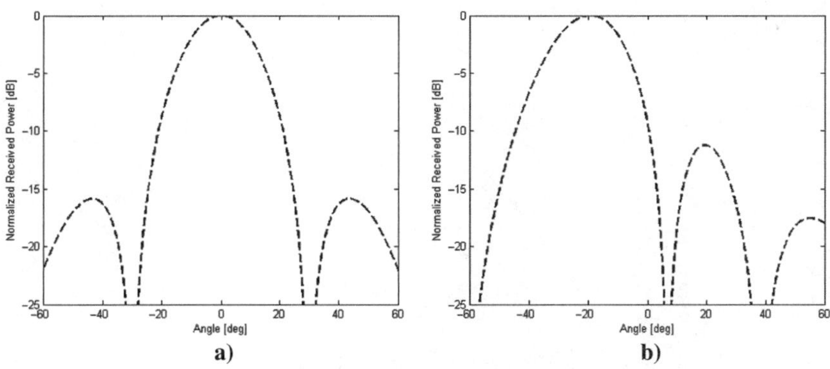

Fig. 17.11 Theoretical bistatic RCSs for a four-element microstrip patch retrodirective antenna array with interrogator at a) 0 deg and b) −20 deg.

In the monostatic case, both the interrogating and receiving antennas are collocated and simultaneously scanned. Because the interrogating and retrodirected signals are both in the same direction, the peak of the array radiation will always be in the direction of the receiving antenna, and thus the monostatic pattern will exhibit a relatively flat pattern, without nulls. The actual monostatic RCS is dependent on the type of antenna element used. The principle of pattern multiplication [17] states that the total radiation pattern of an antenna array consists of the array factor equation multiplied by an individual element's radiation pattern. In the theoretical monostatic RCS, the array factor will remain constant because of the self-steering of the retrodirective system. Therefore, only the individual element's radiation pattern will contribute to the measured monostatic RCS. An example of a monostatic RCS is shown in Fig. 17.12.

Because the interrogating and retrodirected signals share the same frequency, there is always unavoidable coupling between the interrogating and receiving horns. In a phase-conjugating array or modified Van Atta array, this problem is overcome by slightly offsetting the frequencies so that the two signals can be resolved on a spectrum analyzer. For instance, the following frequencies could be used: an interrogating signal of 4.99 GHz, an LO signal of 10.00 GHz, and a retrodirected signal of 5.01 GHz. Phase conjugation is achieved even if the interrogating and retrodirected frequencies are not identical, but the deviation between these two frequencies leads to a pointing error in the return beam. The amount of the error depends on this frequency difference as well as the incoming beam angle because the amount of progressive phase shift needed for array beam steering is a function of antenna element spacing in terms of wavelength. Retrodirective arrays can be either one dimensional, as in the examples presented before, or two

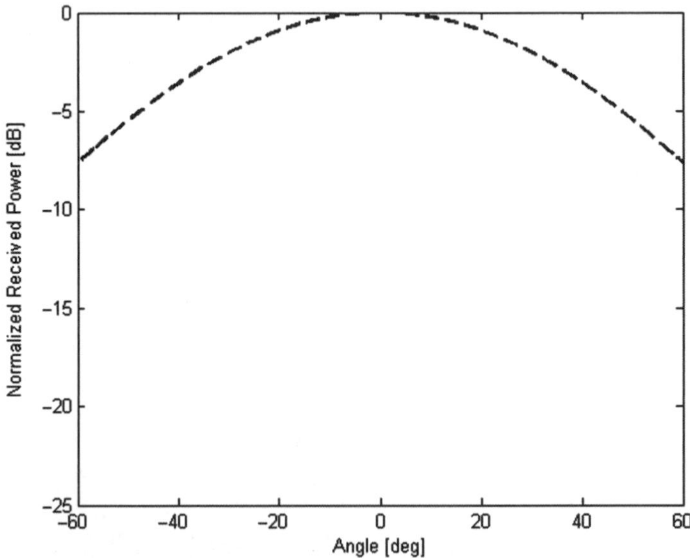

Fig. 17.12 Theoretical monostatic RCS for a four-element microstrip patch retrodirective antenna array.

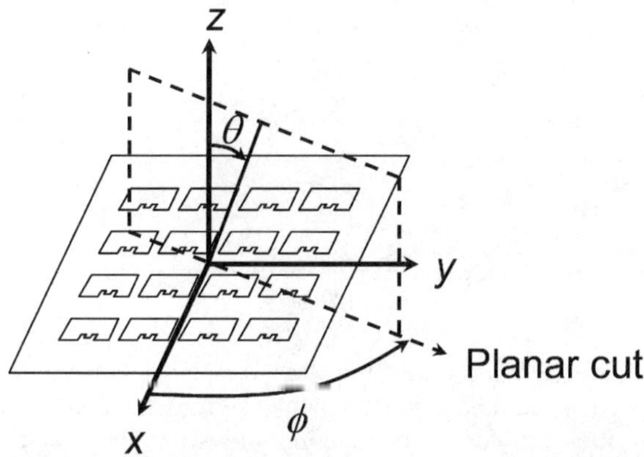

Fig. 17.13 Coordinate system for RCS measurements (reprinted with permission by Grant S. Shiroma [11; Fig. 4.8, p. 43]).

dimensional, as presented in the following sections. Bistatic and monostatic RCS measurements can therefore be taken over several planar cuts to show two-dimensional retrodirectivity, as Fig. 17.13 shows.

V. Preparing Retrodirective Arrays for Space

In wireless communications systems a link budget is created to ensure adequate links are established between nodes. A similar link budget is used for satellite communications in space. The Friis Transmission Formula is used to calculate the received signal power P_{received} based on the transmitted signal power $P_{\text{transmitted}}$, gain of receiving antenna $G_{\text{transmitter}}$, gain of transmitting antenna G_{receiver}, and the frequency-dependent path loss incurred by the transmitted signal *Path Loss*:

$$P_{\text{received}}\,(\text{dBm}) = P_{\text{transmitted}}\,(\text{dBm}) + G_{\text{transmitter}}\,(\text{dB}) + G_{\text{receiver}}\,(\text{dB}) - \textit{Path Loss}\,(\text{dB}) \tag{17.2}$$

$$\textit{Path Loss}\,(\text{dB}) = \textit{Free-space Loss}\,(\text{dB}) + \textit{Other Losses}\,(\text{dB}) \tag{17.3}$$

$$\textit{Free-Space Loss}\,(\text{dB}) = -10\log\left[\frac{\lambda^2}{(4\pi d)^2}\right] \tag{17.4}$$

The frequency-dependent path loss term (17.3) includes free-space losses and other losses encountered along the signal's path. Equation (17.4) shows the calculation of free-space losses, which is a function of signal wavelength λ and path distance d. The *Other Losses* term includes transmission line losses, connector losses, etc., which depend on the specific wireless system.

Table 17.1 illustrates the effect of frequency on free-space path loss. It is shown that at higher frequencies a greater path loss is incurred, but higher frequencies

Table 17.1 Effect of Frequency of Free-Space Path Loss

Frequency	Free-space loss,[a] dB
450 MHz	65.5
1 GHz	72.4
2.4 GHz	80.0
10 GHz	92.4

[a]Based on separation distance of 100 m.

allow greater bandwidth and transfer speeds. A balance between free-space loss and transfer speeds should be reached when designing the communication link.

The unit of measure for wireless communication links is called the link margin:

$$\textit{Link Margin}\ (\text{dB}) = P_{\text{received}}\,(\text{dBm}) - \textit{MDS}\ (\text{dBm}) \tag{17.5}$$

$$P_{\text{received}}\,(\text{dBm}) \geq \textit{MDS}\ (\text{dBm}) \tag{17.6}$$

The link margin is calculated by taking the amount of received power P_{received} and subtracting the minimum detectable signal (MDS) of the intended receiver. A good wireless link has a positive link margin, whereas the minimum requirement to establish a link would require a link margin equal to zero. If the link margin is negative, it would be impossible for a link to be established. Equation (17.6) illustrates the required relationship between received power and minimum detectable signal to have a successful wireless link.

Although retrodirective technology has been around since the 1960s, designing for space applications presents new challenges. Conventional retrodirective arrays must be modified from their normal configurations to enable adequate performance in space. First, the zero-gravity, free-floating nature of the satellites necessitates two-dimensional tracking, and therefore a two-dimensional retrodirective array. Secondly, because the satellites are too small to have attitude control systems, it is impossible to know the orientation of each satellite. The antennas will therefore have to provide circular polarization to allow signal reception and prevent polarization mismatch regardless of each satellite's orientation with respect to the other. The antennas would also need to be on all sides of the satellite to ensure that crosslinks can be established.

Like a two-dimensional corner reflector made up of two metal plates, a one-dimensional antenna array can only steer in one plane. Small satellites would not be able to benefit from one-dimensional steering because nodes will rarely be oriented on the same plane as the antenna steering plane. To enable two-dimensional steering, a two-dimensional antenna array, synonymous to a three-dimensional corner reflector, must be used. A two-dimensional array is capable of steering in both azimuth and elevation directions forward of the antenna. This two-dimensional steering capability would allow the satellites to create crosslinks with other satellites even though their antennas are not on the same plane.

Most antennas used for space communications employ circular polarization. Circular polarization allows antennas to be misaligned and still maintain adequate reception. Linearly polarized antennas require that the antennas be aligned to get reception; misalignment can lead to little, or the worst case, no reception at all. The disadvantage of circularly polarized antennas is that the maximum amount of reception is lessened. The advantage is that there are no situations where communication will be lost, meaning that there will be less variation of reception with changing antenna alignment.

If two-dimensional, circularly polarized, planar retrodirective arrays are placed on each side of a cubesat, it is possible for this single satellite to sustain secure communication links with up to six other satellites in a distributed network, with each satellite randomly moving about in space. Friis' Transmission Formula should be used to determine link parameters, and every effort should be made to maintain a positive link margin even in the worst-case scenario.

VI. Hokulua: Cubesats with Retrodirective Crosslinks

The UH Small-Satellite Program demonstrated the first cubesat-class satellite equipped with retrodirective crosslinks [10], pictured in Fig. 17.14. Hokulua (Twin Stars) is a pair of picosatellites that demonstrated the use of retrodirective crosslinks in cubesats. Hokulua consisted of a pair of one-and-a-half "cube" ($10 \times 10 \times 15$ cm) satellites connected with a semi-rigid 1-m-long tether. One satellite served as the interrogator, and the other satellite housed the retrodirective antenna system; the tether ensured that the satellites were correctly oriented towards each other because of their lack of attitude control. Figure 17.15 shows the deployment sequence of Hokulua.

The retrodirective system used on Hokulua consisted of a 10.5-GHz quadruple subharmonic phase-conjugating array. This type of phase-conjugating array uses the same idea of reversing the phase gradient to achieve retrodirectivity as a conventional phase-conjugating array. However, a quadruple subharmonic uses the fifth-order mixing product to achieve proper output frequency and phase conjugation. A local oscillator (LO) at half the incoming frequency (RF) supplies the mixer instead of a local oscillator at twice the incoming frequency. Using this

Fig. 17.14 The Hokulua (Twin Stars) interrogator satellite (right) and retrodirective antenna satellite (left).

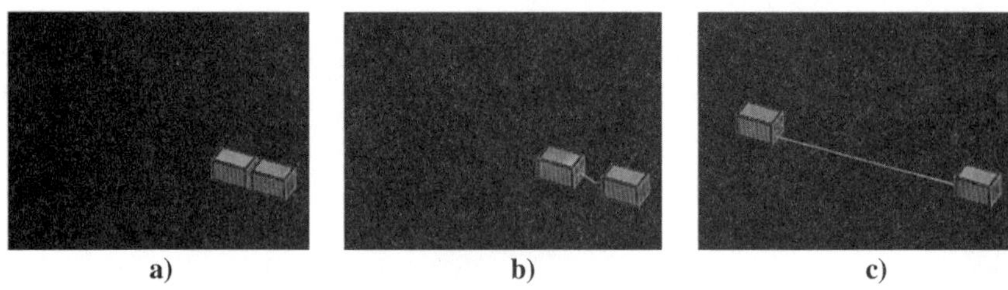

a) b) c)

Fig. 17.15 Deployment sequence: a) deployment of Hokulua from PPOD, b) shape memory tether deploys, and c) final position of Hokulua.

method to achieve phase conjugation is more efficient, which proves advantageous for cubesats because of their limited power generation capabilities.

A cross-shaped circularly polarized microstrip patch antenna array, shown in Fig. 17.16, was used for Hokulua's retrodirective antenna system. This unique antenna array was chosen because of constraints encountered when designing for this particular cubesat application. The antenna array needed to achieve two-dimensional retrodirectivity, but there was a limited number of antenna elements that could fit on the designated end of the cubesat structure. In addition, the internal payload volume available to the retrodirective circuitry was limited. This eight-element sparse (reduced number of elements) array met all requirements and was the final antenna array of choice for Hokulua. The microstrip patch antennas were rotated to allow the antenna element feed lines to fit on the antenna array substrate.

Bistatic RCS measurements were taken in order to characterize this two-dimensional retrodirective array. However, the bistatic RCSs needed to be taken in a different plane from the linear arrangements of elements. If a bistatic RCS measurement were taken along a plane coinciding with a linear array, the results would be indicative of one-dimensional steering, but true two-dimensional steering

Fig. 17.16 Retrodirective circuitry and antenna array in connected flight configuration.

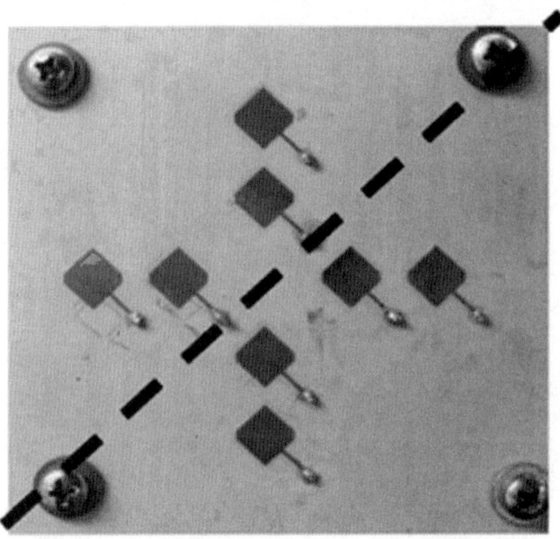

Fig. 17.17 Alignment of the bistatic RCS measurements along the diagonal of the eight-element cross array.

cannot be determined. The alignment of linear elements along the RCS plane could exhibit steering, although the other linear array perpendicular to the plane could not be steering at all. A 45-deg diagonal cut, alignment pictured in Fig. 17.17, was used to take the bistatic RCSs. The use of this plane would give good indication of two-dimensional steering, as horizontal, vertical, and a combination of the two diagonal, steering contributions would be needed to produce good results. The bistatic RCSs of Hokulua are shown in Fig. 17.18.

VII. Drawbacks to Van Atta and Phase-Conjugating Arrays

The retrodirected signal strength at the output of the Van Atta and phase-conjugating arrays are wholly dependent on the distance from the interrogator,

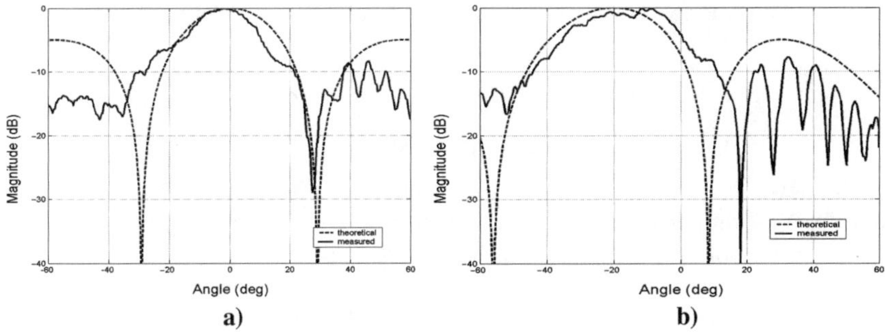

Fig. 17.18 Measured and theoretical bistatic RCSs (plane of Fig. 17.17) of quadruple subharmonic phase-conjugating array with interrogator positioned at a) 0 deg and b) –20 deg (reprinted with permission by Grant S. Shiroma [11; Fig. 5.4, pp. 49–50]).

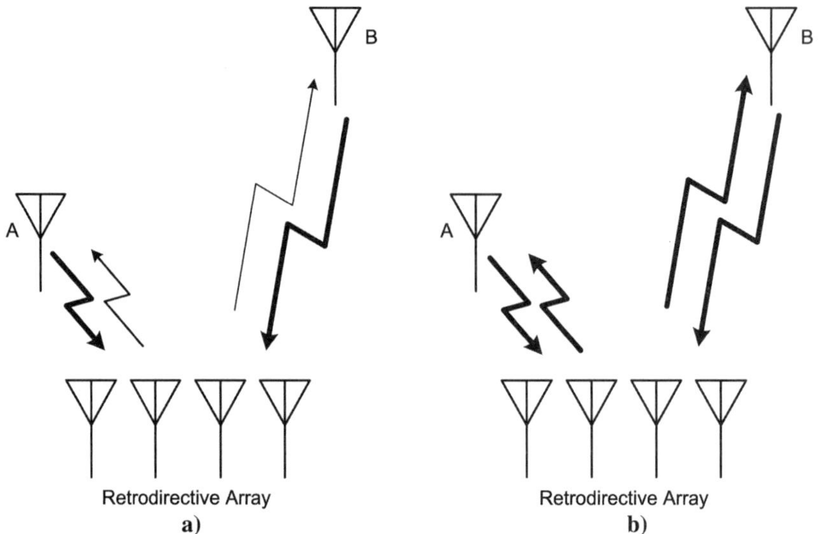

Fig. 17.19 Relative power levels, represented by thickness of arrows (i.e., thicker arrow represents stronger signal), of interrogator and retrodirected signal for a retrodirective array: a) retrodirected power dependent on interrogator power and b) constant retrodirected power, independent of interrogator power.

making them susceptible to a R^4 roundtrip path loss. This is because the signal must travel a roundtrip distance of $2R$, resulting in the power of the signal being reduced by a factor of $1/R^4$. As a result, the retrodirected power transmitted back to source B, which is farther away than source A, will be less than that transmitted to source A. The interrogator signal is reradiated from the antenna element pair as the retrodirected response, and this interrogator signal takes a roundtrip path, which is why the path loss is proportional to R^4 and the distance between A and B impact the received power levels, which is illustrated in Fig. 17.19. In the Van Atta array, bidirectional amplifiers can be placed between elements to increase transmit power, but this adds additional space and power requirements that cannot be afforded by small satellites. Phase-conjugating arrays could use mixers designed with conversion gain, but still does not ensure constant transmit power. However, a signal not dependent on interrogator distance is only going to encounter an R^2 path loss.

VIII. Phase Detection/Phase Shifting

The newest type of demonstrated retrodirective antenna uses phase detection and phase shifting. In the phase-detection and phase-shifting retrodirective system, the incoming interrogator wavefront's phase is sampled using a phase detector. Using this knowledge about the detected interrogator phase, the system is able to phase its response to point back towards the interrogator. Usually the control circuit will tell phase shifters, placed at each radiating element, to create a specified delay that will steer the beam in the proper direction. A block diagram of the general phase-detecting and phase-shifting retrodirective system is shown in Fig. 17.20.

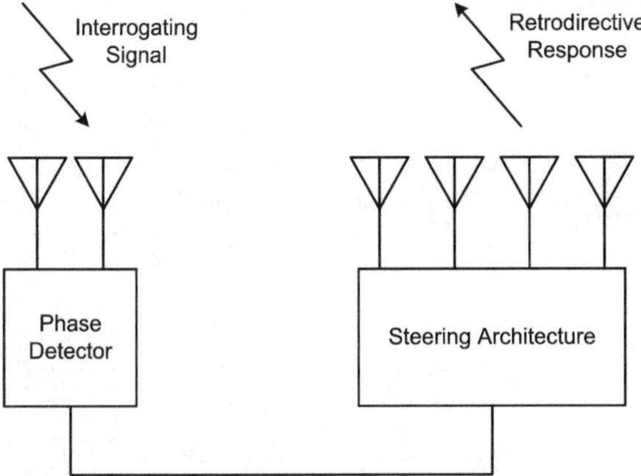

Fig. 17.20 Generic block diagram of a phase-detection and phase-shifting retrodirective array.

The phase-detection and phase-shifting technique for achieving retrodirectivity has one significant benefit when compared with the previously mentioned retrodirective systems. Both the heterodyne and Van Atta retrodirective system suffer from an unpredictable crosslink because both of these systems transmit signal power depend on the interrogator distance. Realistically, the power of the retrodirected signal must be made independent of the power received from the interrogator to ensure efficient system performance over the specified network link distances. Transmit power that depends on interrogator distance limits the allowed separation between satellite nodes because the minimum transmit power is unknown and the link budget cannot be accurately calculated.

Reference [15] presents a novel approach for implementing a retrodirective array that is capable of both full-duplex communication and maintaining a constant retrodirected power level. By using a phase detector to detect the incoming angle of the interrogating source, phase shifters on a transmitting array are automatically controlled to retrodirect a beam back in the direction of the source. In addition to generating a retrodirected signal in the transmitting array, phase shifters in the receiving array ensure coherent combination of the interrogating signal for full-duplex communication. The transmitter and receiver arrays were RF decoupled to increase system efficiency by ensuring a constant transmit power. The entire system is shown in Fig. 17.21.

Because the transmitter and receiver arrays are RF decoupled, the frequency of the interrogator signal can be distinct from the retrodirected frequency. This has the potential of greatly increasing the efficiency of the system requiring a high-frequency retrodirected signal by using an interrogator signal at a much lower frequency. For cubesat applications, a higher transmit and receive frequency could be used to maximize network bandwidth. This allows satellite nodes to be spaced farther using the same amount of interrogator power, while allowing high-speed data exchanges.

Another advantage of this decoupled approach is that increasing directivity requires adding elements only on the transmitting and receiving array because the

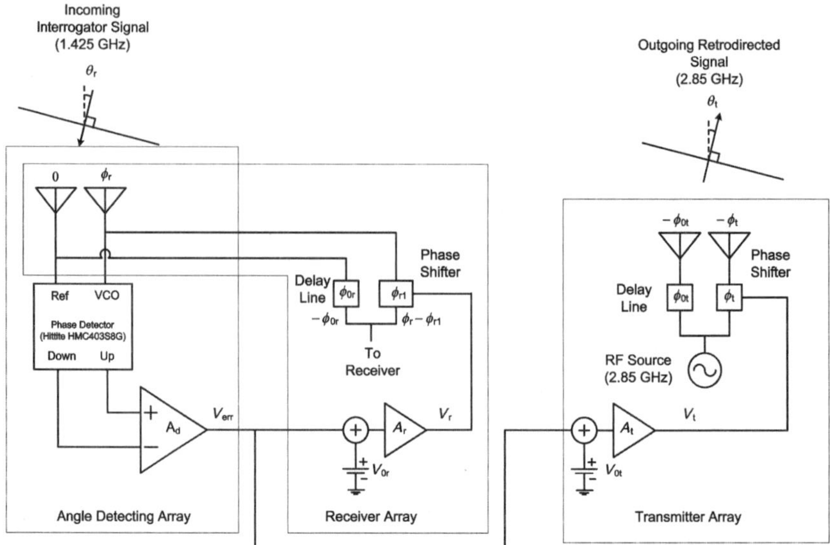

Fig. 17.21 Schematic of retrodirective array using phased detection and phase shifting (reprinted with permission by Grant S. Shiroma [15; Fig. 4.7, p. 64]).

two-element angle detecting array is sufficient for detecting the direction of the source.

Because the source angle is determined by measuring the phase difference between two elements, the angle detecting array is insensitive to the type of modulation used for the incoming signal. A modulated interrogator would be required for network crosslinks to indicate the intended receiver.

Figure 17.22 illustrates the bistatic RCSs of the novel phase-detecting and phase-shifting retrodirective antenna array presented in [15]. Although the

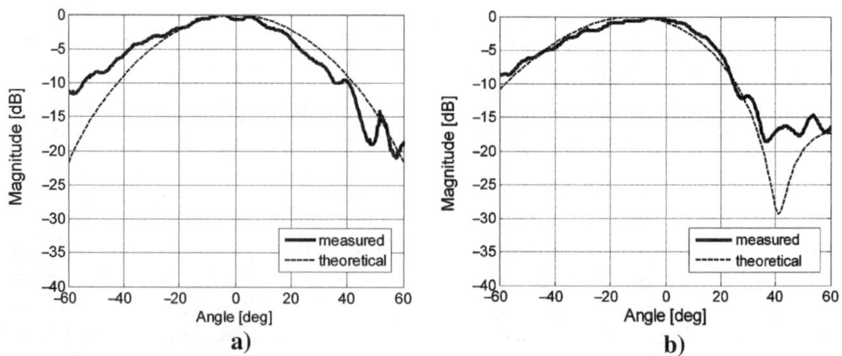

Fig. 17.22 Measured and theoretical bistatic RCSs of retrodirective array using phase detection and phase shifting with interrogator positioned at a) 0 deg and b) –15 deg (reprinted with permission by Grant S. Shiroma [15; Fig. 4.12, p. 74]).

demonstrated array consisted of two elements, sufficient retrodirectivity was demonstrated. The addition of more elements to the antenna array would result in higher directivity and better efficiency.

Although phase detecting and phase shifting solves the large problem of constant transmit signal power, there are downsides to using phase detection and phase shifting. First, the retrodirective system will not work for all interrogating sources. This is because of the design of the phase-detecting array and steering architecture because the entire system design is based on planar wavefronts. Only plane-wave incidence will result in true retrodirective behavior. Unlike the phase-conjugating array, nonplanar wavefronts cannot be recreated using this approach. However, in the envisioned cubesat network there are only line-of-sight links, and all signals are assumed to be plane waves. The second, more complicated problem is that a two-dimensional space-ready retrodirective array would require another phase-detecting array, one for the x and y components and a more complicated phase-shifter control circuit.

IX. Summary

With the emergence of cubesat-class satellites that, when networked, have the capability to replace larger satellites, there comes a need for secure efficient wireless crosslinks. Retrodirective antenna systems show promise of providing these small-satellite network crosslinks thanks to their relative simplicity and autonomous nature. Many of the problems such as security, power efficiency, and processor efficiency associated with other technologies are addressed by retrodirective antenna systems, making them the clear choice for the new generation of cubesat networks.

Conventional retrodirective arrays were not meant for small-satellite space applications, and design challenges need to be addressed accordingly. Miniaturization, beam-steering range, and modulation schemes are some of the numerous issues that must be overcome before retrodirective antennas could be practical for cubesats. The demonstration of retrodirective crosslinks for small satellites by the University of Hawaii shows that overcoming these design challenges are possible.

Novel retrodirective arrays, which are improvements over preceding systems, show promise for cubesat networks. These new developments are simpler, yield higher efficiency, and are more flexible. Future work will consist of further developing and preparing these new retrodirective technologies for space and cubesat networks.

Network reconfigurability is an area that needs to be addressed when using retrodirective antennas for network crosslinks. Normal network architectures will not be able to handle the randomness of the distributed network. New methods for handling the numerous ad hoc connections need to be developed to make retrodirective crosslinked networks possible.

Retrodirective technology for small satellites has not been tested in space, and this is the next step in making small-satellite retrodirective crosslinks a reality. Ideally, a small network of satellites would be launched to test the retrodirective crosslink performance.

References

[1] Ah Yo, D. M. K., Forsyth, W. E., and Shiroma, W. A., "A 360° Retrodirective Self-Oscillating Mixer Array," *IEEE MTT-S International Microwave Symposium Digest*, Vol. 2, June 2000, pp. 813–816.

[2] Forsyth, W. E., and Shiroma, W. A., "A Retrodirective Antenna Array Utilizing a Quasi-Optical Feed for LO Delivery," *USNC/URSI 2001 National Radio Science Meeting*, Paper B/D3-6, Jan. 2001.

[3] Forsyth, W. E., and Shiroma, W. A., "A Retrodirective Antenna Array Using a Spatially Fed Local Oscillator," *IEEE Transactions on Antennas and Propagation, Special Issue on Wireless Information Technology and Networks*, Vol. 50, May 2002, pp. 638–640.

[4] Murakami, B. T., Tamamoto, M. A., Ohta, A. T., Shiroma, G. S., Miyamoto, R. Y., and Shiroma, W. A., "Self-Steering Antenna Arrays for Distributed Picosatellite Networks," *17th Annual AIAA/Utah State University Conference on Small Satellites*, Paper SSC03-XI-1, Aug. 2003.

[5] Shiroma, G. S., Song, C., Miyamoto, R. Y., and Shiroma, W. A., "An Active Self-Steering Array Using Self-Oscillating Mixers," *IEEE MTT-S International Microwave Symposium Digest*, Vol. 3, June 2003, pp. 1685–1688.

[6] Tuovinen, J., Shiroma, G. S., Forsyth, W. E., and Shiroma, W. A., "Multipath Communications Using a Phase-Conjugate Array," *IEEE MTT-S International Microwave Symposium Digest*, Vol. 3, June 2003, pp. 1681–1684.

[7] Sung, S. S., Roque, J. D., Murakami, B. T., Shiroma, G. S., Miyamoto, R. Y., and Shiroma, W. A., "Retrodirective Antenna Technology for CubeSat Networks," *IEEE Topical Conference on Wireless Communication Technology*, IEEE Publications, Piscataway, NJ, 2003, pp. 220–221.

[8] Roque, J. D., Murakami, B. T., Sung, S. S., Shiroma, G. S., Miyamoto, R. Y., and Shiroma, W. A., "Progress in Self-Steering Antennas for Small-Satellite Networks," *Proceedings of the AIAA SPACE 2004 Conference and Exposition*, AIAA Paper 2004-5943, AIAA, Reston, VA, Aug. 2004.

[9] Miyamoto, R. Y., Shiroma, G. S., Murakami, B. T., and Shiroma, W. A., "A High-Directivity Transponder Using Self-Steering Arrays," *IEEE MTT-S International Microwave Symposium Digest*, Vol. 3, June 2004, pp. 1683–1686.

[10] Murakami, B. T., Roque, J. D., Sung, S. S., Shiroma, G. S., Miyamoto, R. Y., and Shiroma W. A., "A Quadruple Subharmonic Phase-Conjugating Array for Secure Picosatellite Crosslinks," *IEEE MTT-S International Microwave Symposium Digest*, 2004, pp. 1687–1690.

[11] Shiroma, G. S., "Self-Steering Arrays for Wireless Communication Networks," M.S. Thesis, Univ. of Hawaii, Honolulu, HI, Aug. 2004.

[12] Mizuno, T. J., Roque, J. D., Murakami, B. T., Yoneshige, L. K., Shiroma, G. S., Miyamoto, R. Y., and Shiroma, W. A., "Antennas for Distributed Nanosatellite Networks," *IEEE/ACES International Conference on Wireless Communications and Applied Computational Electromagnetics*, IEEE Publications, Piscataway, NJ, 2005, pp. 606–609.

[13] Roque, J. D., Sung, S. S., Murakami, B. T., Shiroma, G. S., Miyamoto, R. Y., and Shiroma, W. A., "A Coupled-Antenna Interrogator/Receiver for Retrodirective Crosslinks in a Distributed Nanosatellite Network," *IEEE/ACES International Conference on Wireless Communications and Applied Computational Electromagnetics*, IEEE Publications, Piscataway, NJ, 2005, pp. 610–613.

[14] Takase, B. O., Wong, K. S. Y., Kawakami, C. R., Shiroma, G. S., and Shiroma, W. A., "Modeling of Phase-Conjugating Arrays for Multipath Compensation in a Maritime Environment," *USNC/URSI National Radio Science Meeting*, Paper B5-6, Jan. 2006, p. 68.

[15] Shiroma, G. S., "Security Enhancement and Path Loss Minimization of Retrodirective Arrays for Wireless Communication Networks," Ph.D. Dissertation, Univ. of Hawaii, Honolulu, HI, Dec. 2007.

[16] Shiroma, W. A., Miyamoto, R. Y., Shiroma, G. S., Ohta, A. T., Tamamoto, M. A., and Murakami, B. T., "Retrodirective Systems," *Wiley Encyclopedia of RF and Microwave Engineering*, Wiley, New York, 2005, pp. 4493–4507.

[17] Balanis, C. A., "Arrays: Linear, Planar, and Circular," *Antenna Theory Analysis and Design*, 3rd ed., Wiley, New York, 2005, pp. 283–384.

[18] Van Atta, L. C., "Electromagnetic Reflector," U.S. Patent 2,908,002, 1959.

[19] Pon, C. Y., "Retrodirective Array Using the Heterodyne Technique," *IEEE Transactions on Antennas and Propagation*, Vol. AP-12, 1964, pp. 176–180.

[20] Miyamoto, R. Y., Qian, Y., and Itoh, T., "Active Retrodirective Array for Remote Tagging and Wireless Sensor Applications," *IEEE Transactions on Microwave Theory and Techniques*, Vol. 49, Sept. 2001, pp. 1658–1662.

[21] Miyamoto, R. Y., and Itoh, T., "Retrodirective Arrays for Wireless Communications," *IEEE Microwave Magazine*, Vol. 3, No. 1, 2002, pp. 71–79.

Appendix

Case Study: Overview of ION as Applied to Atmospheric Research and Technology Testing Problems

Purvesh Thakker,* Dustin Ames,† Leon Arber,‡ Mike Dabrowski,§ Aaron Dufrene,¶ Andrew Pukniel,** Alex Rein,†† Victoria Coverstone,‡‡ and Gary Swenson§§

University of Illinois at Urbana-Champaign, Urbana, Illinois

I. Introduction

THIS work presents the Illinois Observing Nanosatellite (ION) from the University of Illinois at Urbana-Champaign (UIUC) as an example of the utility of cubesats (1 to 3 kg). From 1999 to 2006, over 80 organizations registered intentions to develop cubesat-class satellites, and more than two dozen have launched. Typically, deployers [1] hold three 10 × 10 × 10-cm cubesats, which conform to a simple interface defined in the CubeSat Spec (data available online at http://cubesat.atl.calpoly.edu/pages/documents/developers.php) by Stanford and Cal-Poly Universities. These deployers are then carried to orbit in unutilized space on existing launches. Thanks to this growing secondary payload launch infrastructure, these tiny spacecraft provide unique opportunities to access space for education [2], atmospheric science research [3–8], and space technology testing [9,10].

*Program Manager and Graduate Student.

†Graduate Student.

‡Graduate Student.

§Program Manager and Graduate Student.

¶Graduate Student.

**Graduate Student.

††Undergraduate Student.

‡‡Professor and Associate Dean for Graduate and Professional Programs.

§§Professor.

In addition to the educational aspect that cubesat programs typically emphasize, ION provides an effective demonstration of the ability of cubesats to perform remote-sensing and technology test missions. This paper illustrates the feasibility of these missions by providing an end-to-end discussion of the ION project, including recommendations for each system element based on the development experience. Unfortunately, all 18 satellites including ION were lost when the launch vehicle crashed back to Earth during launch. Nevertheless, there are many important results from the ION project that should be preserved. The most important lesson is the demonstration that these satellites can support scientific remote-sensing payloads. This opens up new measurement opportunities because these tiny satellites are well suited to multipoint or multi-angle measurements for both remote-sensing and in situ experiments. Furthermore, these low-cost satellites can be developed in as little as one to two years by small teams of less than ten people.

There are five primary missions for ION:

1. Measure molecular oxygen airglow emissions from the Earth's mesosphere using a 760-nm photometer.
2. Test microvacuum arc thruster (μVAT) and associated power processing unit (PPU).
3. Test small-satellite processor called small intelligent datalogger (SID).
4. Image Earth using CMOS camera.
5. Demonstrate ground-based attitude stabilization.

There are three secondary experiments: a) establish contact with first student satellite completely developed at UIUC, b) survive the first two weeks, and c) demonstrate use of polyethylene shielding.

Notable features of the missions include the following:

- Primary missions demonstrate cubesat utility
- Custom operating system
- Custom communications protocol
- Dual redundant batteries
- Solar-cell powerpoint tracking
- Software upgradeability
- Support for distributed ground stations
- Web dashboard for monitoring telemetry

The first mission involves measuring oxygen airglow emissions from the Earth's mesosphere [3,8,11]. This helps scientists understand how energy transfers across large regions contribute to our knowledge of atmospheric dynamics. Second, ION tests a new microvacuum arc thruster (μVAT) [9,12], advancing ion propulsion technology for small satellites. Third, ION tests a new small intelligent datalogger (SID) processor board that is radiation hardened through system design techniques to allow small satellites in low Earth orbits (LEO) to take advantage of recent advances in processor technology with increased reliability. Fourth, ION performs Earth imaging using a small CMOS camera. Finally, ION demonstrates attitude stabilization [13] showing how reusable ground software can allow attitude control capabilities for simple satellites or a backup capability for satellites with flight attitude control systems. In addition to these primary missions, the preceding list gives some additional educational and technology experiments and some of ION's notable features.

A. Organization

ION is the first project of the Illinois Tiny Satellite Initiative (ITSI), which is organized through an interdisciplinary senior design course [8]. The course objectives include training students to identify, formulate, and solve engineering problems as part of a large multiteam project. Over the course of five years, over 100 students across seven engineering disciplines have participated in the ION project. Most students participate in their senior year with some starting earlier or participating as graduate students. As shown in Fig. A.1, one or two graduate teaching assistants provide day-to-day program management. They also provide continuity across semesters as students in the course change. In addition, three faculty advisors provide mentorship, and numerous other faculty provide technical advice as required. Typically, five to six teams of three to five people each participate in the project.

Cubesat primarily attracts electrical and computer engineering as well as aerospace engineering students (approximately 80% of the class), with minor involvements by computer science, theoretical and applied mechanics, general, and mechanical engineering students. The interdisciplinary design class developed for cubesat has been expanded to incorporate other interdisciplinary topics as well. The course was formulated through the College of Engineering Design Council, which has the charter to foster design innovation and recommendations across the college. The class is structured to incorporate projects that are large (>6 students), with at least two semesters' involvement.

Student developed satellites present many challenges beyond technical issues because of limited expertise, limited equipment, limited budgets, and the students' short-term involvement in the project. As a two-semester course, half of the students change every semester, and all must balance participation with other classes. In addition, like many cubesat development teams, the organization faced many startup challenges since ION is its first satellite. The organizational format proves an advantage in some respects as well. As a student project with limited financial

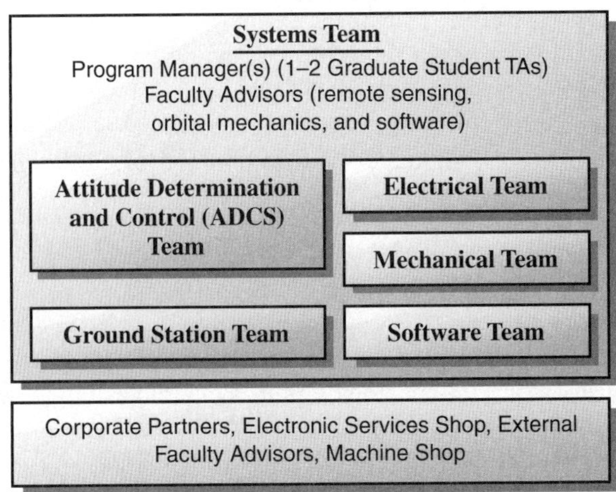

Fig. A.1 Typical team structure for ION.

investment, the organization can tolerate a high degree of risk allowing it to do things that a more expensive project cannot. Even if the satellite does not work, students earn invaluable real hands-on engineering experience.

B. Overview

Section II of this work describes ION's mission objectives. Next, Section III describes the system's requirements and provides an overview of all system elements. Then, Sections IV, V, and VI provide the bulk of the content with discussions of hardware, software, and the integrated system. Finally, Section VII discusses some of the final project results before the conclusion in Section VIII.

II. Mission Objectives

A. Satellite Mission Objectives

ION supports the five mission objectives with five types of additional mission requirements:

1) *Photometer requirements:* Measure oxygen airglow at 89 km from 500 km.
   S/N: >50
   Wavelength: 760 nm
   Airglow intensity range: 6000 Rayleighs
   Wave size: 15 km to 2500 km
   Measurement duration: >2 weeks
   Attitude accuracy: <20 deg of nadir
   Sampling rate: 1 s
   Integration time: <1 s
   Data rate: 58 kB/day
2) *Thruster requirements:* Measure thrust, lifetime, and material fallback.
3) *SID requirements:* Integrate with system, and operate satellite for six months.
4) *Camera requirements:* Photograph coastlines.
   Field of view: 45 deg
   Ground resolution: <1 km
   Imaging frequency: occasionally on demand
   Data rate: <25 kB/image
5) *Attitude requirements:* Attitude determination and control without onboard computer.
   Attitude determination: 5 deg
   Attitude: <5 deg of nadir
   Attitude rates: approximately 0

ION's first mission consists of photometric measurements [3,8] of molecular oxygen airglow emissions from the Earth's mesosphere [11], as shown in Fig. A.2. Oxygen chemistry at this 94-km altitude emits a dim glow of light, which ION's photometer is designed to measure. Measurements should show variations in this airglow caused by various effects such as thunderstorm convection or the presence of mountains. These variations ripple through the atmosphere in 15 km to >2500 km waves carried by wind. By studying these waves, atmospheric

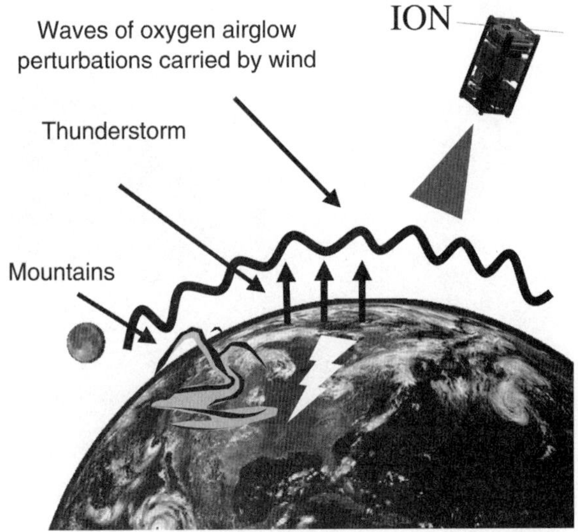

Fig. A.2 ION's photometer studies dynamic drivers of the upper mesosphere (Mission 1).

scientists learn how energy transfers across large spatial regions contribute to knowledge of upper-atmospheric dynamics. This airglow emission is absorbed by the Earth's lower atmosphere preventing study with Earth-based sensors, and satellite detection offers the added benefit of global coverage. In addition, ION's orbit gives it the opportunity to gather these data around the polar regions, which provide special cases that are of extra interest to atmospheric scientists.

The second objective involves testing a new μVAT [9, 12] and associated power processing unit (PPU) from Alameda (Fig. A.3). Small satellites, including cubesats, do not yet have the maneuverability of larger spacecraft. Satellites require lateral movement with high thrust for orbital maneuvers and low thrust for

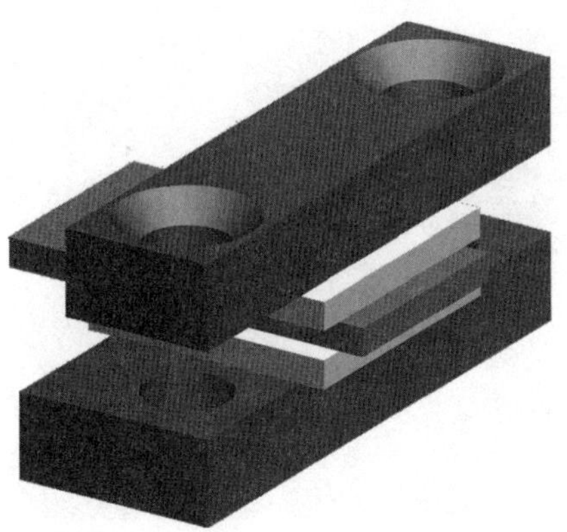

Fig. A.3 ION tests μVAT thrusters (Mission 2).

Fig. A.4 ION tests SID processor board (Mission 3).

stationkeeping. This thruster technology has the potential to provide a high dynamic range without efficiency loss. ION's μVAT experiment serves as a stepping stone towards a versatile low-mass satellite propulsion system capable of lateral movement and fine attitude control, a key enabling technology for small satellites. Once fully developed, such a capability might allow greater interaction with other spacecraft.

The third objective involves testing a processor called the SID from Tether Applications (Fig. A.4). By utilizing a commercial-off-the-shelf (COTS) processor radiation hardened through system design techniques, it allows small satellites to take advantage of the latest in small, low-power, high-performance processor technology with increased reliability. The SID includes latch-up protection circuitry throughout the board as well as heat dissipation and electromagnetic interference (EMI) control techniques. It has been designed specifically to support small satellites, and it has been tested for radiation exposure in LEO orbit

The fourth objective involves using a small black-and-white CMOS camera to provide Earth-imaging capabilities (Fig. A.5). It provides a 640×480 pixel image with a 45-deg field of view covering a 414-km diam on the Earth with slightly better than 1-km resolution.

Finally, ION demonstrates ground-based attitude control [13], as shown in Fig. A.6. Active attitude stabilization is an important enabling capability for the

Fig. A.5 ION tests B/W CMOS camera (Mission 4).

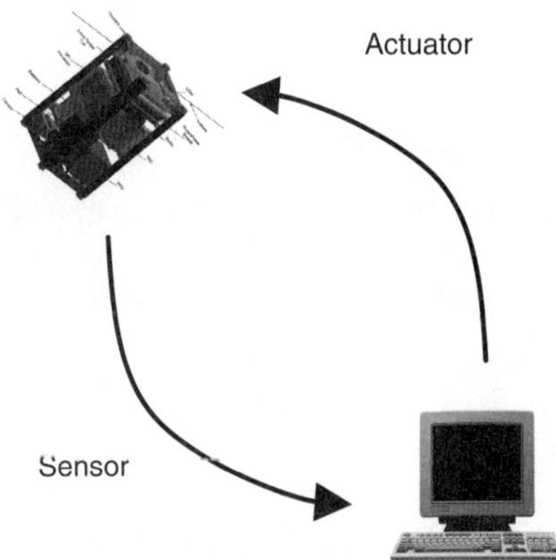

Fig. A.6 ION demonstrates ground-based attitude control (Mission 5).

future of cubesats, and ION's ground-based approach greatly simplifies flight software, processor, and integration requirements. Instead of performing attitude stabilization in real time onboard the satellite, ION transfers sensor data to the ground where simulators determine torque coil firing sequences. These torque coil-firing sequences are then uploaded to the satellite for execution. Attitude stabilization can also be performed in real time while the satellite is in direct contact with the ground station.

B. Educational Objectives

Designing and operating space systems requires close interaction between various fields of engineering and science. Currently, engineering curriculums contain largely theoretical courses in science, physics, mathematics, and engineering. Courses that provide opportunities for students to have hands-on experience with hardware are fewer in number. Those that promote a system engineering experience are even rarer, although academic research groups in space science and the aerospace industry have expressed the need for individuals who have practical systems engineering training. The interdisciplinary design aspects contained within the ION project serve as an incubator for developing system engineers with hands-on expertise.

The educational experience includes project specification, design, simulation, fabrication, assembly, functional testing, and environmental testing. In addition, students develop communication skills through informal and formal, written, and oral communication activities. It is a challenge to operate a long-term project within a semester-based course structure. The project must be organized around specific time schedules with individual and group tasks defined around specific end of semester objectives. Since the course began, a number of students have participated as leaders and continuity providers for the technologies and processes that have been developed. The practical system engineering, communication, and leadership training obtained is highly valued by academia and industry alike, as is illustrated by the heavy recruitment of participants by graduate schools and industry.

C. Research Objectives

In addition to educational objectives and specific mission objectives, the project also has some broader research objectives spanning about a dozen unique publications. These objectives include an expansion of the perceptions of small university satellite capabilities. By providing a concrete example of the utility of these small satellites, it is anticipated that more remote-sensing and technology test mission opportunities will open up. These small satellites can provide a great deal of value for the dollar even if the student engineers are compensated as full-time employees for their time. Some simple remote-sensing measurements do not require the state-of-the-art equipment that is typically employed on larger spacecraft. Similarly, technologies tested on these small satellites might never make it past the idea stage without a university satellite as a stepping stone. In summary, it is hoped that the ION project will help answer the long-standing utility question surrounding small university satellites, particularly including cubesats.

III. System Overview

A. System Requirements

Several of requirements in Table A.1 are imposed directly by the cubesat specification (data available online at http://cubesat.atl.calpoly.edu/pages/documents/developers.php), whereas others are derived internally within the team. Because ION is a double cubesat, it can have a mass up to 2 kg and can occupy two 10-cm cubes with some additional space between the cubes ($10 \times 10 \times 21.5$ cm). The spec includes separation springs on top of the satellite to separate it from its neighbors upon release from the PPOD, and anodized rails to prevent them from sticking to the PPOD as it waits for launch. For wiring, the spec includes kill switches in the satellite's feet, which cut power while the satellite awaits deployment in the PPOD. In addition, a pull pin cuts power while the satellite is outside the PPOD. The spec includes a recommendation for a data port for recharging batteries and performing diagnostics while the satellite is in the PPOD. On ION, this data port also feeds external power sources directly to the satellite in order to avoid wear on the flight batteries during testing. For environmental testing, the spec includes a requirement that the satellite undergo vibration testing beginning with a low-level sine sweep on all three axes to identify resonance frequencies. In addition, the satellite goes through a high-level random vibration test on all three axes at 125% of the launch vehicle's vibration profile. Finally, the satellite goes through a bake-out procedure to prevent outgassing during launch. During this bake-out procedure, the satellite goes through two full heating and cooling cycles in a high vacuum. Finally, because the launch is outside of the United States, hardware exports must be approved through the ITAR process. Note that the design of the Cubesat Spec specifically minimizes risk to the rest of the launch vehicle and payloads by cutting power to the satellite as it awaits deployment, completely enclosing the cubesats and performing vibration and bake-out procedures prior to launch.

Internally, all missions were specifically selected such that they could be completed within a six-month satellite lifetime. This short lifetime allows the team to focus on making the system functional on the first day, without getting excessively

Table A.1 System Requirements

Requirement	Description
	Cubesat spec requirements[a]
Size	10 × 10 × 21.5 cm (double cubesat)
Mass	2.0 kgw
Structure	Separation springs, anodized rails
Wiring	Kill switches, pull pin
Dataport (optional)	Battery recharging, diagnostics, external power
Vibration	Low-level sine sweep on all three axes, 35-s high-level random test at 125% of launch profile for all three axes, 13.85-min low level random test on all three axes
Bakeout	1-h ramp-up from room temp to 70°C followed by 1-h hot soak and 1-h ramp-down to room temperature. The process is repeated. Both soaks occurred at pressures between 0.00021 and 0.00005 torr.
	Internal requirements
Lifetime	Six months
Radiation	<5 krad of radiation exposure
Operating temperature	–20°C to +85°C
Power	50% power margin normal operation, 90% power margin sleep mode, 15% maximum expected discharge depth, no power tracking requirements
Data rate	58-kB payload/day + 12-kB telemetry data/day = 70-kB with 50% data margin
Antenna	Broad pattern, no pointing requirement
Attitude	Determination within 5 deg, stabilization with 5 deg of nadir (20-deg maximum)
Software	Adjustable operating schedules and configurations, activity logging, in-flight debugging tools, upgradeable software
Fault tolerance	Processor and system watchdog timers, "safe mode" after certain failures, backup software system

[a]Data available online at http://cubesat.atl.calpoly.edu/pages/documents/developers.php.

distracted by lifetime concerns. Performing lifetime analysis and testing, and adding fault tolerance, adds tremendously to the complexity of a project that already suffers from limited space, mass, power, time, and expertise. The six-month lifetime requirement limits these lifetime efforts to the simplest, most effective techniques. General component research shows that CMOS components are typically the first to fail after receiving 100 krad to 10 Mrads of accumulated radiation exposure [14]. As a result, ION is required to allow no more than 5 krads of exposure during its expected lifetime. The 5-krad requirement is less than the 10 krads of exposure that CMOS components are often expected to tolerate [10]. The required operating temperature of the satellite is set to a range of –20° to +85°C roughly corresponding to industrial grade components. Military and space grade components offer a wider temperature range, but availability and cost make them much more prohibitive to use. Industrial-grade components offer wide availability, so that there is no need to step down to the narrower commercial temperature range.

The power system is required to achieve a 50% margin under normal operating conditions in order to allow for attitude uncertainty, system characterization error, increased power consumption with radiation exposure, and operating errors. In addition, there should be a 90% sleep mode margin requirement, which allows the satellite to recharge a discharged battery. Although ION is not designed to meet this requirement, it is being included as a recommendation for other programs. Because attitude stabilization is performed from the ground and is a high-risk system, the power system is also required to sustain the satellite with any attitude. A 15% maximum discharge depth goal is required for normal operation. This requirement ensures that the satellite can make it through its 35-min maximum eclipse time, 10-min communication passes, and other burst usage (cameras, thrusters, antenna deployment, initial stabilization with torque coils, etc.) even as the batteries begin to wear. If the system is operated properly, the batteries should remain largely full all the time. See [15] for more information about power system requirements.

The communication system is required to download 58 kB/day of photometer data in addition to 12 kB of system data (20 telemetry sensors at 2 bytes each sampled every 5 min). This should be supported with a 50% data margin to allow for upload traffic and communication difficulties. On-demand camera images (25 kB each) and portions of log files (megabytes) can be downloaded with leftover capacity and in place of photometer data. Because attitude stabilization is performed from the ground, the flight antenna is required to have the broadest pattern possible so that a communication link can be established before attitude stabilization. Attitude determination and stabilization requirements are set to 5 deg each primarily because of mission requirements set earlier.

B. Orbit¶¶

ION's target orbit is a sun-synchronous polar orbit with a period of 95.0065 minutes and an inclination of 97.43 deg (Table A.2). Correct insertion will put the spacecraft into a slightly eccentric (0.0022) orbit with perigee and apogee altitudes of 504.7 and 518.4 km, respectively. The main advantage of this type of orbit is

Table A.2 Orbit Parameters

Parameter	Description
Launch	DNEPR based on SS-18 ICBM from Baikonur, Kazakhstan
Orbit	500-km altitude, 98 inclination, 0.002 eccentricity (near circular), sun synchronous
Period	95-min, 35-min eclipse
Ground station	Urbana, Illinois, United States of America
Passes	Pass sets around 10a and 10p, <10 min each pass
Lifetime	Drag coefficient, $C_D = 2.2$, <4.0-yr orbit

¶¶This section was contributed primarily by Andy Pukniel.

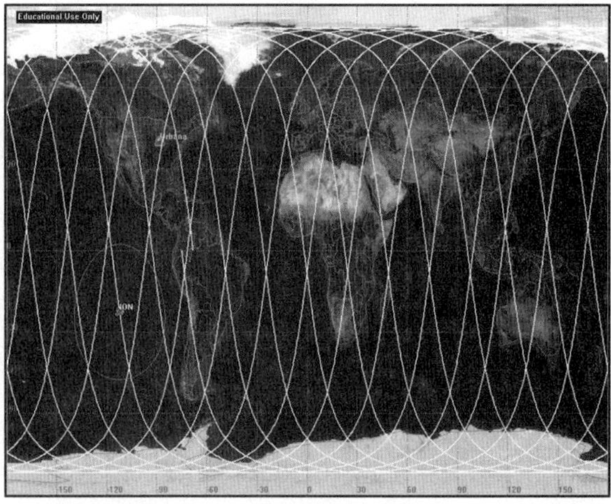

Fig. A.7 ION's Earth ground track.

the global coverage shown by ION's ground track in Fig. A.7. The 97.43-deg inclination ensures that high-latitude and polar regions, which are of special interest to scientists, are available for investigation. An added benefit of ION's orbit is the maximum eclipse time of only 35.38 min per orbit. This allows for sufficient recharging of the onboard batteries and ensures uninterrupted data collection opportunities.

Detailed STK scenarios show that there are five communication opportunities every 24 hours with individual durations ranging between 1.2 and 10.5 min. STK was also used for ION's orbital lifetime analysis by assuming a coefficient of drag C_D of 2.2 and reflection coefficient between 1.0 and 1.3. For ION in its nominal, nadir-facing orientation, the lifetime was estimated at 4.0 years. The use of magnetic actuators in highly inclined orbits can be challenging because it is impossible to generate torque in the direction of the magnetic field, resulting in an uncontrollable subspace. Luckily, because of the time-varying position of ION as the Earth rotates underneath it, this problem can be addressed and controlled.

C. System Elements

Figure A.8 provides a logical layout of the entire ION system including satellite and ground station, which the paper is designed around. Sections I through III have already been discussed and are included for completeness. Each of the remaining sections (IV through VIII) delves into blocks 1 through 11 in order. In addition to the logical layout shown in Fig. A.8, Fig. A.9 shows the physical layout of the satellite, and Fig. A.10 shows the physical layout of the ground station with each item labeled with its logical block number.

The payload block (1) in Fig. A.8 contains items required for some of the primary experiments. In particular, the photometer makes the oxygen airglow measurements for experiment 1. The photometer shutdown diodes detect ambient

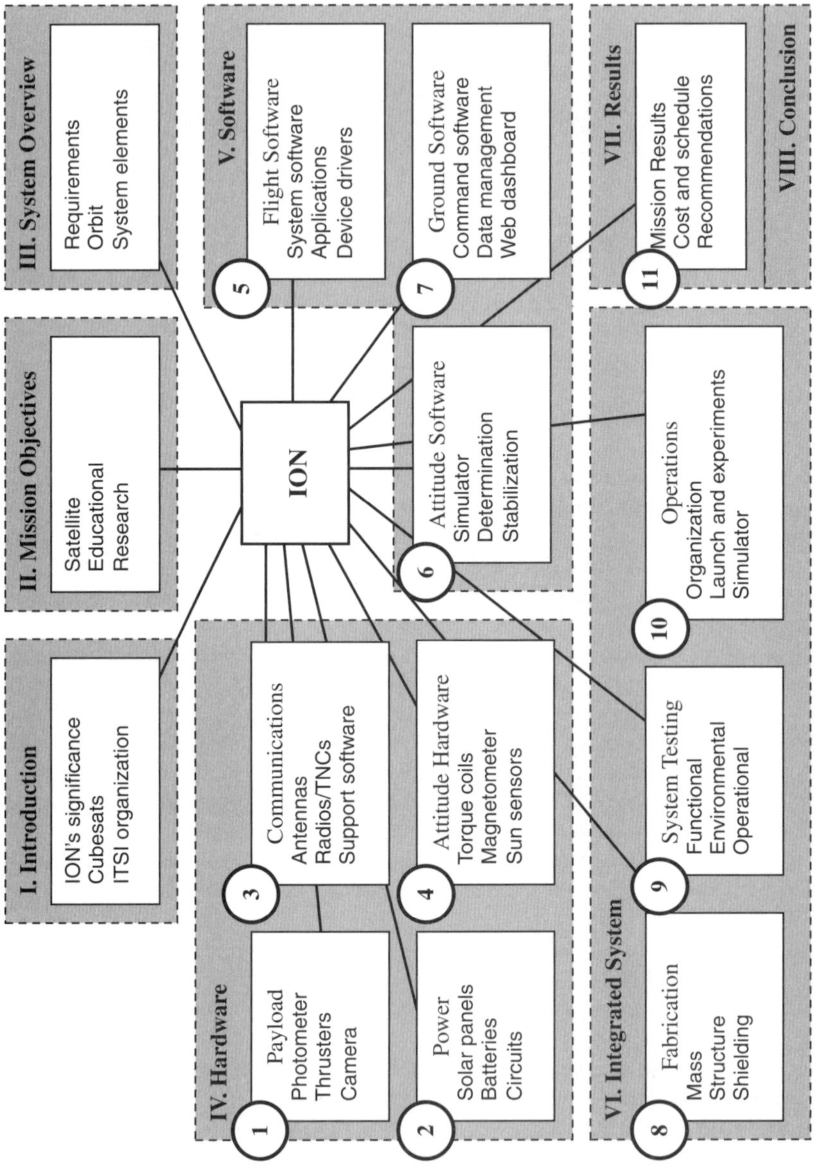

Fig. A.8 Paper overview.

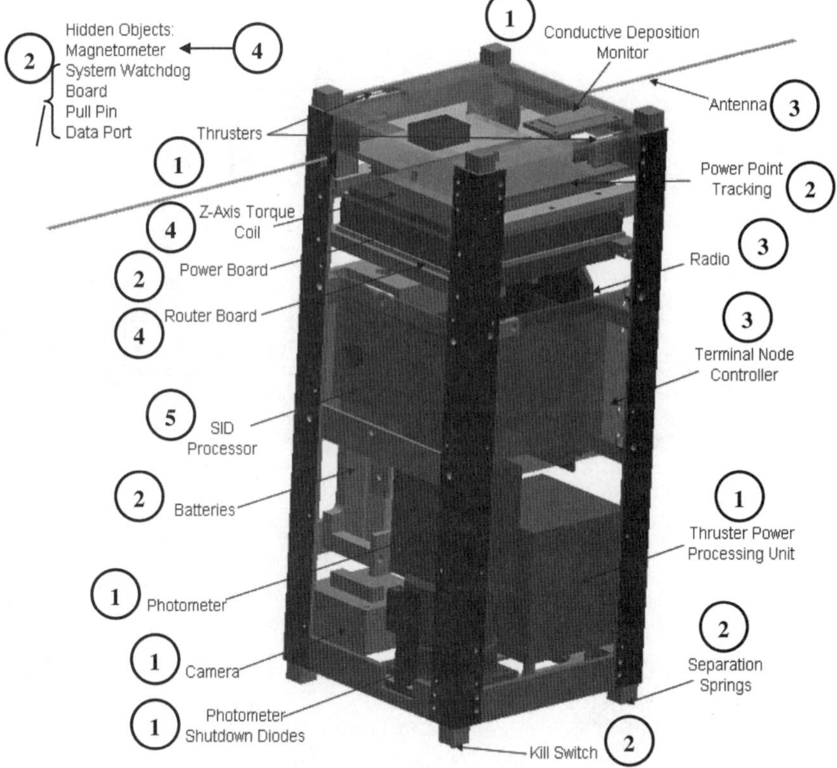

Fig. A.9 ION hardware layout.

light and automatically cut power to the photometer if too much light is present. This provides a last line of defense against photometer burnout. The thrusters and associated PPU are tested in experiment 2 with the help of the conductive deposition monitor (CDM), which is designed to detect thruster material fallback onto the satellite. The camera takes pictures of the Earth for experiment 4.

The power block (2) in Fig. A.8 is responsible for generation, storage, and distribution of power throughout the system. It comprises a number of hardware elements including solar panels, batteries, power board, power point tracking board, system watchdog board, pull pin, kill switches, and data port. In addition, power policy software on the processor manages the system.

The communications block (3) in Fig. A.8 contains items for establishing a link between the satellite and ground station. It includes an antenna, radio, and terminal node controller (TNC) (modem) on both the satellite and ground station. The ground antenna can actively track the location of the satellite as it passes overhead. The communications computer calculates these pass times and controls the direction of the antenna. It also performs Doppler shift calculations and adjusts the radio communication frequency.

The attitude hardware block (4) in Fig. A.8 includes sensors for determining the satellite's attitude and actuators for controlling it. The primary sensor is the

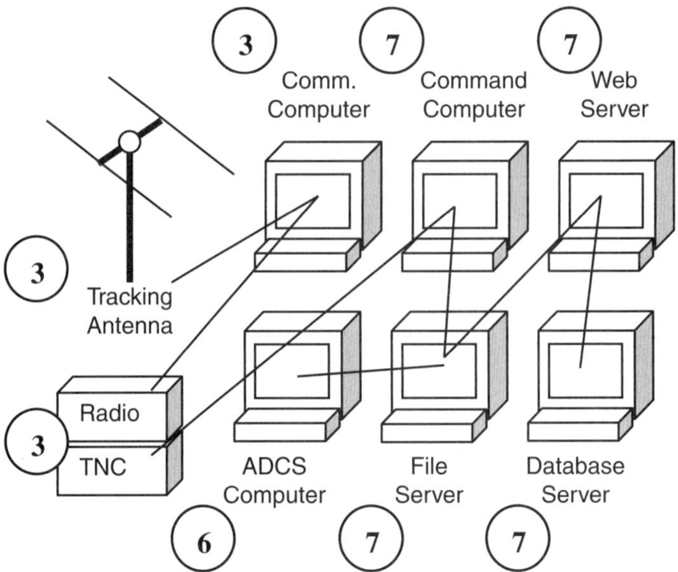

Fig. A.10 Ground station layout.

magnetometer, which provides a three-axis measure of the Earth's magnetic field. A small solar panel on top of the satellite and the four main solar panels also act as sun sensors. With ION's ground-based attitude control scheme, no complex attitude control software is required onboard the satellite. Sensor data are downloaded to the ground computer, where torque-coil firing sequences are calculated and uploaded back to the satellite. The router board contains the torque-coil driver circuitry in addition to assisting with wiring routing. The magnetometer board also contains the photometer shutdown circuitry for the payload block because it is physically located close to the photometer.

The flight software block (5) in Fig. A.8 includes system software such a custom file system and communications protocol. It also contains device drivers for operating all hardware and application managers that schedule processing time for applications. Schedules are controlled from the ground with "config files" containing "work units." These work units specify a date and time, an action to perform, and any device-specific configuration options. The software also includes a great deal of fault tolerance including a simplified "reset mode."

The attitude software block (6) in Fig. A.8 is responsible for attitude determination and control calculations. The ground-based attitude control software includes a simulator that propagates the satellite's attitude forward allowing for a long-duration feedback system. Sensor data are received from the satellite, and torque-coil sequences are uploaded back to the satellite on the same pass or the next pass.

The ground software block (7) in Fig. A.8 contains command software for downloading data and controlling the satellite. Raw data are maintained on the ground computer and published to the team via a separate file server. Here, data are directly accessible in text format, and data are synchronized into a database.

Data analysis calculations are performed on the fly within the database and are accessible to the Web server, which publishes data including graphs to the Web. A public dashboard makes data available to the public, whereas private data, such as file listings, are accessible only to the team.

The fabrication block (8) in Fig. A.8 contains ION's mass budget and describes the satellite's structure, component layout, and fabrication techniques. It also discusses fabrication of some key structural items including solar panels, switches, and the antenna deployment mechanism. Finally, this section discusses radiation shielding including radiation dosage calculations.

The system testing block (9) in Fig. A.8 describes system testing techniques used on ION starting with functional testing. It also shows environmental test results including thermal testing, vibration testing, and thermal-vacuum testing. Finally, the block describes some operational testing procedures and discusses the analysis of flight telemetry data.

The operations block (10) in Fig. A.8 discusses the organization plan for coordination among team members as well as the initial launch priorities and experiment plans.

Finally, the results block (11) in Fig. A.8 reviews the accomplishments and uncertainties from the project. It also reviews the project's actual cost and schedule. Most importantly, this block contains high-level recommendations for future projects.

IV. Hardware

A. Payload

1. Photometer

A simple remote-sensing photometer (Fig. A.11) was designed to observe the Earth's chemiluminescent emission from O_2 atmospheric band (0,0), at 762 nm [3,8] (Table A.3). This bright airglow emission layer has peak brightness at 94 km above the Earth, and it has a thickness of approximately 8 km. The photometer measures brightness perturbations between 0.5% and 50% of the mean layer brightness. This emission is bright and self-absorbed in the lower atmosphere, so that a downward-viewing instrument viewing at this wavelength is not contaminated by city lights, for example. Liu and Swenson [11] have modeled the response

Fig. A.11 Photometer with lens and filter housing.

Table A.3 Photometer Parameters

Parameter	Value
Wavelength	762 nm
Intensity	6000 Rayleighs, 0.5 to 50% perturbation
Wave size	>16 km
Field of view	4 deg
Sampling	1-s Integration time, 1-s sampling interval
Data rate	52 kB/day
S/N	3296 @ 23°C

of the layer brightness perturbations to atmospheric waves propagating through the layer. Small-scale atmospheric gravity waves (AGWs) are a major source of momentum and energy in this altitude region of the atmosphere, and little observational morphology exists. The objective is to perform the first global assessment of these waves.

As shown in Fig. A.12, an optical filter isolates the 762-nm emission, and a lens focuses the light onto the photomultiplier tube (PMT). The photometer samples the target at 1 Hz with a 1-s integration time. Because the satellite orbits the Earth at a speed of 8 km/s, it can resolve waves as small as 16 km without aliasing. As the photometer receives photons from the target signal, its output pulses are counted by the processor. In addition to pulses triggered by the reception of photons, heat triggers pulses creating dark current. This undesired dark current can be characterized over temperature (Fig. A.13) and then removed with the help of onboard temperature measurements. Dark current removal is not as critical for this measurement because the interest is in signal variation and temperature varies slowly. Also, the photometer optical design shows that dark current levels do not have much effect on the instrument's SNR. As a result, the instrument can be operated across the full anticipated temperature range of the satellite (–20°C to 20°C). The final optical design has a field of view of 4 deg with an SNR of 3296 at 23°C. See [3] for more information about the photometer's optical design.

Photometers can burn out if exposed to too much light, and so the photometer is only scheduled to operate at night. In addition, a backup hardware shutdown circuit uses photodiodes to automatically cut power to the photometer when it detects too much ambient light. The shutdown circuit compares an exposed photodiode with a covered photodiode eliminating systematic effects such as

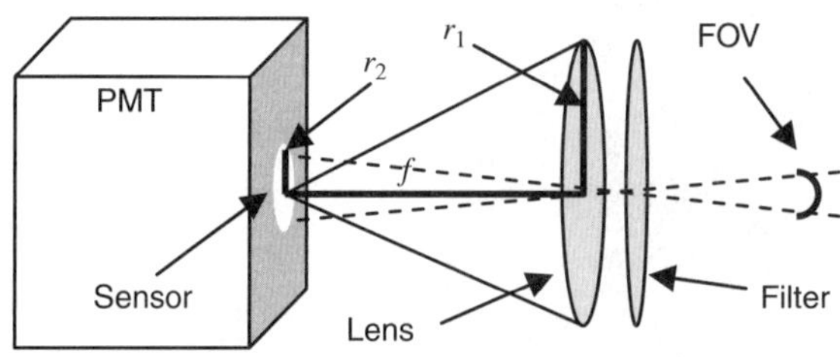

Fig. A.12 Photometer optical design.

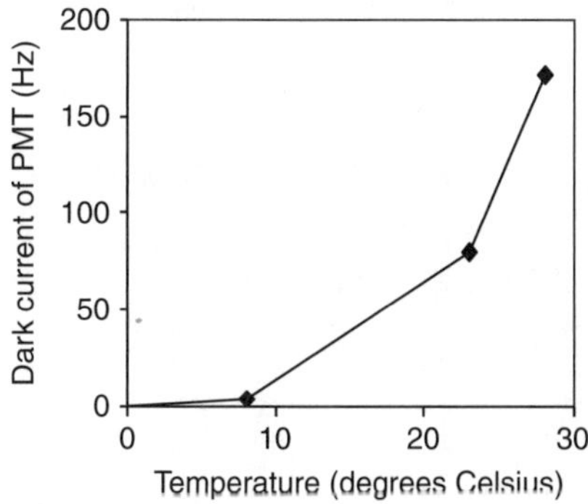

Fig. A.13 Photometer dark current measured by temperature.

temperature. The exposed photodiode also uses a lens that provides a field of view slightly larger than the photometer. The processor can override the automatic shutdown, if needed.

2. *MicroVacuum Arc Thrusters (µVAT)*

The µVAT thrusters [9,12] and associated PPU electronics come from Alameda Corporation with some development contributions from University of Illinois students. On ION the µVAT thrusters serve as an experiment while torque coils act as the primary attitude adjustment actuators (Table A.4). The thrusters are placed in a way that allows translation and two-axis rotation (Fig. A.14).

As an electric arc forms across an anode and a cathode, the cathode material ejects at a high velocity producing thrust (Fig. A.15). ION's frame serves as the cathode material and therefore also the propellant. ION uses a sandwich "BLT" geometry that proves more reliable and easier to manufacture then the cylindrical geometry, but provides less directionality. The layers consist of copper, ceramic, and titanium. When they operate, an arc jumps across the center titanium layer and the outer electrodes. Two high-alumina ceramic plates separate the anode from the structure (cathode). The system produces ion velocities of up to 30,000 m/s driven mostly by local pressure gradients. This gives an excellent thrust/power ratio of approximately 10 µN/W and an individual thrust of 1 µ-s/pulse. It is estimated that a pair of thrusters firing together to rotate the satellite

Table A.4 µVAT Parameters

Parameter	Value
Ion velocities	30,000 m/s
Thrust/power ratio	10 N/W
Individual thrust	1 µ-s/pulse
Acceleration	6.56×10^{-4} rad/s^2 (pair of thrusters producing 5.4×10^{6} N-m of torque)
Rotation	4-s fire turns satellite 90 deg in 10 min
Size	1-cm-wide thrusters, 150-g electronics

Fig. A.14 Thrusters provide lateral translation and two-axis rotation.

should produce 5.4×10^6 N-m of torque resulting in an angular acceleration of 6.56×10^{-4} rad/s^2. With this acceleration, firing the thrusters for 4 s should turn the satellite 90 deg in 10 min. The thrusters measure about 1 cm across, and the associated electronics weigh in at 150 g making for a very small, low mass thruster system. The thrust also scales with power availability making it viable for cubesats, microsats, and larger satellites.

One concern with the use of these thrusters, as with others, involves exhaust depositing on the solar panels, blurring optical experiments, and shorting internal

Fig. A.15 Thruster (top) and thruster firing (bottom).

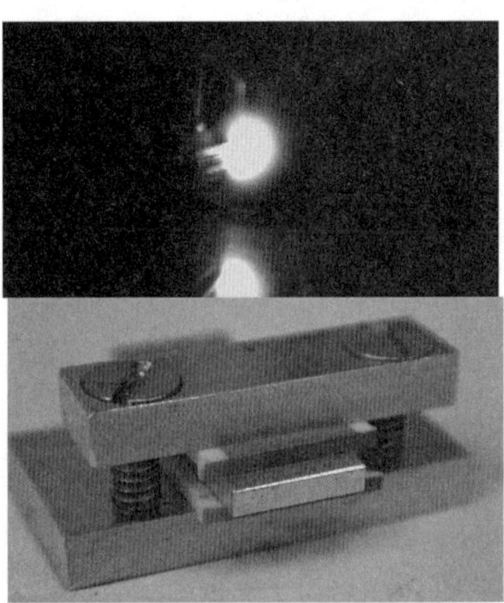

Table A.5 CMOS Camera Parameters

Parameter	Value
Pixels	640 × 480
FOV	45 deg
Ground resolution	~1 km
Compression	90% to 23 kB
Download time	5-min/image @ 1200 baud (50% overhead assumed)

electronics. ION has a CDM near one of the thrusters to help understand this effect. This sensor shorts when excessive conductive material falls back onto the satellite, and it also can detect when that particular thruster fires. The primary method for understanding if the thrusters fire correctly involves detecting attitude changes. ION can also potentially compensate for the thrusters with the onboard torque coils preventing the satellite from losing its target attitude during thruster testing.

3. *CMOS Camera*[***]

ION is equipped with a Photobit PB-300 black-and-white CMOS camera for Earth- imaging applications. The onboard SID computer is specifically designed to work with this camera and has a specialized interface that connects directly to the camera. Although the camera can capture 640 × 480 video at 24 fps (Table A.5), there is not enough computing power onboard to process this much data. Consequently, taking a picture consists of extracting a single 640 × 480 frame from the video stream (Fig. A.16). This image then undergoes JPEG compression, reducing the amount of space required to store the image by 90% to 23 kB, which takes 3 min to download. The onboard JPEG implementation is adapted from the PICOCAM satellite.

Fig. A.16 Photograph taken by a CMOS camera.

[***]This section was contributed primarily by Leon Arber.

In addition to simply capturing images, the PB-300 has numerous built-in camera registers controlled by I^2C that can be used to control parameters such as auto exposure, integration time, signal gain, and brightness controls from the ground. The lens attached to the camera is a wide-angle "fish-eye" lens with a field of view of approximately 45 deg. It is mounted on the side of a metal enclosure built specifically for the camera. To ensure good focus in space, the lens was focused to infinity on the ground and then secured to prevent launch vibrations from changing the focus. The camera is extremely sensitive to bright light making it difficult to optimize for the expected conditions in space. Fortunately, the camera exposure parameters can be adjusted in orbit.

B. Power

1. Overview

A brief discussion of ION's power system is included here. See [15] for a more detailed discussion. Figure A.17 provides an overview of the power system. Solar panels with power point tracking (PPT) circuitry generate power, which recharge the batteries or power the system directly (Table A.6). The data port allows an external power source (see Sec. III.A) to operate the system by feeding lines directly into the PPT like a solar panel. The data port can also directly access the batteries while the satellite sits in the PPOD, so that the batteries can be charged with an external charger. All solar panels and batteries feed through the kill switches (see Sec. III.A), pull pin (see Sec. III.A), and watchdog timer switches so that power can be cut. The watchdog timer provides fault tolerance by using relays to cut power to the entire system for two minutes if the processor does not reset it in time. This resets the whole system if the processor freezes up or the 5-V regulator shuts off due to a fault condition. The two-minute timeframe ensures that all power capacitors have enough time to discharge. Power-path control circuitry switches paths on and off to control power consumption and direct power into or out of each battery. Sensors monitor voltages and currents throughout the power system providing information for automated software control and manual ground monitoring. Finally, the power budget must be carefully characterized

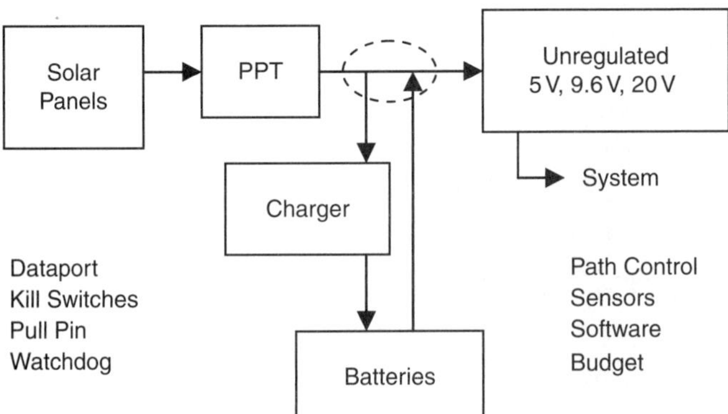

Fig. A.17 System overview.

Table A.6 Power System Parameters

Parameter	Value
Solar panels	Four side panels with five cells each @ 27.5% efficiency, 4.96 W peak power per panel
Batteries	Two Li-ion batteries (12.6 V, 1950 mA-h each)
Ave. power generation	1.954 W (uniform sun angles), 2.561 W (with Earth albedo)
Ave. power consumption	1.12 W (normal), 0.88 W (minimum)
Power margin	42% (56% with Earth albedo)
System watchdog	Cuts power to entire system for 2 min if not reset within 11 min

through analysis and testing of generation and consumption in order to ensure the system does not run short of power.

2. Power Generation and Storage

Power generation is a major constraint on the system (see Sec. IV.B.3), and so it is critical that the solar cell and PPT design maximize power generated. ION uses high efficiency (27.5%) advanced triple junction (ATJ) cells from Emcore. Figure A.18 shows the characteristic I-V curve test data for ION's panels. Maximum power is clearly generated near the knee of a curve, and PPT circuitry ensures that the panels operate near this peak power point. ION utilizes a simple circuit with a comparator and boost regulator to keep the panels' operating voltage at around 10 V. See [15] for more information about this design. Table A.7 shows test results in bright sunlight for the four completed solar panels with PPT circuit. Each panel was tested individually normal to the sun, and pairs of panels around the satellite were tested at 45-deg angles. Because power tracking is approximately 95% on average at room temperature and sunlight intensity is approximately 41% brighter in space without atmospheric losses, expected values from the datasheet are scaled by 0.95/1.41. Results show that the panels are generating approximately

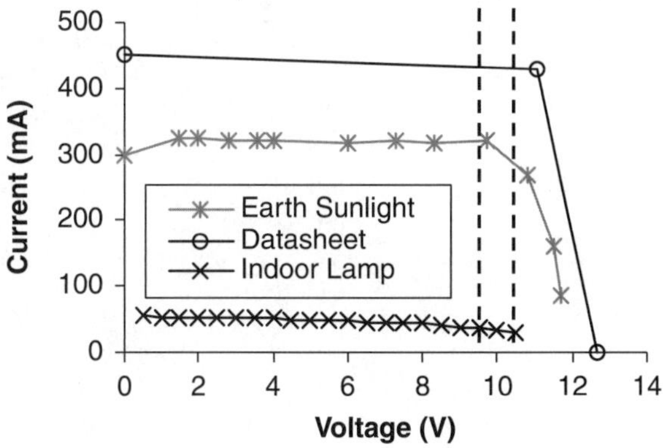

Fig. A.18 Solar-panel I–V curve.

Table A.7 Solar-Panel Test Results

Parameter	Description
Setup	Bright sunlight, onboard sensors
Normal panels	2.8 to 3.4 W (tested) 4.96 W * 0.95/1.41 = ~3.3 W (expected)
Two panels @ 45 deg	4.0 to 4.6 W (tested) 4.96 W * 2 * sin 45 deg * (0.95/1.41) = ~4.7 W (expected)

the correct amount of power. Errors in panel assembly and testing are accounted for by including a large power margin in the power budget (see Sec. IV.B.3). ION uses Li-ion batteries because mass and space are at a premium, and Li-ion offers excellent storage for a given mass. Each battery can hold 1950 mA-hrs of charge for a total of 3900 mA-hrs. In normal operating mode, the satellite consumes an average of about 95 mA (1.2 W at 12.6 V, see Sec. IV.B.3). As a result, the batteries store 41 h of charge.

3. *Power Budget*

As discussed in Section III.A, the satellite must generate more power than it consumes with any attitude. Power consumption of each subsystem was carefully characterized and adjusted for the efficiency losses of the regulator that supports it (Table A.8). A uniformly distributed sunlight scenario was calculated, while low- and high-power generation attitude scenarios were simulated to determine power generation. As discussed in Section III.A, a sleep mode with minimal power consumption is important to allow the batteries to recharge from a low capacity

Table A.8 Power Budget (Consumption)

Location	Raw power, mW	Source, V	Efficiency, %	Duty cycle, %	Normal power, mW	Duty cycle, %	Sleep power, mW
Power and router boards	250	Unreg.	100	100	250	100	250
SID	383	5	90	100	426	100	426
PMT shutdown circuit	40	5	90	100	44	100	44
Magnetometer	180	9.6	85	5	11	0	0
PMT	85	5	90	30	28	0	0
TNC and radio receiver	467	9.6	85	12.0	66	11.7	64
TNC and radio transmitter	8455	9.6	85	2.40	239	1.00	99
Torque coils	1143	Unreg.	100	5	57	0	0
Average consumption					**1120**		**883**

Table A.9 Power Budget (Generation and Margin)

	Generated power, mW					Margin, %	
	Generated power (incl. cell efficiency, eclipse, & angle)	PPT board power consumption	PPT tracking effectiveness, 95%	PPT regulator efficiency, 85%	Charger efficiency, 88%	Normal power margin	Sleep power margin
	Calculated/ simulated, mW	minus 150 mW * 66% sunlight	mW	mW	mW	1120 mW Cons.	883 mW Cons.
Low (w/albedo)	2200 (2860)	2100 (2760)	1995 (2622)	1696 (2229)	1492 (1961)	25% (43%)	41% (55%)
Uniform (w/albedo)	2850 (3705)	2750 (3605)	2612 (3424)	2220 (2910)	1954 (2561)	42%(56%)	55% (66%)
High (w/albedo)	3000 (3900)	2900 (3800)	2755 (3610)	2342 (3069)	2061 (2700)	46% (59%)	57% (67%)

situation. ION's sleep mode consists of turning off optional systems, such as the payload items and torque coils, while minimizing communication requests. Power margins are calculated with and without a 30% Earth albedo (reflection) bonus (Table A.9). The resulting power margin under uniform power generation and normal power consumption is 42% (56% with albedo), which is near the 50% goal. A sleep mode with 90% margin was not incorporated in the original design, but is included as a recommended requirement for future designs in Section III.A. ION's margin increases to 55% (66% with albedo) when all of its optional systems are turned off (sleep mode). With the worst random attitude case simulated, ION still maintains a 41% (55% with albedo) margin in its sleep mode.

4. *Recommendations*

- *Include a sleep mode that allows the satellite to recover from discharged batteries*: When batteries fall below a certain level, the system should automatically go into a sleep mode until the batteries charge back up. The batteries self-discharge while the satellite is in storage waiting for launch, and batteries might discharge in orbit due to orientation, power budget, or operational errors. For extra fault tolerance, this sleep mode should be implemented independently of the main processor with hardware or a simple power processor.
- *Use power diodes to connect power sources*: The design and testing would have been greatly simplified if the two batteries and solar panels were connected through power diodes instead of being dynamically switched. The switching requires extra hardware, extra software, careful timing, and a great deal of integrated system testing. Battery charge level, line voltage, and line currents can be challenging to use as decision inputs because of transient effects and noise. Accounting and testing for all scenarios in the integrated system can also be very challenging, especially as software is developed on a central processor by a separate team. With diodes, all three sources are backing each other up all the time, and the one with the highest voltage provides current first. See [15] for a more detailed discussion.
- *Include a watchdog timer or ground reset switch for fault tolerance*: The power system should assume that the main system can freeze up. ION's watchdog timer was frequently used a backup plan to deal with various anomaly scenarios including hardware, software, and radiation errors. As with a home PC, sometimes the only solution is to cut power and reset the whole system from scratch. It is also possible to directly trigger the watchdog timer from a reset switch on the ground. See [15] for more information.

C. Communications

1. *Satellite Hardware*

A brief discussion of ION's communication system is included here; see [16] for a more detailed discussion. The communication system onboard the satellite is implemented by three pieces of hardware: a baseband modem, a radio-frequency transceiver, and a dipole antenna with 180-deg phase shifter (Table A.10). The components are connected together as shown in Fig. A.19. The central processor implements the internal communication protocol.

Table A.10 Communication System Parameters

Parameter	Description
Data transmission	Baseband AFSK 1200 bps, Bell 202 Standard (1.2-kHz and 2.2-kHz tones)
Radio transmission	Frequency modulation onto carrier at 437.505 MHz with 11.5-kHz signal bandwidth
Protocol	Half-duplex channel, custom protocol implemented over AX.25
Beacon	Every 5 min; plain ASCII text using AX.25 UI frame
Flight antenna	Linear half-wave dipole, deployed after orbital insertion
Ground antenna	Circularly polarized Yagi with 21-deg beam width and 19-dBi directional gain and pre-amp
Ground station	Satellite tracking, antenna positioning, Doppler tuning, Web data publication

2. *Ground Hardware and Software*

The complete ground station system is shown in Fig. A.20. Essentially three independent systems function together to provide satellite communication support: a radio transceiver, Doppler shift tracking and tuning, and orbital tracking and antenna positioning. The central system is the radio system. When the other supporting systems are stripped away, the radio system is nearly identical to the satellite side radio system. It consists of a baseband modem, radio-frequency transceiver, and antenna. These devices perform the same function as their counterparts on the satellite. However, two additional components are in line between the transceiver and antenna: a lightning arrestor and preamplifier.

3. *Baseband Modems*

The communication hardware is highly symmetric and with both the satellite and ground station systems performing identical duties. The outermost portions of the communication systems are the baseband modems that serve to modulate/demodulate digital data into/from an analog waveform and implement the media layers (1–3) of the OSI network model. The modems in the ION communication system use the Bell 202 audio-frequency shift keying modulation standard. This modulation scheme maps the binary data from the serial connection onto distinct

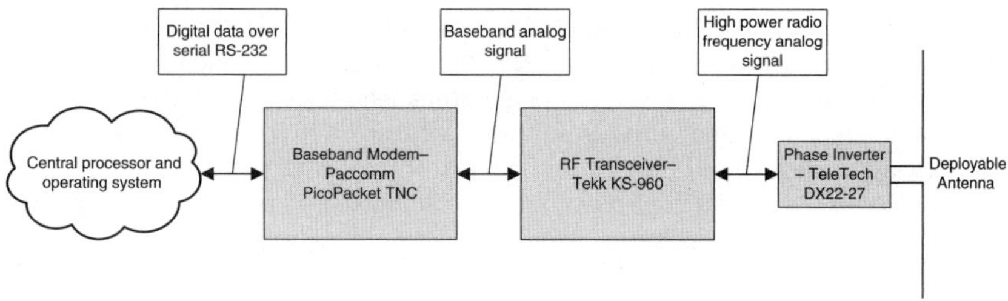

Fig. A.19 Communication system block diagram.

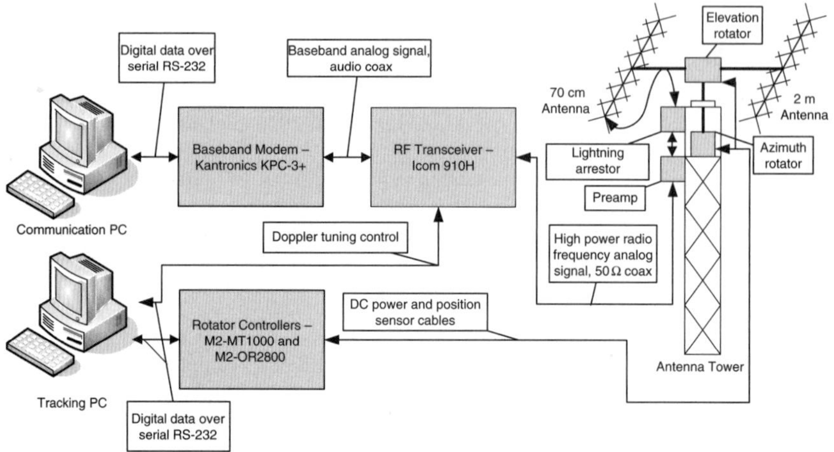

Fig. A.20 Ground station block diagram.

audio tones of 1.2 and 2.2 kHz. The tones are produced for a period of 0.83 μs for each bit enabling 1200-bpstransmission.

The modems implement the AX.25 protocol, which is very popular for amateur data radio purposes. The AX.25 protocol provides error detection, collision avoidance, and bit stuffing, which enable the wireless data link to function efficiently and transparently. The ground station and satellite software can treat the communication link as serial connection, which can drop packets occasionally under marginal link conditions. AX.25 can also provide packet retransmission for dropped packets using a send-and-acknowledge scheme. Because of the limited pass time and lag time of the low Earth orbit, this is inefficient. For ION we chose to implement a higher level protocol (on top of AX.25), which requests retransmission of lost packets in a more efficient manner.

4. *Transceivers*

The next layer of the communication systems are the radio transceivers. These modulate/demodulate the baseband signal onto/from a high-power radio-frequency signals suitable for communication between a LEO satellite and ground station. ION is severely power constrained like most cubesats and transmits only a 2-W signal, which is still a significant draw on the power system. The ground station is not power limited, and the Icom radio can transmit up to 75 W. Both transceivers use frequency modulation to modulate the baseband signals, which has two benefits for cubesats: by using more radio spectrum than necessary, a boost in SNR performance is achievable; FM modulation is insensitive to small tuning errors due to Doppler shift.

The 70-cm (420–450 MHz) amateur band is used because COTS equipment is readily available for this band, and it is relatively easy to obtain a permit to transmit in this band. The wavelengths of this frequency range allow antennas to be a reasonable size for pico-class satellites.

5. *Antennas*

The last layer of the communication systems is the antennas. ION uses a common antenna configuration for cubesats. A linear dipole is used on the satellite, and a circularly polarized high-gain Yagi antenna is used on the ground station. This configuration is robust with respect satellite orientation. A linear dipole antenna provides a relatively unidirectional radiation pattern and is relatively simple to construct. The ideal radiation patter is shown in Fig. A.21.

6. *Recommendations*

- *A 2-W radio is not needed for cubesats*: Many amateur satellites operate with radios that output 1-W power or less. Furthermore, theoretical link budget analysis shows that 1 W should suffice to close the link budget even at 10-deg elevation even at 9600 bps. Reducing the transmitter power by 50% yields numerous benefits for cubesat systems.
- *The 2-m band might be better suited to cubesats*: The Doppler tuning requirements are relaxed at lower carrier frequencies. At 140 MHz the maximum Doppler shift is only 3.3 kHz. Using frequency modulation and a radio with a sufficiently wide channel selection filter would avoid the need to use real-time Doppler correction. The 2-m band also has a lower free-space loss. The obvious drawback is that the satellite antenna would need to be larger.

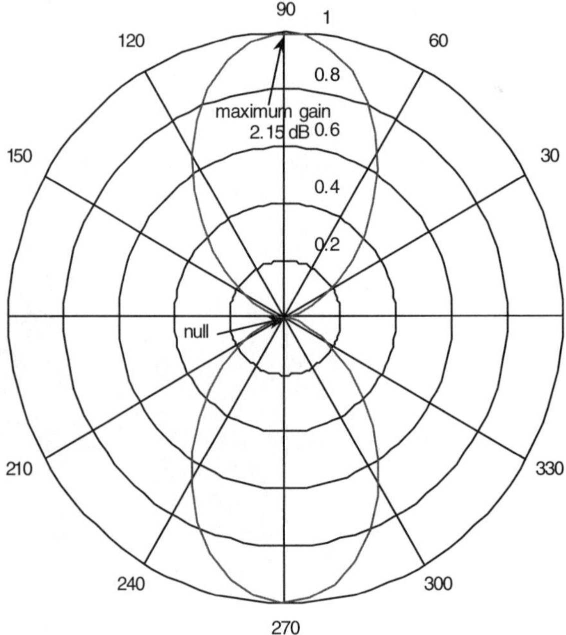

Fig. A.21 Dipole radiation pattern.

Table A.11 Attitude Hardware Parameters

Parameter	Description
Actuators	Three torque coils with one on each axis; 5 Gauss maximum magnetic field
Turns	30 turns per layer × 4 layers = 120 turns
Resistance	99.4 Ω coil + 47 Ω resistor
Current	84 mA @ 12.6 V (1.06 W)
Magnetic dipole moment	0.0222 Am^2
Magnetic sensors	3-axis magnetometer
Sun sensors	Four solar-panel currents, solar-cell sun sensor on top

D. Attitude Control System

1. Design

ION utilizes six separate sensors to estimate its attitude. The first one is the HMC2003 Honeywell magnetometer that provides magnetic measurements on each of the three axes of the satellite (Table A.11). The recommended set/reset circuit removes magnetic history and temperature effects. Also, after installation into the satellite, the magnetometer was calibrated to eliminate hard and soft metal effects from the system. The four solar panels serve also serve as sun sensors. In conjunction with a fifth solar-cell sun sensor located on top of ION, they provide the necessary information for attitude determination.

The primary actuators consist of three torque coils, one along each of the body axes of the satellite. The torque coils are manufactured by running wire traces around a standard four-layer printed circuit board, creating a light, easy-to-mount design (Fig. A.22). With 30 loops per layer, the total resistance of a single torque coil is 99.3 Ω, and the system can apply a maximum of 84 mA of current. With these specifications, the coils produce a maximum magnetic dipole moment of 0.0222 Am^2. The processor can control magnetic field strength via a duty-cycle

Fig. A.22 Torque coil on a PC board.

waveform, and it can control current direction with the help of support electronics.

2. *Recommendations*

- *PC board torque coils simplify mechanical integration*: Traditional wire loops are hard to mount, snag, and are delicate. The pc board design is durable and easy to mount, but has fewer turns and requires more current/power. Nonetheless, the overall power requirements are still low enough, especially in tracking mode.

V. Software[†††]

A. Flight Software

A brief discussion of ION's software is included here (Table A.12); see [17] for a more detailed discussion.

Computational facilities onboard the satellite are provided by a central computer, dubbed the SID (small integrated datalogger). The SID, developed by Tether Applications, is about the size of a credit card and is a prototype of an extremely compact, embedded computer to be used for space-based applications. The SID provides a unique computing platform that requires custom-built software. ION's onboard software includes drivers for all hardware, a custom operating system, a custom file system, a custom communications protocol, an application scheduler, power management software, and other application software. In total, ION's flight software contains 35,000 lines of code and 15,000 lines of in-code documentation spread across 300 files. The software is also remotely upgradeable from the ground station.

1. *Software Architecture*

All devices onboard ION have custom device drivers, which are controlled by applications that users interact with from the ground. These applications are, in turn, integrated into a software component called the "application manager,"

Table A.12 Software Parameters

Parameter	Description
Processor	SID from Tether Applications, based on Hitachi SH2 7045
Memory	1 MB external RAM, 256 kB EEPROM, 8 MB flash storage
Timekeeping	Three onboard real time clocks
Software	Custom operating system; custom file system based on FAT12; custom communications protocol

[†††]This section was contributed primarily by Leon Arber.

which schedules and allocates processing time for all of the applications. Control of the applications is provided from the ground through the use of "config files," which are binary data files, specific to each application, that contain a set of "work units." Work units specify a date and time, an action to perform, and any device-specific configuration options. Multiple work units are packaged into a config file, and this file is then uploaded to the satellite from the ground station. In this manner, ION's software provides for on-the-fly scheduling, sampling, and configuration of satellite hardware. This flexibility added to the complexity of the software. One of the dominant themes of the design was an attempt to isolate the various system components from one another in software. This allows it to handle revisions in device drivers as wiring and functionality requirements change.

2. *File System*

All data collected by the instruments onboard ION, as well as the software system itself, and any configuration files, are stored in a custom-developed file system inspired by FAT12 (Fig. A.23). The file system supports up to 128 files of unlimited size, with 11 character filenames. The 8 MB of flash storage is separated into two logically separate file systems. "Filesystem A" is 7 MB in size and stores all of the data files and some of the config files created while the satellite is running. "Filesystem B" is 1 MB in size and can be thought of as a boot partition that holds the software system itself, as well as the initial startup configuration files.

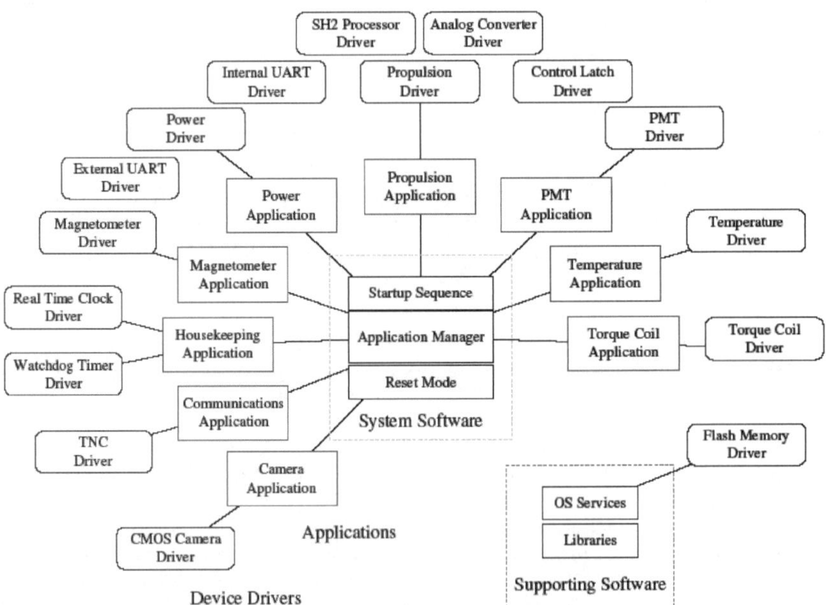

Fig. A.23 "Application manager" plays the central role in controlling and integrating all of the software interfaces for the various hardware devices onboard.

Upon bootup, the system searches for a valid software system on filesystem A, with a backup copy always existing on filesystem B. This provides the ability to upload new system software from the ground if new features are necessary or bugs are discovered. In the event of failure, filesystem A can be safely formatted, with a backup copy of the software system always in place on filesystem B.

3. *Fault Tolerance*

Throughout the software, as much fault tolerance as possible was built in. Where available, this fault tolerance relies on hardware devices, such as multiple real-time clocks, watchdog timers, and dual redundant batteries. There was also a great deal of software fault tolerance built into handle system crashes or other unforeseen anomalies. This includes software-based watchdog timers, a NOP sled, automatic termination of misbehaving applications, and a special communications mode called "reset mode." The system enters reset mode upon certain system failures. Reset mode is essentially a software system within a software system. It contains a separate copy of all necessary device drivers, a rudimentary power management system, and a completely distinct communications protocol that allows for reading, writing, and execution of arbitrary memory locations to aid in debugging.

4. *Communications Software*

Communications between ION and the ground station are facilitated by a custom communications protocol, dubbed SRTP (semireliable transport protocol). The protocol supports file upload/download, supports various filesystem and satellite control commands, and provides the ability to read and write arbitrary memory locations for debugging purposes. SRTP is akin to reliable UDP and runs on top of AX.25, which in turn runs on a 1200-baud AFSK modulated signal at 437.5 Mhz. The communications system has three distinct layers. The lower-level layer includes devices drivers that control the radio and the TNC. A middle layer provides all of the protocol functionality, which includes state machines, packet handling, and error handling. The high-level communications application is responsible for scheduling communications windows, as well as broadcasting a periodic beacon. The beacon contains constantly updated system information, including vital power and temperature readings.

5. *Recommendations*

- *Follow the COTS philosophy, and use open source libraries when available*: Because of the uniqueness of the SID and an inherent distrust and dislike of prebuilt software libraries, ION's developers chose to write almost all of the software onboard the satellite from scratch. In retrospect, there were many components, ranging from simple things like a queue, to complex components like the file system, which could have been obtained from open-source libraries and integrated into ION. This approach would no doubt have saved much development and debugging time. On the other hand, because all of the software onboard ION was written by ION's developers, there is a great deal of trust and understanding of how the system works, almost no unnecessary functionality, and very tight integration among all of the system components.

- *Consider utilizing "smart" hardware devices*: On ION, software developers needed to have a large amount of in-depth knowledge about how the system hardware and software components operate. This resulted in a handful of experienced developers who did most of the work on the satellite and a large number of peripheral developers who, although willing to contribute, simply did not have enough of an understanding of the system to be useful participants. Giving the hardware devices more smarts with local processors would make for a more modular design that would spread the development workload.

B. Attitude Control System[‡‡‡]

A brief discussion of ION's attitude software is included here (Table A.13); see [13] for a more detailed discussion.

The attitude determination and control subsystem is responsible for calculating the satellite's orientation at any given time and stabilizing it into a nadir-pointing direction. The attitude is found using current readings from four solar panels, a voltage reading from the top cell, and a three-axis magnetometer reading. The full three-axis control is achieved using three magnetic torquers that use a linear quadratic regulator as the control law. The satellite's optical payload constrains the attitude to be within ±20 deg of nadir, but the system is capable of 5 deg or better pointing accuracy.

Because of computational limitations and ease of software integration, the final design of ION's ADCS is a "long-duration feedback system" in which only data sampling and torque actuation occurs onboard, whereas the remainder of the subsystem's workload is done on the ground. The entire software for attitude estimation and control is written in MATLAB®, and Simulink and is run on the ground. This design greatly reduces the amount of hardware and software that needs to be successfully integrated with all other systems. It also offers unparalleled flexibility when updating or changing the code, and it ensures much quicker adjustment time because there is no need to upload the changes to ION.

Table A.13 Attitude Software Parameters

Parameter	Description
Method	Combination of onboard sensors for data acquisition and ground-based simulator for optimal torque profile calculation
Actuation	3 independent, variable-strength magnetic torque coils
Pointing accuracy	<5 deg

[‡‡‡]This section was contributed primarily by Andy Pukniel.

The most notable disadvantage of this system is the lack of real-time attitude data. With the current design, the team was forced to develop several attitude propagators that not only increase computational complexity, but are also less accurate than real-time attitude sampling methods. This method can also suffer from time synchronization problems. Spacecraft running a real-time operating system experience inherent clock drift, resulting in a time differential between onboard and ground clocks. This slippage can cause the torque coils to fire at a time when the spacecraft's orientation is no longer what it was predicted by the propagators. Although the effects of this error are minimal for a slow spinning or stabilized satellite, they can be very substantial if the spacecraft is tumbling or spinning at an appreciable rate. To prevent this problem, clock synchronization should be performed during communication opportunities.

1. Ground Simulator

Although the ground simulator is written specifically for the "long-duration feedback system" that is incorporated on ION, with minor changes it can be adapted to a wide family of nanosatellites with magnetic torque actuation. The simulator begins by acquiring the two line element (TLE) from Celestrak (www.celestrak.com) and the sensor data downloaded from the satellite. Next, using a custom-built orbit propagator, ION's ephemeris is propagated to the time of data acquisition, allowing calculation of the Earth's magnetic field at those locations. Using this information as well as the current, voltage, and magnetic field readings, the initial attitude of the spacecraft is found and then propagated until the desired time of actuation. It is important to note that this and all other propagators take into account the disturbance torques generated by the aerodynamic drag (AD) and the gravity gradient (GG). This attitude is then fed into a separate simulator that includes the asymptotic periodic linear quadratic regulator (LQR) similar to one proposed by Psiaki [18]. Based on the attitude and position along the orbital path, the simulator computes the optimal torquing profile needed to despin and orient the satellite. As seen in Fig. A.24, this simulator is a simple feedback system in which the applied torques are summed up and fed into the satellite dynamics block that finds the response of the system. The new state is then fed into LQR to find a new optimum torque, which in turn is fed back into system dynamics block. Once the satellite is stabilized and oriented correctly, the simulator saves the torquing profile to be uploaded to ION for execution during the next pass.

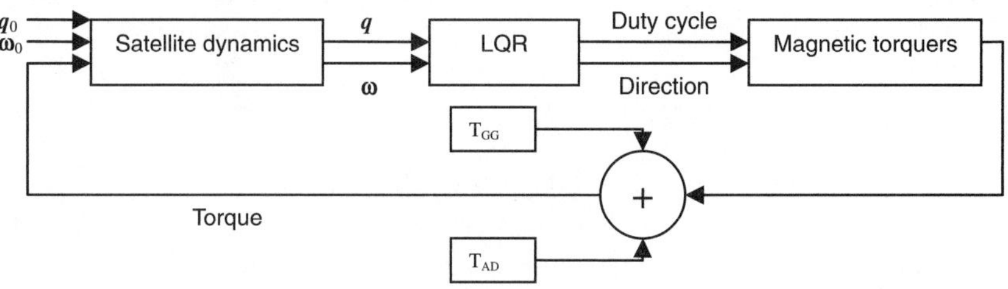

Fig. A.24 Simplified simulator diagram.

2. ADCS Operational Plan

Immediately after the team establishes first contact and performs necessary system checks, the ADCS will begin its operational procedures. This process will begin by first calibrating the torque-free attitude propagators. This will be done by sampling the attitude at various points along the orbit and comparing these values to the output produced by our simulator and then adjusting the propagator until desired accuracy is achieved. The second part of calibration will check how well the simulated response of the system to an input torque corresponds to the actual system response. A series of simple torquing profiles will be uploaded to the satellite for execution at various points along the orbit, and the sensor data will be acquired and downloaded. The computed attitude after each torque was applied will be compared to the attitude obtained using our simulators, and if the results are satisfactory, the team will move on to a more complicated torquing profile.

Once the desired accuracy of all the simulators is verified, the team will proceed to despin and stabilize the satellite. This process is composed of two parts. The first uses an LQR that emphasizes slowing down the rotational rates in minimum time at the expense of power and pointing accuracy. In contrast, the second mode puts more emphasis on the final nadir-pointing state using more efficient power profile, but at a penalty of increased stabilization time. The threshold between the "detumbling" and "tracking" modes is user defined and is currently set to 0.1 deg/s. The different objectives of each mode are accomplished by assigning different weighting matrices, Q and R in the LQR. One method of obtaining the desired system response is to pick Q and R, study and closed-loop stability, and adjust R accordingly—usually by increasing R if the system is unstable. It has been noted by Psiaki [18] that Floquet analysis is a good method for tuning Q and R. Another method, adopted by the authors, uses genetic algorithms to compute near-optimal sets of the weighting matrices for both detumbling and tracking modes. Details of this approach can be found in http://cubesat.atl.calpoly.edu/pages/documents/developers.php. The two modes are incorporated into one simulator that computes the entire stabilizing torquing profile that includes duty cycle and direction of current applied to each coil. A sample duty cycle output can be seen in Fig. A.25. The simulator also outputs the evolution of ION's rotational rates and offset from the nadir-pointing direction taken from a sample scenario.

It is worth noting that, in the current state, the highly accurate magnetic field model, orbit and attitude propagators, and long-duration LQR algorithm require large processing capabilities. However, if the propagation time is decreased to either instantaneous or very short duration (approximately 20 s) the code itself runs in real time. As a result, with few modifications, this code can be adapted for other nano-satellites that use magnetic actuation as an active onboard-control method.

3. Recommendations

- *Implement Kalman filter for attitude determination*: Attitude determination can be greatly simplified by removing its dependence on solar panels. Instead, we strongly recommend implementing a Kalman filter that finds the attitude solely from magnetometer readings. Extensive work on such systems has shown superior accuracy to the implemented method.

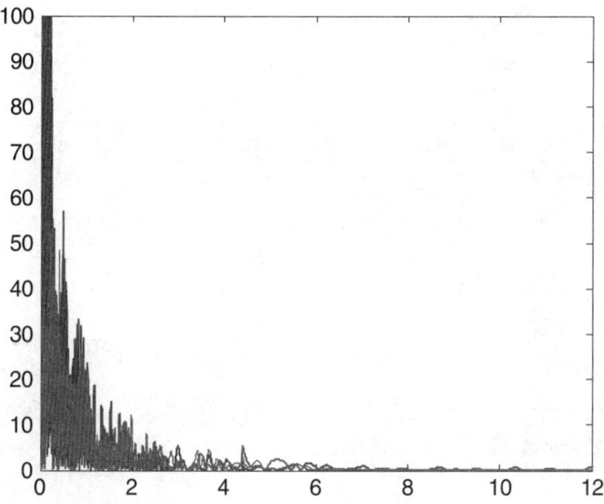

Fig. A.25 Sample duty cycle profile.

- *Develop ADCS software specifically to run on hardware and vice versa*: The a priori decision of using a ground-based ADCS allowed utilization of MATLAB® and Simulink. Onboard ADCS will not have the advantage of built-in libraries and toolboxes and, in the interest of shortened development time, should be programmed in hardware-specific language. Recent advances in PIC technology make them a good choice as ADCS-specific processor.

C. Ground Station

In addition to the communication system software discussed in Section IV.C, the ground station has some software for controlling the satellite and managing data related to the satellite. A user interface exists for issuing commands to the satellite including the upload of config files and the download of data files. A simulator of the satellite software was developed for training ground personnel in the operation of the satellite. This simulator consists of a modified version of the actual satellite software making for a realistic interaction. Simulations are conducted over a wireless link to simulate actual radio transmissions.

Once data are downloaded to the ground, a copy of the files is published to a shared drive accessible to the entire team. The data are also synchronized into a database, which is accessible via the Web (Fig. A.26). This "ION dashboard" also contains other useful data for operating the satellite including orbit data from Satellite Tool Kit, by Analytical Graphics, Inc. A public version of the dashboard is available for the general public to browse ION's telemetry data. A private dashboard contains additional internal data such as file listings. The starting time and ending time of the desired data is entered at the top of the page. Then for each data set, the minimum, maximum, and average value is displayed in a table, while the data are graphed next to the table. In addition, below the table there are notes with information about how to analyze the data. Nearly 30 data sets such as the one

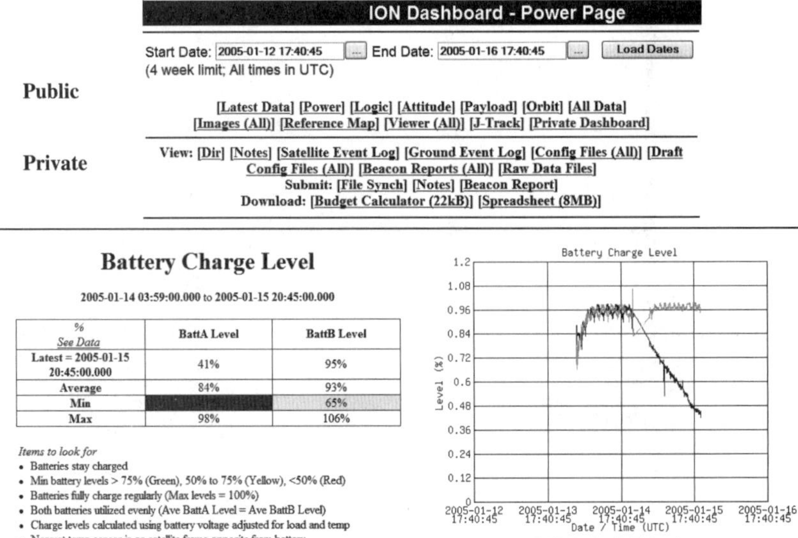

Fig. A.26 ION dashboard.

displayed for battery charge level exist. The dashboard contains all of the information required to make command decisions by the geographically distributed members of the team.

VI. Integrated System

A. Fabrication

This section was contributed primarily by Andy Pukniel. The fabrication of any satellite, regardless of size, is complex. Size and weight constraints must be met to a high degree of accuracy. Additionally, the payload and supporting hardware must be securely mounted to withstand launch loads, and all epoxies and materials must be capable of surviving the harsh environment of space. To support its ambitious goals, ION uses 24 mounts, 30 standoffs, 6 different epoxies, and 187 screws. A detailed three-dimensional drawing consisting of 108 individual components, keeps all of the tolerances in check, and helps prevent wiring and assembly difficulties. This comprehensive model is also used to determine center of mass and moments of inertia. At several points during integration, fit checks of the entire system were performed to identify mechanical mounting problems and update the actual mass budget. The final mass budget of ION is listed in Table A.14.

1. Frame, Mounts, and Component Layout

The frame is light, and the design is simple, making it easy to machine, assemble, and disassemble. The design, however, is very component specific, so that if boards change in the future, the entire layout of the satellite could change.

Table A.14 Mass Budget

Item	Cost
Power (solar cells, batteries and mount)	393.5
Structure/mounts/fasteners	584.2
Controls (torque coils, thrusters, and PPU)	252.3
Sensors (camera, PMT, CDM, magnetometer)	208.4
Circuit boards (antenna, power, radio, router, TNC, SID)	294.0
Wires	59.3
Misc (padding, shielding, epoxy)	235.3
Satellite total	**2,027.0**

Assembly and disassembly of the entire satellite is also challenging. The payload and the batteries are the only components that need to be placed in a specific location. ION's payload is located at the bottom, or the Earth-facing side of the satellite. The battery is centered along one of the sides to help keep the center of mass in a central location. The remaining boards are positioned to maximize internal space and minimize wire lengths. As seen in Fig. A.27, maximizing the internal space lead to complicated wiring procedures. These complicated wiring procedures slowed down integration time. Because of the complicated wiring procedures, wires were bundled with lightweight, wax-coated, flat Dacron tie cord to minimize clutter.

To survive the launch and space environment, several fabrication considerations are necessary. A solid frame is required with rubber washers to protect sensitive equipment from vibrations. The washers also provide electrical isolation.

Fig. A.27 ION internal wiring.

Aluminum 6061 is the material of choice for the frame and the majority of the mounts, as a result of its thermal expansion properties, cost, availability, and strength. Smaller mounts are constructed from Delrin, which is a lightweight polymer with good strength and thermal properties. All boards are covered with conformal coating to strengthen the solder contacts and prevent movement. Larger components and connectors are staked down using staking epoxy. Electrically conductive epoxy ensures strong connections where necessary. Additionally, a thermally conductive epoxy secures the solar cells to the side panels, and another thermally conductive epoxy secures the thermal sensors. Limited quick-drying epoxy also helps secure external wiring.

2. *Solar Panels, Switches, and Antenna Deployment*

Our solar panels are very difficult to assemble, and each side of ION is slightly different. The first generation of solar cells had no glass covering or built-in diodes making them extremely sensitive to cracking during assembly. These panels were eventually discarded and replaced with newer glass coated cells with built-in diodes. Five individual solar cells are epoxied to each fiberglass panel and wired in series to an external connector.

There are three external components to ION excluding the antenna. The first of these is a pair of Delrin plungers located at opposite feet that act as the kill switches. When depressed in the PPOD, they close a set of lever switches. ION also has a rectangular pull pin. This pin depresses two sets of switches both above and below the pin when it is inserted into the satellite. The final external component is the data port. This port as well as the pull pin can be accessed through windows on the PPOD.

ION uses a simple but effective technique for antenna deployment. The antenna is constructed out of rolled steel, the same material used for retractable measuring tapes. It folds down two sides of the satellite away from the PPOD's ejection opening, and it is tied down using fishing wire wrapped around special NiCr wire. The NiCr wire heats up when current is run through it cutting the fishing wire thereby releasing the antenna.

3. *Shielding*

This section was contributed primarily by Purvesh Thakker. Unfortunately, space-grade components resistant to radiation cost too much for the ION project and do not typically get designed for satellites as small as ION. They also tend be older, and they have limited availability. For these reasons, ION primarily uses commercial hardware not specifically designed to withstand radiation. Although these components present some extra risk, the benefits outweigh the risks for small satellites in LEO orbits. The SID processor from Tether Applications is the only part specifically designed for operation in a radiation environment. It also consists of commercial parts, but it includes latch-up protection circuitry throughout the board as well as heat dissipation and EMI control techniques. It has been tested for radiation exposure in a LEO orbit.

As discussed in Section III.A ION is required to allow no more than 5 krads of radiation exposure during the satellite's six-month lifetime. The online tool,

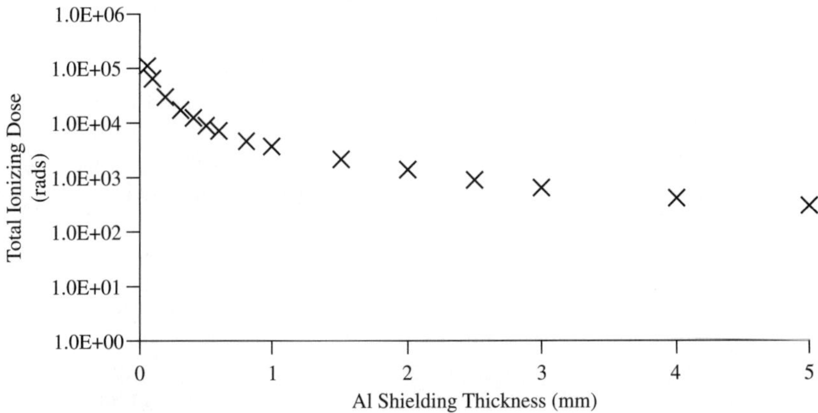

Fig. A.28 ION's six-month radiation dosage vs shielding level from SPENVIS (http://www.spenvis.oma.be/spenvis/intro.html).

SPENVIS (data available online at http://www.spenvis.oma.be/spenvis/intro.html), is used to help understand the total radiation dosage for ION's orbit and shielding level (Fig. A.28) [19]. Because of strict mass restrictions, ION could only support approximately 1 mm of shielding in the form of one 0.2-mm aluminum plate and four 0.2-mm polyethylene sheets. The polyethylene sheets are expected to provide protection similar to the aluminum at a cost of less mass. This level of shielding should result in 2.7 krads of radiation exposure for the six-month mission. The system is not expected to have problems reaching the two-week minimum lifetime requirement and should satisfy the longer six-month operational goal.

4. *Recommendations*

- *Utilize staking epoxy and conformal coating*: It is a good idea to stake down large components and conformal coat all boards to make the entire system more rugged.
- *Keep the inside of the satellite neat and organized*: It is easy to ignore wiring and mounting of boards until later, but a lack of planning will result in mechanical integration problems and spaghetti wiring. Everything eventually came together, but wiring crossed the satellite taking up most of the space in the center, and every mount was custom designed. This design makes tracking wires and mounts challenging. This design also makes sharing the hardware and working with the integrated system challenging. Minimize wiring in the system by using a backplane or stacking boards. Design boards specifically to fit into the space available, and have a plan for mounting them. Leave extra space for unexpected electronics.
- *Develop a comprehensive three-dimensional model*: This model prevents mechanical integration problems later, and should be updated constantly throughout the design phase with actual parts and their associated weights.

- *Purchase solar cells with built-in diodes and interconnects*: High-efficiency solar cells are very fragile and expensive. Develop a panel assembly procedure that minimizes risk and practice assembly before using the real cells.
- *Design the solar panels to be identical with traces and internal connectors*: This way an extra panel can be built ahead of time that can replace any panel with a broken cell. The panels can be designed with traces to eliminate external wiring, and connectors can be placed on the inside of the panel with the help of vias.
- *Clean rooms are not always necessary for cubesats*: Many cubesat teams do not have or do not use clean rooms. They can protect the flight unit and keep optics and solar panels clean, but most cubesats, including ION, do not face design requirements that require high cleanliness.
- *Plan to install fresh batteries just before delivery*: Batteries will likely fail first, and launch dates might slip. Make it easy to replace batteries at the latest date possible.

B. System Testing

1. *Functional Testing*

Before electronics integration, each custom board was functionally tested and tested for hot and cold operation. Cold tests going down to –20°C utilized a can of cold spray, and hot tests going up to +60°C utilized a heat gun. Individual boards were gradually tested and debugged together until a full-system flat board test could be run outside of the satellite frame. In addition to more detailed tests, at this time a standard a set of simple diagnostic tests was developed, which could be run over the data port. The system could also be tested over the communications system using the ground station. This diagnostic test suite was run regularly especially around environmental tests that can damage previously working hardware. System-wide mechanical fit checks earlier in the project ensured that mechanical integration would proceed with a minimum of difficulty. The components were installed into the frame, and detailed diagnostic tests were run. Then the satellite was disassembled in order to fix some newly discovered bugs. At this time, voltage and current sensors were fully characterized, and readouts were adjusted in software. The satellite was then reassembled a second and final time approximately six months prior to delivery. Finally, items such as magnetometer readings, camera focus, and self-reported power measurements were characterized and calibrated.

2. *Environmental Testing*

A number of environmental tests were run on the satellite and its components to better simulate conditions in space. As discussed earlier, individual custom boards were tested at hot and cold temperature extremes using a heat gun and cold spray. The vacuum chamber was modified to include a coil surrounding the satellite, which could be filled with liquid nitrogen to cool the satellite while in a vacuum. For hot tests in a vacuum, several halogen lamps were placed around the satellite to heat it. A sample solar panel was attached to the satellite frame early

Table A.15 Vibration Profile

DNEPR high-level qualification profile—Test on each axis for 35 s							
Lower frequency	20	40	80	160	320	640	1280
Higher frequency	40	80	160	320	640	1280	2000
Spectral density	0.011	0.011	0.033	0.053	0.053	0.053	0.026
DNEPR low-level qualification profile—Test on each axis for 831 s							
Lower frequency	20	40	80	160	320	640	1280
Higher frequency	40	80	160	320	640	1280	2000
Spectral density	0.011	0.011	0.011	0.014	0.014	0.007	0.007

on in order to understand its heating and cooling characteristics in a vacuum. From these tests, it was estimated that the ION's internal electronics would experience temperatures of approximately −20°C to +20°C in orbit. A sample solar panel was also vibration tested early on to verify that the delicate solar cells would survive launch.

Vibration information in this section was provided primarily by Dustin Ames. After integration, the assembled satellite was tested in a room-temperature vacuum, in a cold vacuum, and in a hot vacuum. The standard diagnostic test suite was run over the satellite's data port while it was in each of these conditions, and it was also operated from the ground station. An external company specializing in running vibration tests was used to satisfy vibration-testing requirements set by CalPoly (Table A.15). There, low-level sine sweep vibration tests were conducted on all three axes to better understand the satellite's resonance frequencies. Next, low- and high-level random vibration tests were run on all three axes at 125% of the launch-vehicle's vibration profile. The data for the high-level test on the

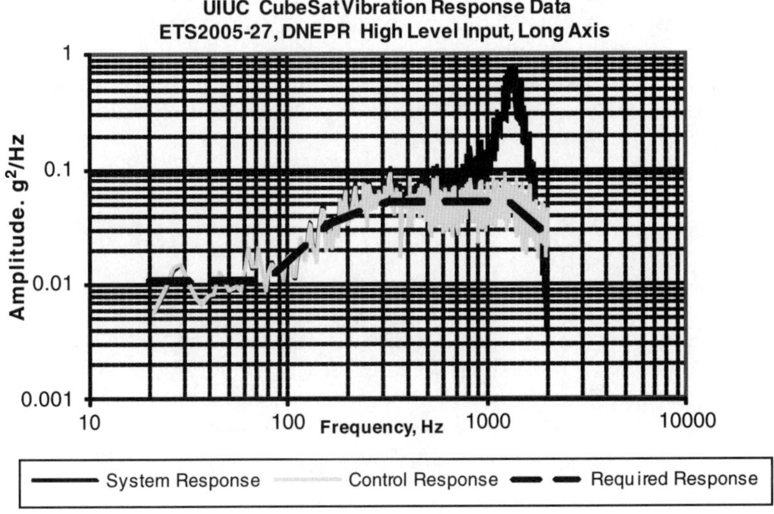

Fig. A.29 ION vibration response.

satellite's longitudinal axis are shown in Fig. A.29. The figure shows the vibration system's control response (control response) along with the combined satellite and PPOD response (system response) and the required response as dictated by the DNEPR launch vehicle and CalPoly (required response). Because no new problems arose, the satellite was permanently sealed by gluing or tying down all major joints, screws, wires, and connectors. Next the satellite went through a bakeout procedure defined by Cal-Poly to prevent outgassing in space. This entailed heating and cooling the satellite for two full cycles while in a high vacuum:

1. Bring vacuum level to at least 5×10^{-4} torr.
2. Raise temperature of heating element to 70°C from room temperature.
3. As temperature increases, record pressure levels every 20 min.
4. Let hardware bake at 70°C for 1 h.
5. Record pressure and temperature every 10 min.
6. Bring chamber back to room temperature.
7. As temperature decreases, record pressure levels every 20 min.
8. Keep hardware at room temperature for 1 h.
9. Raise temperature 70°C for final bakeout.
10. As temperature increases, record pressure levels every 20 min.
11. Let hardware bake at 70°C for 1 h.
12. Record pressure and temperature every 10 min.
13. Bakeout is complete.

3. *Operational Testing*

Many full system operational tests were run after integration and in parallel with environmental tests. These tests included comprehensive automated tests on the whole system simulating operation in orbit. First, the system was run for an extended period of 24 hours gathering telemetry data as if the system were in orbit (Fig. A.30). These tests checked charging cycles, power consumption, the absence of system resets, correct data file downloads, and correct command file uploads. In addition, anomaly situations were simulated for various custom configurations. In general, the last four months were spent running any tests that would assist in finding new bugs, learning how to operate the satellite, and characterizing performance of various items.

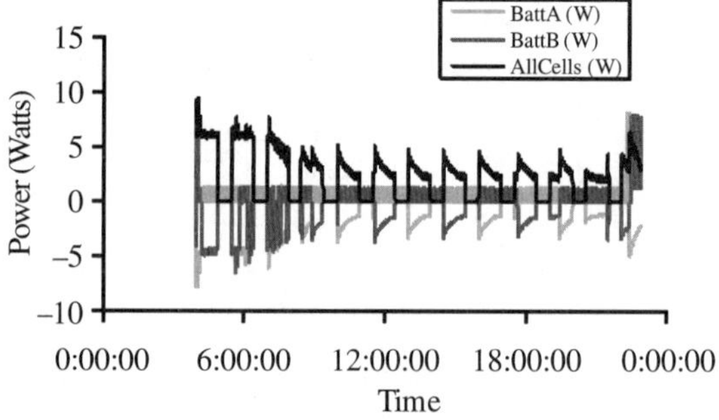

Fig. A.30 Power during 24-h operational test.

4. Recommendations

- *Outsource vibration testing*: A year of effort was spent developing limited vibration capability internally, while a very simple, effective, and reasonably priced solution was a phone call away.
- *Establish a simple test suite for the hardware*: All hardware needs at least a minimal functionality test after every environmental test and assembly of the satellite. This test suite is run many times, and so automate it to the greatest extent possible.
- *Generate new test ideas continually*: A repeatable test suite is important, but it tests things that were working in the past. Finding new bugs requires new tests.
- *Know that bugs always remain*: Avoid skipping tests because something is "already working," or because "there isn't anything that can be done anyways." The purpose of testing is to find bugs, and there are often creative solutions to these unknown problems. New bugs frequently popped up in items that were thought to be completed. Sometimes these bugs could be tolerated, sometimes a simple work around was available, and sometimes a major fix was required to keep the project viable.

C. Operations

Of course, the satellite didn't quite make it to the operations phase because of the launch failure, but a great deal of planning was put into place prior to launch. The operations work is divided between two teams, a command team and a ground team. The command team consists of ION veterans, most of whom were no longer involved with the team because the launch was delayed for one year after satellite delivery. Some were not even on campus. The command team is responsible for nonroutine satellite communication, all operational decisions, and ground team training. The ground team consists of newer students, some of whom helped upgrade the ground station during the previous year. Command team members would perform most operations during the first few weeks and during nonroutine operations later. The ground team would handle routine uploads and downloads after the first few weeks. The ground team is required because a ground operator is needed at the ground station during each pass, which occurs in two pass sets per day consisting of two or three passes 97 min apart. This ground team downloads data, performs a preliminary analysis, and synchronizes it to the Web site where remote command team members can access it. The ION simulator helps improve the ground team members' skill with operating the system according to well-documented procedures.

The launch day plans include a sequence of actions with two command team members at the ground station. A prelaunch announcement is sent to the world-wide ham radio community, seeking assistance with listening for beacons using standard AX.25 equipment. The beacons contain some key measurements from the satellite such as battery voltages. The first 100 ham radio operators that submit these measurements via the Web site receive a QSL card confirming contact. Ham radio operators would also send contact announcements to their email lists allowing the team to quickly ascertain the status of the satellite. When the satellite passes over the UIUC ground station for the first time, the ground operators would download key data including recent power measurements as well as initial and recent log files. If the satellite is not generating adequate power, then it should be placed into a sleep mode until a course of action can be determined, possibly including changes in the satellite's attitude. If the satellite is healthy, then hardware can be tested, and predefined experiments can begin including camera, photometer, and thruster operation. As the satellites separate, the universities cooperate to identify satellites and can help each other find missing satellites if communication equipment is compatible.

VII. Results

A. Cost and Schedule

The actual development time line for ION appears in Fig. A.31. Originally, the team planned to spend the first year designing the satellite and the second constructing it. Because of the complexity of the project and the team's inexperience in developing satellites, it took four years to complete the satellite and the fifth year to complete the ground station. Fortunately, numerous launch delays provided extra time allowing the team time to complete the satellite without losing its launch deposit. Estimated costs appear in Table A.16 and give insight into the value of the satellite if it were to be constructed in industry. Specialized equipment costs include the initial cost of acquiring a thermal-vacuum chamber. Regular educational expenses are not included because these would normally be provided as part of a lab course. The low cost and short development time make cubesats appealing as a technology platform because they can support high-risk experiments with fast turnaround.

B. Mission Results

Although the satellite did not make it to orbit because of a launch failure, the project offers many achievements that should be recognized and preserved. The project provided a unique opportunity to allow higher risk, nonflight heritage, or otherwise difficult to fund ideas to move forward in a manner that would be very difficult to do outside of a university environment. Until recently, it was not clear that cubesats could perform missions of value because of their small size. ION's oxygen airglow mission shows that these satellites can, in fact, support photometers for atmospheric science remote sensing experiments. ION's successor, ION 2, also supports a dual photometer for measuring hydrogen and off-channel background airglow. These experiments likely provide just a small sample of the remote-sensing and in situ measurements that cubesats can perform in support of atmospheric science research. For technology missions, ION developers provided

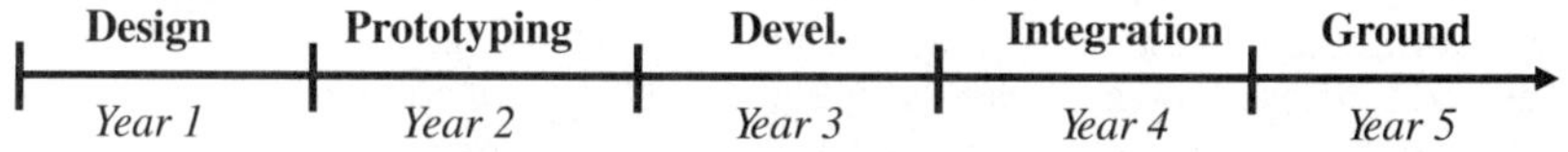

Fig. A.31 Actual project time line.

Table A.16 Estimated Costs[a]

Item	Cost
Parts	\$35 k
Specialized equipment	\$20 k
Mechanical shop	\$30 k (700 hrs)
Launch	\$85 k
Subtotal	**\$170 k**
Estimated student labor time	4 full time people × 4 years = 16 man-years
Estimated student labor value	× \$150,000 per man-year = ***\$2.4 M***

[a] Normal educational expenses include lab, test equipment, teaching assistants, and faculty.

a great deal of customer feedback to the manufacturers of the prototype μVAT thrusters and SID processor board. Although the hardware was not demonstrated in space, it was integrated and tested as part of a completed spacecraft by a third party. Finally, the ION project demonstrated the feasibility of ground attitude stabilization with extensive simulation results. Overall, ION provided a stepping stone to larger goals for some objectives and enabled new work to proceed for other low-cost applications.

In addition to these satellite objectives, the project achieved many of its research objectives. The work is being preserved in about a dozen publications spanning the entire project. Some of these publications are being written about specific novel elements of ION, such as the science experiment [3,8], thrusters [9,12], and ground attitude control system [13]. Others provide system studies including power [15], communications [16], and software [17] topics. In addition, some students from the project went on to utilize ION as their application example for their work in other research groups. These include a number of hardware [19] and software [20] dependability studies. If flight data had been obtained, many of these publications would have been further strengthened or would have had follow-up publications. As time goes on, there will likely be future publications that continue to draw on the ION experience in the form of comparisons and references. Finally, it should be noted that many of these publications offer relevance in their own specialty research areas in addition to small satellites.

For educational objectives, the project trained over 100 engineers and laid the foundation for future spacecraft developments at UIUC. These student engineers learned to operate as part of an interdisciplinary team developing valuable hands-on technical, leadership, and teamwork skills as part of an ongoing multi-team project. They learned from more experienced graduate and undergraduate students who often stayed around as advisors even after their direct involvement ended. The project also is credited with opening up new doors for students because employers value the experience as much as internship experience. For some of these students, the ION project provided the only work experience opportunity they could discuss beyond their primary academic work. Some students from the project went on to careers within aerospace, whereas others carried their technical skills and experience into other fields. Roughly ¼ of the students gained experience as team leads, and roughly ¼ stayed involved for extended periods allowing them to develop strong leadership and teamwork skills. In some cases, these strong foundations lead to leadership positions in industry. The organization as a whole has moved on to develop ION 2, a balloon launch program, and the primary optical payload for the Taylor University's TEST nanosat.

C. Recommendations

- *Select a mission idea carefully, one that has real scientific significance beyond educational value*: The first question people outside the team always ask is, "What does the satellite do?" Satellites that perform real missions are better respected and more motivating to work on.
- *Select one good mission idea, and keep the rest of the satellite simple*: During project definition, people do not understand the full complexity of the project. The experiment with the photometer is the team's most important one, but looking at it critically illustrates this point. The photometer can be purchased off the shelf making it look simple, but its design must maintain adequate signal to noise, it must be mechanically packaged with the lens and filter, it requires multiple revisions on boards with support circuits, all wiring must be piped to the proper locations, the data must be managed, software must learn to interface with it, and it must be operated from the ground. Extra missions take up precious resources like mechanical space, board space, mass, power, a specific regulator's current, memory, communications link time, and processing power, putting pressures on the rest of the system. Most importantly the design, development, testing, and integration all take up the time of key people who are always spread too thin delaying the delivery. What seems like a simple additional mission objective adds more complexity than anyone realizes causing the project to be completed late. The people who define the project have a responsibility to keep it manageable for later team members who will carry the burden of developing and integrating all of these ideas. One good well-executed idea delivered on time and within budget is best.
- *Reuse as much of the organization's previous satellite designs as possible*: There likely will not be enough capable and committed people to design and develop an entire satellite from scratch in the time available. Like many cubesat projects, ION had only five to eight people at any given time that could handle major tasks, and the rest were helping with smaller tasks. Even if more of these people are available, they often cannot help because the system is very tightly bundled and interdependent. Focus key people on the most important pieces and live with old designs for the rest. These older systems can sometimes be upgraded once higher-priority items have been completed.
- *Design the mechanical, software, and electrical systems to be as modular as possible (even though it is a small package)*: This way the teams can work independently, and no one person or group ends up having to know everything. Also, having more people does not help if the system is not designed to be modular, as there are too many interdependencies. Consider using simple microcontrollers to handle unique hardware requirements of power, ADCS, health monitoring, and payload subsystems.
- *Purchase extra components and make extra boards*: With two of a component, people can compare their performance to determine if the primary is functioning

properly or if there is a common design or testing problem. Also, hardware breaks during testing or because of mistakes. Progress on the project, especially during integration, requires a series of many small activities to occur in a certain order. Having backups can shave a great deal of time off of the project schedule. Finally, sometimes multiple people need to work with the same hardware and giving them each their own unit eliminates one more interdependency.

- *Avoid signing up for a delivery date until the end is in sight*: Students have many outside commitments to juggle and do not have the experience to predict how long things will take. Hopefully, launch opportunities for cubesats will increase in the coming years, so that launch opportunities are available on shorter notice.
- *Make simple radiation immunity improvements for low-Earth-orbit cubesats, but do not get overly distracted by it*: A few things can be done to improve radiation immunity with minimal effort. These include adding shielding, using tantalum tape on commercial components, and using radiation-tolerant types of components like ceramic capacitors. Latch-up circuitry can also help protect boards with commercial components. However, radiation primarily affects the lifetime of the satellite, and cubesats tend to have short mission lifetime requirements. They also have less than perfect first contact success, and so it is best to keep the priority on making the satellite functional on the first day.

VIII. Conclusion

This work illustrated the utility of cubesats with a discussion of the ION project. The work showed how cubesats can perform remote-sensing and technology test missions within the context of a complete spacecraft project. The work further provided an end-to-end feasibility discussion for these spacecraft including recommendations for each system element based on the development experience.

As a whole, the objectives for UIUC's first completely student-developed satellite were set much higher than what most experts would consider achievable. As discussed in the preceding section, many of the project's objectives were successfully accomplished including the education of over 100 students and the publication of about a dozen papers. The successful completion of all objectives was ultimately limited by bad luck with the launch rather than the student's capabilities and efforts.

In conclusion, these tiny spacecraft offer unique opportunities to access space for education, atmospheric science research, and space technology testing thanks to a growing secondary payload launch infrastructure. ION's successfully completed stretch development goals should help expand the perceptions of what these tiny university satellites can do further growing their opportunities, and the detailed preservation of this work should help contribute to the success of future university satellites.

References

[1] Nason, I., Puig-Suari, J., and Twiggs, R., "Development of a Family of Picosatellite Deployers Based on the CubeSat Standard," *IEEE Aerospace Conference Proceedings*, Vol. 1, IEEE, Piscataway, NJ, 2002, pp. 1-457–1-464.

[2] Lan, W., Brown, J., Toorian, A., Coelho, R., Brooks, L., Suari, J. P., and Twiggs, R., "CubeSat Development in Education and into Industry," *AIAA Space*, AIAA Paper 2006-7296, Sept. 2006.

[3] Thakker, P., Swenson, G., and Waldrop, L., "ION-1 and -2 Cubesat Optical Remote Sensing Instruments," *Emergence of Pico- and Nanosatellites for Atmospheric Research and Technology Testing*, edited by P. Thakker and W. A. Shiroma, Progress in Astronautics and Aeronautics, AIAA, Reston, VA, 2010.

[4] Holmes, W. C., Bryson, J., Gerig, B., Oehrig, J., Rodriguez, J., Schea, J., Schutt, N., Voss, D., Voss, J., Whittington, D., Bennett, A., Fennig, C., Brandle, S., Dailey, J., and Voss, H. D., "TU Sat 1—A Novel Communications and Scientific Satellite," *16th Annual AIAA/USU Conference on Small Satellites*, Paper SSC02-I-1, Aug. 2002.

[5] Obland, M., Hunyadi, G., Jepsen, S., Larsen, B., Klumpar, D. M., Kankelborg, C., and Hiscock, W. A., "The Montana State University NASA Space Grant Explorer-1 Science Reflight Commemorative Mission," *15th Annual AIAA/USU Conference on Small Satellites*, Paper SSC01-III-2, Aug. 2001.

[6] Flagg, S., Bleier, T., Dunson, C., Doering, J., DeMartini, L., Clarke, P., Franklin, L., Seelback, J., Flagg, J., Klenk, M., Safradin, V., Cutler, J., Lorenz, A., and Tapio, E., "Using Nanosats as a Proof of Concept for Space Science Missions: QuakeSat as an Operational Example," *18th Annual AIAA/USU Conference on Small Satellites*, Paper SSC04-IX-4, 2004.

[7] Gregorio, A., Bernardi, T., Carrato, S., Kostadinov, I., Messerotti, M., and Stalio, R., "AtmoCube: Observation of the Earth Atmosphere from the Space to Study 'Space Weather' Effects," *Proceedings of Recent Advances in Space Technologies*, Aeronautics and Space Technologies Inst., Istanbul, 2003, pp. 188–193.

[8] Swenson, G., Thakker, P., Kamalabadi, F., Frank, M., Coverstone, V., and Voss, H., "Optical Sensing Atmospheric Emissions from Cubesats and Nanosats," *SPIE Sensors and Systems for Space Applications*, Vol. 6555, April 2007, p. 655506.

[9] Rysanek, F., and Hartmann, J. W., "MicroVacuum Arc Thruster Design for a CubeSat Class Satellite," *16th Annual AIAA/USU Conference on Small Satellites*, Paper SSC02-I-2, Aug. 2002.

[10] Simburger, E. J., Liu, S., Halpine, J., Hinkley, D., Srour, J. R., Rumsey, D., and Yoo, H., "Pico Satellite Solar Cell Testbed (PSSC Testbed)," *Conference Record of the 2006 IEEE 4th World Conference on Photovoltaic Energy Conversion*, IEEE, Piscataway, NJ, 2006, pp. 1961–1963.

[11] Liu, A. Z., and Swenson, G. R., "A Modeling Study of O_2 and OH Airglow Perturbations Induced by Atmospheric Gravity Waves," *Journal of Geophysical Research*, Vol. 108(D4), 2003, pp. ACH11.1–ACH 11.9, doi:10.1029/2002JD002474.

[12] Rysanek, F., Burton, R. L., and Keidar, M., "Macroparticle Charging in a Pulsed Vacuum Arc Thruster Discharge," *42nd Joint Propulsion Conference*, AIAA Paper 2006-4499, July 2006.

[13] Pukniel, A., "Attitude Determination, Control, and Related Operations of the Illinois Observing Nanosatellite," *Emergence of Pico- and Nanosatellites for Atmospheric Research and Technology Testing*, edited by P. Thakker and W. A. Shiroma, Progress in Astronautics and Aeronautics, AIAA, Reston, VA, 2010.

[14] Barnes, C., and Selva, L., "Radiation Effects in MMIC Devices," Jet Propulsion Lab., *JPL Publication*, 96-25, Pasadena, CA, 1996, Chap. 10, pp. 203–243.

[15] Thakker, P., and Kimball, J. W., "Power Systems for Cubesat-Class Satellites," *Emergence of Pico- and Nanosatellites for Atmospheric Research and Technology*

Testing, edited by P. Thakker and W. A. Shiroma, Progress in Astronautics and Aeronautics, AIAA, Reston, VA, 2010.

[16] Rein, A., "Cubesat Radio Communication Systems," *Emergence of Pico- and Nanosatellites for Atmospheric Research and Technology Testing*, edited by P. Thakker and W. A. Shiroma, Progress in Astronautics and Aeronautics, Reston, VA (to be published).

[17] Arber, L., "Evolution of the ION Cubesat Software Architecture," *Emergence of Pico- and Nanosatellites for Atmospheric Research and Technology Testing*, edited by P. Thakker and W. A. Shiroma, Progress in Astronautics and Aeronautics, Reston, VA, 2010.

[18] Psiaki, M., "Magnetic Torquer Attitude Control via Asymptotic Periodic Linear Quadratic Regulation," *Journal of Guidance, Control, and Dynamics*, Vol. 24, No. 2, 2001, pp. 386–394.

[19] Athanasopoulou, E., Thakker, P., and Sanders, W. H., "Evaluating the Dependability of a LEO Satellite Network for Scientific Applications," *Second International Conference on the Quantitative Evaluation of Systems (QEST'05)*, 2005, pp. 95–104.

[20] "The Dependency Management Framework: A Case Study of the ION CubeSat," *18th Euromicro Conference on Real-Time Systems*, 2006, p. 10.

Index

Supporting Materials

Many of the topics introduced in this book are discussed in more detail in other AIAA publications. For a complete listing of titles in the AIAA Progress Series, as well as other AIAA publications, please visit www.aiaa.org.

AIAA is committed to devoting resources to the education of both practicing and future aerospace professionals. In 1996, the AIAA Foundation was founded. Its programs enhance scientific literacy and advance the arts and sciences of aerospace. For more information, please visit www.aiaafoundation.org.